Advanced Signal Processing for Industry 4.0, Volume 1

Evolution, communication protocols, and applications in manufacturing systems

Online at: https://doi.org/10.1088/978-0-7503-5247-5

Advanced Signal Processing for Industry 4.0, Volume 1

Evolution, communication protocols, and applications in manufacturing systems

Edited by

Irshad Ahmad Ansari and Varun Bajaj

Electronics and Communication Engineering, PDPM Indian Institute of Information Technology, Design and Manufacturing Jabalpur, Jabalpur, India

IOP Publishing, Bristol, UK

ISBN   978-0-7503-5247-5 (ebook)
ISBN   978-0-7503-5245-1 (print)
ISBN   978-0-7503-5248-2 (myPrint)
ISBN   978-0-7503-5246-8 (mobi)

DOI   10.1088/978-0-7503-5247-5

Version: 20230601

IOP ebooks

British Library Cataloguing-in-Publication Data: A catalogue record for this book is available from the British Library.

Published by IOP Publishing, wholly owned by The Institute of Physics, London

IOP Publishing, No.2 The Distillery, Glassfields, Avon Street, Bristol, BS2 0GR, UK

US Office: IOP Publishing, Inc., 190 North Independence Mall West, Suite 601, Philadelphia, PA 19106, USA

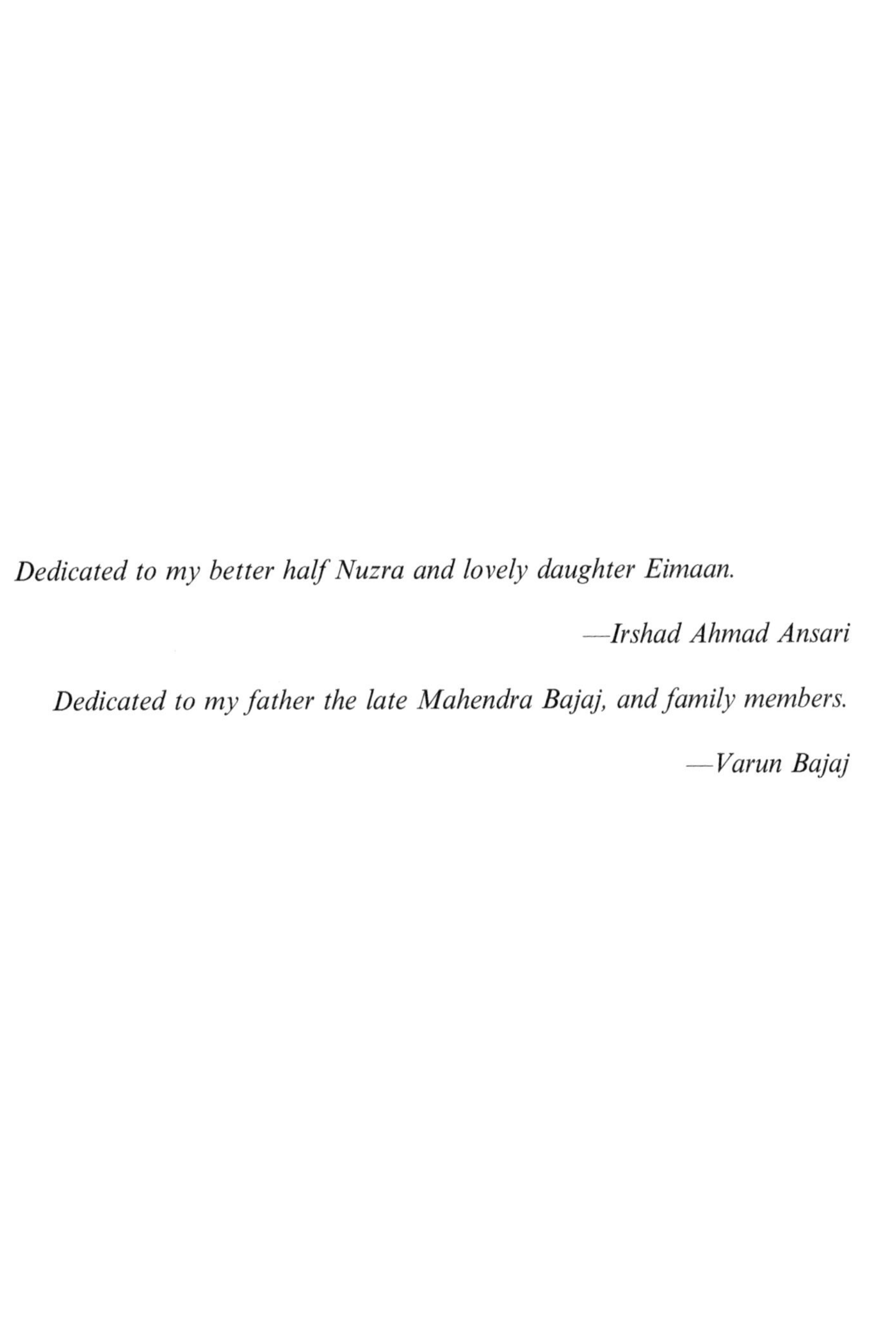

Dedicated to my better half Nuzra and lovely daughter Eimaan.

—Irshad Ahmad Ansari

Dedicated to my father the late Mahendra Bajaj, and family members.

—Varun Bajaj

Contents

4 Applications of infrared imaging for non-destructive testing and evaluation of industrial components — 4-1

Navpreet Kaur, Geetika Dua, Ravibabu Mulaveesala and Vanita Arora

5 Customer-driven healthcare through mission-focused approach in 4IR   5-1

Marjo Rissanen, Antti Rissanen and Kari Rannaste

6 The application of Industry 4.0 technologies for automated health monitoring and surveillance during pandemics and post-pandemic life   6-1

Hamid Reza Marateb, Mónica Rojas-Martínez, Shadi Zamani, Mehdi Shirzadi, Amirhossein Koochekian and Miguel Angel Mañanas

7 A novel computational intelligence approach to making efficient decisions under parametric uncertainty of practical models and its applications to Industry 4.0

N A Nechval, G Berzins and K N Nechval

8 Role of artificial intelligence in industries for advanced applications

Resham Raj Shivwanshi and Neelamshobha Nirala

9 Artificial intelligence based flexible manufacturing system (FMS)   9-1

Ashutosh Mishra

Preface

Industry 4.0 is all about making a drastic change, a revolution in the industrial production, organization, and process in years to come. Industry 4.0 combines concepts like computer vision, machine learning, artificial intelligence (AI), cloud computing, the Internet of Things (IoT), etc., in order to make production, operation, and manufacturing more efficient. Industry 4.0 uses various sensor inputs for operation and IoT for communication. These sensors and fields like data analysis, Big Data, AI, robotics, computer vision, etc., generate a large amount of data; This makes signal processing a major area of research in this revolution. Industry 4.0 is not only about automation, it is about creating intelligent factories and production units. Industry 4.0 should not be seen just as the introduction of some gadgets into already existing industries, but as a space where the process, product, technology, and people are all intertwined. Industry 4.0 is about creating an environment where people can work with machines through seamless connectivity. It is about creating a cyber-physical system for the manufacturing sector. The communication and the field of advanced signal processing form the heart of the inevitable fourth Industrial Revolution.

Industry 4.0 is an amalgamation of digital technologies with the industries; it is required for enhancing production, flexibility, and scalability in industries. Industry 4.0 is a journey towards an integrated environment with human–machine interaction being its important aspect. With devices like cameras and different types of sensors an environment could be created for enhanced man–machine interaction. These sensors and cameras would generate a large amount of signals, making advanced signals processing an important area of research in manufacturing units. The real time/near real time signal processing for these sensors becomes a major research domain.

The main aim of this book is to bring together the frontier of research happening around advance signal processing for Industry 4.0. The main focus of volume 1 is on vision systems for Industry 4.0, communication protocols, human–machine interference and various applications of signal processing in the industrial domain. The chapter-wise description of this book is as follows.

Chapter 1 describes the role and importance of computer vision in industrial applications. Different types of industrial vision systems are discussed in this chapter. It also provides an example of implementing a vision based control mechanism for object sorting. Chapter 2 presents details of the crux of capnography systems for delivering precise capnogram signals in the advancement of Industry 4.0. This chapter also outlines the systematic analyses and prospective tools utilized in literature for incorporating capnography in medical facilities.

Industry 4.0 and the use of private 5G networks are covered in Chapter 3 along with the difficulties and challenges in its implementation. The advantages of private networks for Industry 4.0 are also discussed in this chapter. Chapter 4 proposes an algorithm for automatic defect detection and defect diameter estimation in fiber-reinforced polymer materials using signal and image processing approaches.

Frequency Modulated Thermal Wave Imaging (FMTWI) is discussed in detail for nondestructive testing in industrial applications. Chapter 5 presents a design and evaluation that underlines mission-strategic patterns in customer-centric health technology development that play a significant role in this change. Chapter 6 discusses the contribution of Industry 4.0 (IoT, digitalization, Big Data, AI, and cloud-based computing) for automatic surveillance and health monitoring during pandemics and post-pandemic life. More than 50 papers published in 2020–2022 were analyzed in this chapter.

Chapter 7 proposes a novel computational intelligence approach with applicability to Industry 4.0 in order to make effective judgments under parametric stochastic model uncertainty. In addition, examples from the real world are provided to demonstrate the use of the proposed technique. Chapter 8 aims to highlight some of those important impressions of AI on industries responsible for bringing a significant shift in their timeline and consistently altering the future with unprecedented potential efforts. Chapter 9 covers the use of some of Expert Systems (ES) in various manufacturing processes in flexible manufacturing systems (FMS). Chapter 10 discusses some of the present age applications of deep learning-based algorithms in the domain of the stock market, marketing industries, bioinformatics, and cybersecurity with an emphasis on how modern techniques are revolutionizing these fields.

Chapter 11 discusses the perspectives and application situations of family businesses undergoing digital transformation. In the application part of the study, the digital transformation process of Muratbey Cheese Company, a leading family business in the cheese manufacturing sector in Turkey, is presented as a situation analysis. Chapter 12 provides an analysis of a human–machine interface (HMI) system that can be used as an industrial robotic hand based on electromyogram (EMG) signals. A detailed analysis of figure movement identification is provided with various classification techniques. Chapter 13 focuses on data-driven models for the deployment of energy-efficient wireless networks in the context of Industry 4.0 applications, as Industry 4.0 heavily relies on IoT sensor data or knowledge-based data. Additionally, this chapter provides several automatic opportunistic relay selection methods for multi-input multi-output (MIMO) multi-relay cognitive networks (MRCNs) employing deep neural networks (DNN).

Acknowledgments

Dr Ansari expresses his gratitude and sincere thanks to his wife, family members, and teachers for their constant support and motivation. He really appreciates his wife for all the time that she has given him to work on this book, while she took care of everything.

Dr Bajaj expresses his heartfelt appreciation to his mother Prabha, wife Anuja, and daughter Avadhi, for their wonderful support and encouragement throughout the completion of this important book on Advanced Signal Processing for Industry 4.0, Volume 1: Evolution, communication protocols, and applications in manufacturing systems. His deepest gratitude goes to his mother in law and father in law for their constant motivation. This book is an outcome of sincere efforts that could be given to the book only due to great support of the family.

We sincerely give thanks to Prof. Bhartendu K. Singh, Director of PDPM IIITDM Jabalpur, for his support and encouragement. We would like to thank all our friends, well-wishers and all those who keep us motivated in doing more and more; better and better. We sincerely thank all contributors for writing relevant theoretical background and applications for this book.

We express our humble thanks to Dr John and all editorial staff of IOP for great support, necessary help, appreciation and quick responses. We also wish to thank IOP for giving us this opportunity to contribute on some relevant topic with a reputed publisher. Finally we want to thank everyone, in one way or another, who helped us editing this book.

Dr Bajaj especially thanks his family, who encouraged him throughout the time of editing this book. This book is heartily dedicated to his father who took the lead to heaven before the completion of this book.

Last but not least we would also like to thank God for showering us his blessings and strength to do this type of novel and quality work.

Irshad Ahmad Ansari
Varun Bajaj

Editor biographies

Irshad Ahmad Ansari

Irshad Ahmad Ansari (PhD, SMIEEE20) has been working as a faculty member in the discipline of Electronics and Communication Engineering at PDPM Indian Institute of Information Technology, Design and Manufacturing (IIITDM) Jabalpur, India since 2017. He received his BTech degree in Electronics and Communication Engineering from Gautam Buddh Technical University (formally UPTU), Lucknow, India in 2010, and his MTech degree in Control and Instrumentation from Dr B R Ambedkar National Institute of Technology Jalandhar, Punjab, India in 2012. He completed his PhD from IIT Roorkee with MHRD teaching assistant-ship, and subsequently joined Gwangju Institute of Science and Technology, South Korea as a Postdoctoral fellow. His major research interests include Signal and Image Processing, Brain Computer Interfaces and Machine Learning. He is contributing as an active technical reviewer of leading international publishers such as IEEE, IOP, Elsevier, and Springer. He has more than 57 publications, which includes journal papers (27), conference papers (26), books (3), and book chapters (4). The citation impact of his publications is around 791 citations, with a h-index of 14, and i10 index of 18 (Google Scholar November 2022). He has guided three (2 submitted and 1 in process) PhD Scholars and 13 MTech Scholars.

Varun Bajaj

Varun Bajaj (PhD, SMIEEE20) has been working as an Associate Professor in the discipline of Electronics and Communication Engineering, at the Indian Institute of Information Technology, Design and Manufacturing (IIITDM) Jabalpur, India since July 2021. He worked as Assistant Professor in IIITDM Jabalpur from March 2014 to July 2021. He also worked as a visiting faculty member in IIITDM Jabalpur from September 2013 to March 2014. He worked as Assistant Professor at the Department of Electronics and Instrumentation, Shri Vaishnav Institute of Technology and Science, Indore, India during 2009–2010. He received his PhD degree in the discipline of Electrical Engineering, at the Indian Institute of Technology Indore, India in 2014. He received his MTech degree with Honors in Microelectronics and VLSI design from Shri Govindram Seksaria Institute of Technology and Science, Indore, India in 2009, and his BE degree in Electronics and Communication Engineering from Rajiv Gandhi Technological University, Bhopal, India in 2006.

He is an Associate Editor of the IEEE Sensor Journal and Subject Editor-in-Chief of IET *Electronics Letters*. He served as a Subject Editor of IET *Electronics Letters* from November 2018 to June 2020. He is a Senior Member of IEEE since June 2020,

MIEEE 16-20, and also contributing as an active technical reviewer of leading international journals from IEEE, IET, and Elsevier. He has 145 publications, which include journal papers (93), conference papers (31), books (10), and book chapters (11). The citation impact of his publications is around 4400 citations, with a h-index of 36, and i10 index of 89 (Google Scholar December 2021). He has guided seven (5 completed and 3 in process) PhD Scholars and eight MTech Scholars. He has been listed in the world's top 2% researchers/scientists by Stanford University, USA (October 2020 and October 2021). He has worked on research projects funded by DST and CSIR. He is a recipient of various reputed national and international awards. His research interests include Biomedical Signal Processing, AI in Healthcare, Brain Computer Interface, Pattern Recognition, and ECG signal processing.

List of contributors

Irshad Ahmad Ansari
PDPM IIITDM, Jabalpur, India

Vanita Arora
InfraRed Vision and Automation Pvt Ltd, Punjab, India

Varun Bajaj
PDPM IIITDM, Jabalpur, India

Matadeen Bansal
PDPM IIITDM Jabalpur, M.P., India

Kausik Basak
JIS Institute of Advanced Studies and Research Kolkata, JIS University, Kolkata, India

Gundars Berzins
Latvia University, Riga, Latvia

Geetika Dua
Thapar Institute of Engineering and Technology, Adarsh Nagar, Prem Nagar, Patiala, Punjab, India

Meral Erdirençelebi
Necmettin Erbakan University, Konya, Turkey

V Ganesan
Department of Electronics and Communication Engineering, Bharath Institute of Higher Education, Chennai, T.N., India

Güzide Karakuş
Necmettin Erbakan University, Konya, Turkey

Navpreet Kaur
Thapar Institute of Engineering and Technology

Tapomoy Koley
Capgemini Technology Services India Limited, Kolkata, India

Amirhossein Koochekian
Child Growth and Development Research Center, Research Institute for Primordial Prevention of Non-Communicable Disease, Isfahan University of Medical Sciences, Isfahan, Iran

Anurodh Kumar
PDPM IIITDM Jabalpur, M.P., India

M B Malarvili
School of Biomedical Engineering, Universiti Teknologi Malaysia, UTM Skudai, Johor, Malaysia

Miguel Angel Mañanas
Biomedical Engineering Research Centre (CREB), Automatic Control Department (ESAII), Universitat Politècnica de Catalunya-Barcelona Tech (UPC), Barcelona, Spain

Hamid Reza Marateb
Biomedical Engineering Research Centre (CREB), Automatic Control Department (ESAII), Universitat Politècnica de Catalunya-Barcelona Tech (UPC), Barcelona, Spain

Shikha Maurya
NIT, Agartla, India

Ashutosh Mishra
Thapar Institute of Engineering and Technology, Patiala, India

Ravibabu Mulaveesala
Centre for Sensors, Instrumentation and cyber-physical Systems Engineering (SeNSE), Indian Institute of Technology Delhi

Konstantin Nechval
Transport Institute, Riga, Latvia

Nicholas Nechval
Latvia University, Riga, Latvia

Neelamshobha Nirala
National Institute of Technology Raipur, India

Santheraleka Ramanathan
School of Biomedical Engineering, Universiti Teknologi Malaysia, UTM Skudai, Johor, Malaysia

Kari Rannaste
Aalto University, Espoo, Finland

R Ravindraiah
Department of Electronics and Communication Engineering, Siddharth Institute of Engineering and Technology, Puttur, A.P., India

Antti Rissanen
National Defence University, Helsinki, Finland

Marjo Rissanen
Prime Multimedia Ltd., Helsinki, Finland

Mónica Rojas-Martínez
Biomedical Engineering Research Centre (CREB), Automatic Control Department (ESAII), Universitat Politècnica de Catalunya-Barcelona Tech (UPC), Barcelona, Spain

Rohit Roy Chowdhury
JIS Institute of Advanced Studies and Research Kolkata, JIS University, Kolkata, India

Deepak Sahu
PDPM IIITDM Jabalpur, M.P., India

P D Selvam
Department of Electronics and Communication Engineering, Siddharth Institute of Engineering and Technology, Puttur, A.P., India

Mehdi Shirzadi
Biomedical Engineering Research Centre (CREB), Automatic Control Department (ESAII), Universitat Politècnica de Catalunya-Barcelona Tech (UPC), Barcelona, Spain

Resham Raj Shivwanshi
National Institute of Technology Raipur, India

J Sridhar
Department of Electronics and Communication Engineering, Siddharth Institute of Engineering and Technology, Puttur, A.P., India

Dinesh Kumar V
PDPM IIITDM Jabalpur, M.P., India

Amit Vishwakarma
PDPM IIITDM Jabalpur, M.P., India

Shivam Vyas
PDPM IIITDM, Jabalpur, India

Shadi Zamani
Biomedical Engineering Department, Engineering Faculty, University of Isfahan, Iran

Contributors' biographies

Dr Vanita Arora

Dr Vanita Arora, PhD received her PhD degree from the Indian Institute of Technology Ropar (IIT Ropar), India. She is currently working as a Faculty Member with the Indian Institute of Information Technology Una, Transit Campus-II: Vill. Chandpur, Teh. Haroli, India. Her research interests include the development of infrared imaging technologies for noninvasive/nondestructive testing applications. She serves as a Guest Editor for several refereed journals of the Institute of Physics, Institution of Engineering and Technology, and Springer Nature.

Matadeen Bansal

Matadeen Bansal received his BE degree from the Department of Electronics and Communication Engineering, Madhav Institute of Science and Technology, Gwalior, India, in 2001, and his ME degree in Communication, Control and Networking from Madhav Institute of Technology and Science, Gwalior, India in 2006. He received his PhD degree from the Department of Electronics and Communication Engineering, ABV Indian Institute of Information Technology and Management Gwalior, India, in 2013. He has been working as an Assistant Professor at the PDPM Indian Institute of Information Technology, Design and Manufacturing, Jabalpur, India, since 2013. He has various publications in SCI indexed journals such as Springer *Wireless Networks*, Elsevier *Physical Communication*, IET *Communication Journal*, IEEE *Wireless Communications Letters*, Springer *Wireless Personal Communications*, and Wiley *Transactions on Emerging Telecommunications Technologies* (formerly ETT). He is the reviewer of various SCI indexed journals. His area of interest is wireless communications, cooperative communications, cognitive radios, multiple-input multiple-output (MIMO) systems, non-orthogonal multiple access (NOMA), and optimization of communication systems.

Dr Kausik Basak

Dr Kausik Basak, PhD, received his BTech in electronics and instrumentation engineering from West Bengal University and Technology in 2008, and MTech and PhD degrees in biomedical engineering from Indian Institute of Technology Kharagpur in 2010 and 2015, respectively. Since July 2019, he has been with the Centre for Health Science and Technology, JIS Institute of Advanced Studies and Research (JISIASR) Kolkata, JIS University, where he is currently working as an Assistant Professor. Previously, he was a

Post-doctoral Research Scientist in the Molecular Imaging Engineering division of Technical University Munich (TUM), Germany. Prior to this, he worked as an Assistant Professor at the School of Engineering Sciences, Mahindra Ecole Centrale, India. He has also worked as a Visiting Professor at CentraleSupelec, Paris, France. His research interests focus on superficial and deep-tissue imaging at the visible and near infrared regime, encompassing imaging modalities like laser speckle, multispectral and optoacoustic imaging, and associated analysis using machine learning-based computational algorithms. He has published several papers in journals with reputed publishers e.g., Springer Nature, Elsevier, IEEE, etc. He is a senior member of IEEE, and Engineering in Medicine and Biology Society. He is the recipient of a TUM University Foundation Fellowship (TUFF) (2018), and doctoral (2010) and post-graduate (2008) scholarships from the Ministry of Human Resource and Development (MHRD), Government of India.

Dr Gundars Berzins

Dr Gundars Berzins is the Dean of the Faculty of Business, Management and Economics of the University of Latvia (Doctor's degree in Management), Corresponding Member of the Latvian Academy of Sciences; Senior Researcher, Institute of Economics and Management, Faculty of Business, Management and Economics, University of Latvia; member of the Senate Strategy Commission of the University of Latvia (UL); member of the UL Senate Financial and Budget Commission; Member of UL Senate Academic Staff Group; member of the Council of the Faculty of Business, Management and Economics of the UL; member of the Faculty of Business, Management and Economics Strategy Council; member of the Student Business Incubator Council, UL: Board Member of the Stockholm School of Economics in Riga; consultant of the International Open Innovation Platform 'DEMOLA Latvia' in cooperation with Riga Technical University; member of the Organizing Committee of the International Conference 'New Challenges of Economic and Business Development', Faculty of Business, Management and Economics, UL; member of the Organizing Committee of the 4th World Congress of Latvian Scientists, Leader of the Social Sciences Section: 'Latvia in the Global Supply and Value Chains'; member of The Association to Advance Collegiate Schools of Business (AACSB); member of the Editorial Board of the Faculty of Management and Economics, Journal of Economics and Management Research; Expert Rights in Management. He previously held the position of Director of Finance and Accounting Department of the University of Latvia, later Chancellor of the University of Latvia. For a long time he also worked in the private business sector. Dr Berzins has authored/co-authored 170 publications in various high impact factor peer-reviewed journals. His research interests include strategic management, big data analysis, and product design.

Dr Geetika Dua

Dr Geetika Dua, PhD received her PhD degree from the Indian Institute of Technology Ropar (IIT Ropar), India. She is currently working as a faculty member with the Thapar Institute of Engineering and Technology, Patiala, Punjab, India. Her major areas of research interest are the development of signal, image, and video processing methods for nondestructive testing/noninvasive imaging modalities for the inspection of various industrial and biomedical materials. She serves as a Guest Editor for several refereed journals of Institute of Physics (IOP) and Institution of Engineering and Technology (IET).

Dr Meral Erdirençelebi

Dr Meral Erdirençelebi received her doctorate degree from Selcuk University. Since 2013, she has been with the Faculty of Applied Science in Necmettin Erbakan University. Currently, she is Associate Professor at the Faculty of Applied Science in Necmettin Erbakan University. She is also the Assistant Director at the Sustainable Family Businesses Application and Research Center of Necmettin Erbakan University. She specialized in organizational behavior and human resources management. Her research interests include sustainability, family business, entrepreneurship, and current issues in organizational behavior (work-life balance, presenteeism, burnout, nepotism, organizational loneliness, etc). Dr Erdirençelebi has authored/co-authored publications in 22 peer-reviewed journals. Dr Erdirençelebi has written 12 book chapters and she has presented about 10 papers in international conferences. She is a member of the TOBB Women Entrepreneurs board in Konya-Turkey.

Dr V Ganesan

Dr V Ganesan is currently working as an Associate Professor at the Department of Electronics and Communication Engineering, Bharath Institute of Higher Education and Research, India. He received his Bachelor of Science degree and MSc degree in Physics from the Bharathidasan University Thiruchirapalli in 2002 and 2005, respectively. He received his Master's Degree (ME) in Applied Electronics from Sathyabama University in 2007. He received his doctorate in the Faculty of Electronics Engineering from Sathyabama University in April 2015. His research interest focuses mainly on image processing and VLSI technology. He has published more than 35 Scopes-indexed publication in international level conferences and journals. Presently he is guiding 8 PhD students. He is a member of the IEEE.

Dr Güzide Karakuş

Dr Güzide Karakuş is an Assistant Professor at Necmettin Erbakan University, Department of Aviation Management in Turkey. She holds a Bachelor's degree in Industrial Engineering from Kocaeli University (1999) and an MBA degree from Gebze Institute of Technology (2002). She completed her PhD at Selçuk University in Production Management and Marketing in 2014. Prior to joining academia, she worked in consulting about quality management systems, CE marking, and project management. Her research interests include sustainable operations management, environmentally friendly manufacturing, digital transformation in the perspective of Industry 4.0, and its reflections. She teaches innovation management, technology management, project management, quality management, decision making techniques, and supply chain management, both at undergraduate and graduate level. She has authored/co-authored publications in 14 peer-reviewed journals. Dr Karakuş has written 8 book chapters and she has presented about 15 papers in international conferences. She has undertaken consultancy/manager duties in a total of 9 projects.

Navpreet Kaur

Navpreet Kaur received her BTech (2017), ME (2021), degrees in Electronics and Communication engineering from Punjabi University, Patiala and Thapar Institute of Engineering and Technology, Patiala, Punjab, India respectively. During her Masters she pursued her interest in non-destruction testing and researched the same topic for her final year dissertation.

Mr Tapomoy Koley

Mr Tapomoy Koley received his BTech in electrical engineering from West Bengal University and Technology in 2005, his Executive MBA (specialization—finance) from IGNOU in 2012, and his MTech degree in software systems (specialization—data analytics) from Birla Institute of Technology (Pilani) in 2019. Since November 2020, he has been with Capgemini Technology Services where he is currently working as Senior Manager in the Data Science and Analytics division. He is also pursuing his PhD research work at the JIS Institute of Advanced Studies and Research (JISIASR) Kolkata, JIS University. Previously he worked in Cognizant Technology Solution for more than 15 years. He has served leading clients in different domains and geographies across the world. He has led data science solution builds for large and complex requirements in the capacity of lead data scientist. He is experienced in analytics product development as well. He has published several research papers in reputed journals such as the International Journal of Economics and Finance. His research interest includes federated learning, semantic AI, application of AI in the field of cybersecurity, banking and financial services, operations, and the insurance sector.

Amirhossein Koochekian

Amirhossein Koochekian is currently a BS student of Biomedical Engineering at the Islamic Azad University Central Tehran Branch. He is also affiliated with the Pediatrics Department, Child Growth and Development Research Center, Research Institute for Primordial Prevention of Non-Communicable Disease, Isfahan University of Medical Sciences. His research interests include medical data mining and biomedical signal processing.

Anurodh Kumar

Anurodh Kumar received a BE degree in electronics and communication engineering from the University of RGPV, Bhopal, India, in 2010 and an MTech degree in digital communication from the University of RGPV, Bhopal, in 2013. He worked as an Assistant Professor at the Sagar Institute of Science and Technology, Ratibad campus, Bhopal from 2013 to May 2019. He is currently a PhD Research Scholar at the PDPM Indian Institute of Information Technology, Design, and Manufacturing, Jabalpur (PDPM IIITDMJ), Jabalpur, India. His research interests include image processing applications in biomedical, machine learning, and deep learning.

Associate Professor Ir. Dr Balakrishnan Malarvili

Associate Professor Ir. Dr Balakrishnan Malarvili has acquired more than 15 years of experience in the academics as well as involvements in the biomedical industry. She is the head of the Biosignal Processing Research Group under the Health and Wellness Research Alliance at Universiti Teknologi Malaysia. She obtained a Doctor of Philosophy in Medical Sciences Engineering from The University of Queensland, Australia in 2008. Currently, she is attached to the School of Biomedical and Health Sciences Engineering, Universiti Teknologi of Malaysia. Dr Malarvili is actively involved in research related to digital signal processing, time–frequency signal analysis, pattern recognition, biosensors, and medical monitoring devices. To date, she has published nearly 100 publications consisting of refereed journals (30), books (8), Scopus indexed journals (35), ISI indexed journals (20), book chapters (8) and over 20 refereed conference papers. Dr Malarvili has also produced 13 copyrights and 5 patents. She also received 5 international awards, 12 national awards, and many more appreciation certificates. In addition, she has chaired sessions as chairperson in prestigious conferences, editorial board member and technical reviewer for IEEE, EMBS and medical societies besides being appointed as a scientific and technical committee member for several international conferences. She was recently selected for an 'Outstanding Woman in Health and Medical Sciences' award at the 4th Venus International Women Awards—VIWA 2019, Chennai, India.

Miguel Angel Mañanas

Miguel Angel Mañanas, PhD, received his Telecommunications Engineering and PhD in Biomedical Engineering (BME) degrees from Universitat Politècnica de Catalunya (UPC) in 1993 and 1999, respectively. He is currently an Associate Professor at the Department of Automatic Control (ESAII) at the same university, a member of the Biomedical Engineering Research Center (CREB) at the UPC, and a member of the CIBER-BBN. He is the founder and leader of the multidisciplinary BIOsignal Analysis for Rehabilitation and Therapy Group (BIOART, UPC), composed of engineers and physicians. His research is focused on applying engineering techniques in the health areas. His career has been mainly focused on biomedical signal processing and biological system modeling in three fields: (1) neuromuscular: evaluating neuromuscular disorders and motor rehabilitation by high-density electromyography, (2) neurological: evaluating brain activity by electroencephalography and magnetoencephalography, and (3) respiratory: modeling the respiratory control system and simulating pulmonary diseases and mechanical ventilation. He is the co-author of 65 JCR articles (31 in Q1 and 23 in Q2), has more than 100 peer-reviewed scientific communications, and co-invented two patents. He has co-supervised 8 PhD theses. He has also been PI of 30 national/international projects. He received the Leonardo Award from BBVA Foundation for career recognition in 2016.

Hamid R Marateb

Hamid R Marateb, PhD, received BS and MS degrees from the Shahid Beheshti University of Medical Science and Amirkabir University of Technology, Tehran, Iran, in 2000 and 2003, respectively. He received his PhD and the postdoctoral fellowship from the Laboratory of Engineering of Neuromuscular Systems, Politecnico di Torino, Turin, Italy, in 2011 and 2012, respectively. He was a visiting researcher at Stanford University in 2009 and Aalborg University in 2010. He was a visiting professor at UPC, Barcelona, in 2012 and 2017. His research line is Cognitive Informatics in Health and Biomedicine, mainly focusing on clinical neurophysiology, computational neurosciences, and medical data mining. Dr Marateb is a reviewer in more than 30 international ISI journals and is now on the editorial board of international journals including Frontiers in Integrative Physiology and Frontiers in Cognitive Neuroscience. He also received European Union/Spanish grants (e.g., Marie Curie Fellowship and MYOSLEEVE). He is currently with the Department of Biomedical Engineering, Faculty of Engineering at the University of Isfahan, IRAN, and also BIOsignal Analysis for Rehabilitation and Therapy Research Group (BIOART),

Department of Automatic Control, Biomedical Engineering Research Center, Universitat Politècnica de Catalunya, Barcelona Tech (UPC), Barcelona, Spain.

Shikha Maurya

Shikha Maurya received her BTech (Electronics and Communication Engineering) degree from Uttar Pradesh Technical University in 2012 and MTech (Digital Communication) degree from ABV Indian Institute of Information Technology and Management, Gwalior, India in 2014. She has received a PhD from the Department of Electronics and Communication Engineering at the PDPM Indian Institute of Information Technology, Design and Manufacturing Jabalpur, India in 2019. Presently, she is an Assistant Professor at the National Institute of Technology, Agartla, India, since October 2022. She has various publications in SCI indexed journals such as Elsevier's *Physical Communication*, IET's *Communication Journal*, and IEEE's *Wireless Communications Letters*. Her research interests include energy-efficient resource allocation in wireless networks, MIMO, and cognitive relay networks.

Ashutosh Mishra

Ashutosh Mishra (M'79) received his PhD degrees in computer science and engineering from the Indian Institute of Technology (IIT-BHU), Varanasi, India. He is currently working as an assistant professor in the Department of Computer Science and Engineering, Thapar Institute of Engineering and Technology, Patiala, India. His research interests include software engineering, data mining, semantic web technologies, and cognitive computations. He is a member of various professional bodies and has published research papers in national and international journals and conferences. He is also worked as Visiting Associate at the Department of Computer Science and Statistics, Trinity College Dublin (Ireland) from 20 February 2017 to 13 April 2017.

Dr Ravibabu Mulaveesala

Dr Ravibabu Mulaveesala, PhD received his PhD degree from the Indian Institute of Technology Delhi (IIT Delhi), India in 2006. Presently he is working as Associate Professor in the Centre for Sensors, iNstrumentation and cyber-physical Systems Engineering (SeNSE), Indian Institute of Technology Delhi, Hauz Khas, New Delhi, India. His research interests include thermal, acoustical, and optical methods for noninvasive/nondestructive imaging technologies. He serves on editorial or advisory boards for several refereed journals of the American Institute of Physics (AIP), Institute of Physics (IOP), Institute of Electrical and Electronics Engineers (IEEE), Institution of Engineering and Technology (IET), Springer Nature, and Elsevier, and also several peer-reviewed conferences.

Dr Sc.Ing. Konstantin Nechval

Dr Sc.Ing. Konstantin Nechval After defending my doctoral dissertation I have received a doctoral degree in the field of engineering sciences in 2008 Riga Technical University. A year after on a competitive basis I was elected as an associate professor at the Aviation Department of Transport and Telecommunication Institute. In 2013 I had experience in enterprise of LNK group—company Centre Composite Design Bureau in Latvia, which is engaged in engineering calculations and the development of new types of aircraft. This opportunity allowed me to reconsider aspects of the current aviation professionals training and currently has group of graduates working in the test laboratory in Aviatest, at the air company AME in Estonia, in FL technics in Lithuania and of course at Riga airport. I was involved in some of the projects related to aviation specifying.

Nicholas A Nechval

Nicholas A Nechval is a leading researcher at the BVEF Research Institute and professor at the University of Latvia. PhD in Automatic Control and Systems Engineering (1969), Department of Applied Mathematics, Graduate school: Riga Civil Aviation Engineers Institute (RCAEI), Riga, Latvia. Dr Habil.Sc. in Radio Engineering (1993), Department of Applied Mathematics, Riga Aviation University (RAU), Doctoral studies: Moscow Aviation Institute (MAI), Moscow, Russia. His current research interests are in the following areas: mathematics, econometrics, stochastic processes, pattern recognition, digital radar signal processing, operations research, statistical decision theory, adaptive control, and mechanical engineering. Broadly viewed, his research interests are in models and algorithms for nonlinear optimization (under conditions of incomplete information) and their application to problems arising in science, engineering, operations research, risk management, finance, insurance, economics, hydrology, material sciences, telecommunications, and many other industries. Application areas of interest focus mainly on optimization problems with a special interest in applying the proposed techniques to yield management and decision making in aircraft service as well as to all aspects of business administration and management. The database 'Research Gate' (Selected Publications: 178, Research Interest: 333.0, Reads: 8350, Citations: 537, h-index: 11); the database 'Scopus' (Documents by Author: 129, Citations: 124, h-index: 10); Springer Link (46 Results for 'Nechval'); dblp (34 Results for 'Nechval'). Winner of the Exhibition Committee Award: Silver Medal (No. 13105) of the Exhibition Committee in connection with research on the problem of Prevention of Collisions between Aircraft and Birds, Moscow, Russia, 1991.

Dr Neelamshobha Nirala

Dr Neelamshobha Nirala, PhD, is Assistant Professor at the Department of Biomedical Engineering, NIT Raipur. She has completed her Bachelor of Technology from the National Institute of Technology Raipur in Biomedical Engineering. She has done her Masters at SRM University in Biomedical Engineering. She completed her PhD from NIT Raipur, and her interest of area is signal processing and analysis of type-2 diabetic disease. She is also experienced in medical imaging, image processing wide variety of filters, filter applications, and signals acquisition and analysis.

Dr Santheraleka Ramanathan

Dr Santheraleka Ramanathan obtained her PhD in Nanobiotechnology Engineering in Universiti Malaysia Perlis, Malaysia (2020). Her research field is in the generation of high-performance biosensors. She has gained experiences in synthesizing and characterizing nanomaterials, and its incorporation with biomolecules for detecting disease biomarkers. She is currently continuing her postdoctoral study under Biosignal Processing Research Group in Universiti Teknologi Malaysia. Dr Santheraleka pursuing her interest in research related to signal processing, with regard to medical monitoring devices. Her research works were published (total of 24 with 349 citations) in referred scientific journals. She has contributed to 3 book chapters and 1 edited book.

Kari Rannaste

Kari Rannaste received his MSc degree in Theoretical Physics from Jyväskylä University, Finland, and MBA degree from University of Genève (IMI). He has extensive work experience in the areas of industrial development tasks and business management consulting. He has worked in higher education teaching and development positions at universities in Finland and also taught at several universities of applied sciences, e.g., hospital engineers' studies in the field of measurement and control technology, modeling, and IT systems. In the Finnish Defense Forces, he has been responsible for the management and information security related projects. He has worked as a product manager of the Standardization Association (SFS) in the coordination of measurement unit systems and metrology activities and taken care of organizing the introduction of the SI- system in Finland. He is a member of several steering groups, boards and organizations and has given presentations at seminars in various fields both in Finland and in international contexts.

Dr R Ravindraiah

Dr R Ravindraiah is working as a Professor in the Department of Electronics and Communication Engineering in Siddharth Institute of Engineering and Technology, Puttur. He received his undergraduate degree from S.K. University, Ananthapur in 2008 and a Master's degree from JNT University, Ananthapuramu, Ananthapur, Andhra Pradesh, India in 2011. He was awarded a PhD degree from JNTUA, Ananthapuramu in 2021. His research topics include biomedical image processing and embedded systems. He has published 13 papers in reputed journals, 9 international conferences, and 3 national conference proceedings. He holds a lifetime membership in ISTE, IACSIT, and IAENG. He has guided 5 PG projects and 15 UG projects.

Dr Antti Rissanen

Dr Antti Rissanen, PhD, received his BSc, MA, and PhD degrees in Physics from Helsinki University, Finland in 1981, 1984 and 2011, respectively. He obtained a Licentiate in Technology degree (electric engineering with specialty in biomedical engineering) in 1992 at the University of Technology in Tampere. In 1997 he joined the National Defence University (Finland) as a Senior Lecturer for General Technology where he currently works as a researcher. He is also the leader of the Educational Research Group at the Military Technology Department. His research interests include science education, e-learning, and simulations or artificial intelligence in educational contexts. Previous research interests are biomedical engineering, signal analysis, technology transfer, and thin film technology. Dr Rissanen has authored/co-authored 37 publications in peer-reviewed journals. Dr Rissanen has written several book chapters and has presented about 50 papers in international conferences.

Dr Marjo Rissanen

Dr Marjo Rissanen is a senior consultant, focusing on health informatics design areas. She received her MSc degree from Helsinki University, Faculty of Medicine, Finland and Doctor of Science degree in neuroscience and biomedical engineering from Aalto University, Finland, in 2022. She has worked in Helsinki University, in the Ministry of Social Affairs and Health, and in consultancy companies in research and design activities. She has also worked in international projects e.g., in cost-control information system project for an orthopedic hospital in Colombo, Sri Lanka. She has also lectured for academia, health organizations, and industry in international contexts. Her research interests cover topics connected to translational health technology and design, health informatics, and related quality and design challenges. Dr Rissanen has written 16 peer-reviewed articles and 4 book chapters about timely health technology linked topics.

Dr Mónica Rojas-Martínez

Dr Mónica Rojas-Martínez, PhD, graduated in 1999 as an Electrical Engineer from the Universidad de Los Andes in Bogotá, Colombia, and got her PhD degree in Biomedical Engineering in 2012 from the Polytechnic University of Catalonia (Barcelona Tech), Barcelona, Spain. She has been an Associate Professor and Researcher at the Department of Bioengineering, Faculty of Engineering, Universidad El Bosque, Bogotá, Colombia. She is currently with the BIOsignal Analysis for Rehabilitation and Therapy Research Group (BIOART), Department of Automatic Control, Biomedical Engineering Research Center, Universitat Politècnica de Catalunya, Barcelona Tech (UPC), Barcelona, Spain. Her research interests include signal processing and machine learning techniques with applications in the study of motor rehabilitation, human–machine interfaces, prostheses, and assistive devices. During the last few years, she has been working on several research projects focused on analyzing EMG signals for diagnosing muscular pathologies, monitoring rehabilitation therapies, and using HD-EMG to control external devices.

Mr Rohit Roy Chowdhury

Mr Rohit Roy Chowdhury, received his BTech in Computer Science Engineering from West Bengal University in 2017 and is pursuing his MTech in Computer Science and Engineering with specialization in Data Science from JIS Institute of Advanced Studies and Research (JISIASR) Kolkata, JIS University (to be completed in 2022). He started his career in Wipro Technologies working with HP North America in solving business issues. Currently he is working with BYJU's, an EdTech Company as a data scientist. His core interest lies in machine learning-based implementations to cater solutions for various industry and business problems.

Deepak Sahu

Deepak Sahu received his BE degree in Electronics and Communication Engineering from Rajiv Gandhi Technical University, Bhopal (M.P.), India, in 2012 and MTech in Digital Communication from ABV-India Institute of Information Technology and Management, Gwalior, India in 2014. He is currently pursuing his PhD degree in Electronics and Communication Engineering from PDPM Indian Institute of Information Technology, Design and Manufacturing, Jabalpur,

India. He has qualified for the national level Graduate Aptitude Test in Engineering (GATE). His current research includes deep learning, wireless communication, cognitive relay network, and precoder designing.

Dr P D Selvam

Dr P D Selvam is working as a Professor at the Department of Electronics and Communication Engineering, Siddharth Institute of Engineering and Technology, Puttur, A.P., India. He received a BE in Electronics and Communication Engineering from University of Madras, Chennai, India, an ME in Communication Systems from Anna University, Chennai, India, and a PhD in Information and Communication Engineering from Anna University, Chennai, India in 2002, 2007, and 2020, respectively. He has more than 14 years of teaching experience in teaching for undergraduate students. His research interests include 5G, Massive MIMO, and Device-to-Device Communication.

Mehdi Shirzadi

Mehdi Shirzadi received the BS and MS degrees in Biomedical Engineering from Islamic Azad University of Najaf Abad, Najaf Abad, Isfahan, Iran, and the University of Isfahan, Isfahan, Iran, in 2015 and 2018, respectively (ranked in the top two students). He is a PhD candidate of Biomedical Engineering at the Universitat Polit'ecnica de Catalunya-Barcelona Tech (UPC), Barcelona, Spain. He won a four-year scholarship from the Ministry of Economy and Competitiveness (MINECO) in Spain in 2019. His research interests include electromyographic signal processing, optimization, medical data mining, and machine learning.

Resham Raj Shivwanshi, (PhD Pursuing)

Resham Raj Shivwanshi, (PhD Pursuing) is a research scholar at the Department of Biomedical Engineering, NIT Raipur. He is currently working on medical imaging, CT scan analysis, machine learning and artificial intelligence methodologies. He completed his Masters in embedded system design from the Center for Development of Advanced Computing (CDAC), Mohali, and worked on wearable wireless sensor technologies. He also spent a year as an MTech Trainee in the Department of Biomedical Instrumentation at the Central Scientific Instrument Organization, (CSIO, Chandigarh) and worked upon wireless body-worn sensors for the mobile healthcare monitoring system.

Dr J Sridhar

Dr J Sridhar received a Bachelors Degree in Computer Science and Engineering in 2012 from Kalasalingam University, Krishnankovil, and a Masters Degree in Computer Science and Engineering in 2014 from Anna University, Chennai. He completed a PhD degree in Computer Science and Engineering from Bharath Institute of Higher Education and Research, Chennai, Tamilnadu, India in 2021. Since December 2021, he has been with the Faculty of Engineering, Siddharth Institute of Engineering and Technology, where he is currently an Associate Professor and Researcher in the Computer Science and Engineering Department. He has published articles in various international journal publications and attended conferences from various institutions. He has published a patent. His research interests include chronic liver disease, algorithms of enhanced Grass Hopper optimization, dimensionality reduction, nonlinear correlation parameters, niche mechanisms, and tools using Python.

Dinesh Kumar V

Dinesh Kumar V received a PhD degree in Electronics and Communication Engineering from the Indian Institute of Science (IISc), Bangalore in 2005. He is currently a professor in the department of Electronics and Communications Engineering. He was a visiting foreign researcher in Japan from 2006 to 2009. Prior to this, He was employed in Ministry of Communication and IT Government of India through Indian Engineering Services (IES). His research interests include Analytical and computational electromagnetics, microwave circuits, antenna and wave propagation, applied photonics, and THz and optical communication. He is a senior member of IEEE and a member of the Antenna and Propagation Society, Communication Society, and Photonics Society.

Dr Amit Vishwakarma

Dr Amit Vishwakarma received a BE degree from the Takshila Institute of Engineering and Technology, Jabalpur, India, in 2011, and an MTech degree from PDPM Indian Institute of Information Technology, Design, and Manufacturing, Jabalpur, in 2014. He pursued a PhD degree from IIT Guwahati, Guwahati, India. Currently, he is working as a faculty member in the discipline of Electronics and Communication Engineering, at PDPM Indian Institute of Information Technology, Design and Manufacturing (PDPM IIITDMJ) Jabalpur, India. His current research interests include computer vision, sparse representation, signal processing, pattern recognition, and machine and deep learning.

Shivam Vyas

Shivam Vyas received a BE degree in Electronics and Communication Engineering from the Acropolis Institute of Technology and Research, Indore, India, in 2020, and is currently pursuing an MTech in Mechatronics Engineering from Pandit Dwarka Prasad Mishra Indian Institute of Information Technology Design and Manufacturing, Jabalpur. His research interests include robotics, industrial automation, and computer vision, signal processing applications in biomedical fields, EEG and EMG signals, machine learning and deep learning.

Shadi Zamani

Shadi Zamani received the BS and MS degree in Biomedical Engineering from the University of Isfahan, Isfahan, Iran, in 2019 and 2022, respectively. Her current research interests include machine learning, pattern recognition, data mining, and medical image processing.

Advanced Signal Processing for Industry 4.0, Volume 1
Evolution, communication protocols, and applications in manufacturing systems
Irshad Ahmad Ansari and Varun Bajaj

Chapter 1

Robotics vision for industrial automation

Shivam Vyas, Irshad Ahmad Ansari and Varun Bajaj

This chapter describes the role and importance of computer vision in industrial applications. First, it describes the different aspects and applications of computer vision such as code scanning, machine control and so on, and the way it will be helpful in making industries smarter, more productive, more reliable and more efficient. Then a colour-based sorting system is implemented that is able to detect the colour of objects and sort it out accordingly. Also, a demo of colour-based apple sorting is implemented based on industrial sorting. The tools used are raspberry pi, OpenCV Python.

1.1 Introduction

Industrial automation is the process of controlling mechanical, electrical computational, chemical, or hydraulic machinery to produce a given product and to automate all the processes by means of autonomous systems. This is automated through the integration of measurement, control, and computer software [1].

There are many aspects in the field of industrial automation that are subject to implementation of computer vision like code scanning, robotic guidance, machine control, and sorting. These areas will be described in detail in the upcoming sections.

1.1.1 Code scanning

Code scanning, which is normally carried out by either a camera or a laser scanner, is the process of decoding the information encoded in binary format. There are many other encoding systems, but the two that we use the most frequently in daily life are the QR code and the barcode. The codes are typically used for inventory management, administrative operations, and tracking purposes [2].

1.1.1.1 QR code
A QR code consists of a two-dimensional grid that can store information. The grid is a square matrix and consists of small squares inside which are either black or white,

Figure 1.1. A simple QR code.

Figure 1.2. A linear barcode.

see figure 1.1. The size of the grid defines the amount of data stored in it; a small grid holds less information whereas a large grid holds a considerably high amount of information in it. A QR code is created with the help of an encoder which encodes the supplied data, and a scanner is used to decode the QR code and retrieve the data. Initially it was developed to track the parts of a vehicle in manufacturing but later on with time has found many applications [3].

1.1.1.2 Barcode

A barcode consists of a sequence of white and black stripes. The information is encoded in the width of the stripes. The code comprises of several segments with an equal number of stripes, where different characters are encoded by different segments [4]. The most commonly used barcode is the linear bar, see figure 1.2.

1.1.2 Robotic guidance

A robot can be guided with the help of a camera or computer vision. Considering the present scenario, robots are of two types; the first is the robotic arm and the second one is mobile robots. Presently in automobile manufacturing companies a lot of work like welding, joining parts and painting vehicles is done by robotic arms [5], as these tasks are tedious and require a lot of precision. Robotic arms can do these tasks without any stoppage and with better precision as compared to a human. Mobile robots are also replacing conventional tools as they are far better in terms of consistency, precision, and cost effectiveness. Mobile robots needs sensors like an ultrasonic distance sensor to calculate nearby distances and computer vision to monitor the nearby environment. The information generated by these sensors is processed through a feedback control system and required actions are to be taken. With the help of the feedback system the robot can maintain its position and speed very precisely as it gets closer to an object [6].

1.1.3 Machine control

Machine control is the act of controlling a machine or a process, with the help of some sensor or some feedback mechanism [1]. In industries generally controllers like programmable logic controllers (PLCs) are used for automation purposes. A PLC consists of both inputs and outputs; inputs such as temperature, pressure, and velocity are taken from the sensors and based on the logic defines inside the PLC programming. Certain actions are taken, and PLC output can be connected to a motor, an actuator/valve, or some switch depending upon the application. A practical application of machine control is a control valve fitted on a steam line, which is connected to a reactor, and a temperature sensor is fitted on the reactor, which continuously senses the reactor temperature and the controller accordingly adjusts the valve of the steam line to maintain the temperature.

1.1.4 Sorting

Sorting refers to the process of arranging/ordering data in a defined way, such as in increasing/decreasing order, or arranging it in a more meaningful way such that it can fulfil our requirements. In many industries a sorting process is carried out; a practical example is the food products manufacturing industry. For example, a tomato sauce producing company uses sorting to pick only ripe and red tomatoes and reject unripe tomatoes [7]. Sorting can be done in many ways, such as colour-based sorting, size-based sorting, and product-based sorting. In a waste product management industry wastes like plastic, metal, paper, and recyclable and non-recyclable waste are sorted out and processed separately [8, 9].

1.2 Computer vision system

1.2.1 Image representation

This section explains the technicalities of an image and how to represent one. An image is made up of a 2D array of pixels, each of which describes the light at a given point in the image. A pixel can be represented in a variety of ways, and the three most popular in computer graphics are described below.

1.2.2 RGB colour model

The RGB colour model is a three-channel colour model with red, green, and blue values for each pixel. The three hues are then mixed together in various quantities to create a specific colour. Since a pixel typically consists of a red, green, and blue LED, the RGB colour model is extensively employed in display technology because it is straightforward to alter the intensity of each LED. This model has the disadvantage of being sensitive to non-uniform illumination and having a non-linear relationship between colours [10].

1.2.3 HSV colour model

Researchers created the Hue-Saturation-Value (HSV) colour model in the 1970s with the goal of creating a colour model that interprets colour in a way that is more akin to how humans see colour [11]. A HSV colour model consists of three components: hue, saturation and value. The hue component defines the colour portion, saturation defines amount of grey component present, and value and saturation work together and define the intensity of the colour. The HSV colour model can be represented in the form of a cylindrical model, in which hue varies around the circumference of the cylinder where red is at $0°$, green at $120°$, and blue at $240°$. Saturation ranges from 0 to 100 where 0 implies greyscale and 100 implies pure colour. Value also ranges from 0 to 100, where 0 indicates pure black while a colour with value 100 indicates that no black component is present in the colour [12].

1.2.4 Greyscale image

A greyscale image is made up of only black, white, and grey colours, where the intensity of each pixel varies from 0 to 255. 0 represents pure black, 255 represents pure white, and the intermediate values represent multiple levels of grey. greyscale images are often used for image segmentation and image processing purposes [13, 14].

1.2.5 Image segmentation

Image segmentation refers to the process of transforming a picture into a binary image where pixel's value is either 0 or 1, from any kind of representation, such as a greyscale image. This is done to simplify the image and to highlight the important elements present in the image. Image segmentation can be used to find edges, detect objects, and in further image analysis, since it is one of the most significant process in the area of computer vision [1].

1.2.6 Thresholding

Thresholding is a widely used technique for image segmentation. The key concept is to fix a thresholding value for a pixel, where all pixel values exceeding the threshold can be set either to black or white and pixels below the threshold are set to opposite colours. There are many types of thresholding methods, and the most widely used is Otsu's method [15]. A commonly used thresholding technique in image segmenta-tion is the binary thresholding method, which uses a fixed value to differentiate between an object and its background [16], refer to figure 1.3.

1.2.7 Blurring

Blurring is a very common and extensively used method in image processing, used to smooth an image and reduce the noise present. There are various factors which can cause an image to be noisy, such as sensor malfunction, or environmental factors

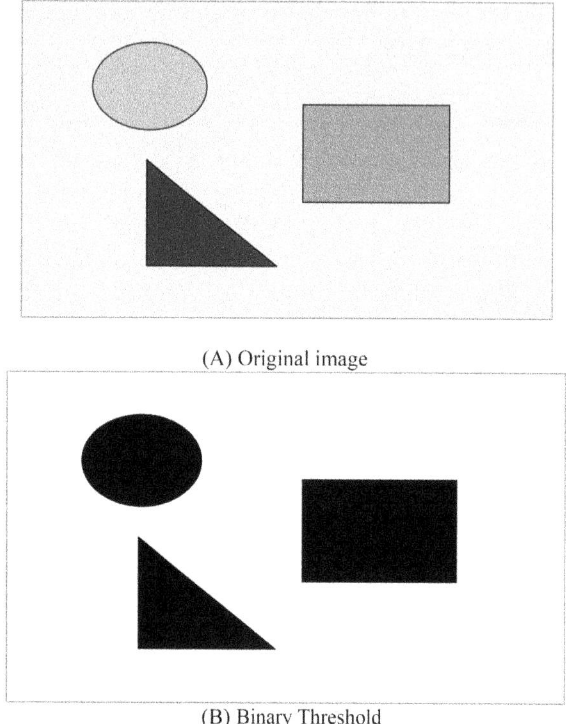

(A) Original image

(B) Binary Threshold

Figure 1.3. Binary thresholding: (A) original image and (B) binary threshold of 150.

like poor lightning, direct lighting, and dust present in the air. Blurring can be related to mixing of a pixel intensity with its neighbouring pixel intensities.

The mostly commonly used filtering methods are Gaussian blurring, median blurring and bilateral blurring.

1.2.7.1 Gaussian blurring

In Gaussian blurring, the neighbouring pixels that are nearer the central pixel give the average more weight. It is preferable to use Gaussian blurring, which utilises a Gaussian kernel rather than a conventional one, when the image contains Gaussian-like noise. The kernel value should be an odd number. As the name implies, noise that roughly follows a Gaussian distribution is removed using the technique known as 'Gaussian smoothing' [17]. An example of Gaussian blurring can be seen in figure 1.4.

1.2.7.2 Median blurring

With median blurring, the centre element is changed to the median value obtained from all the pixels within the kernel area. Median blurring finds its use case mostly in removing salt-and-pepper type noise. Salt-and-pepper noise is a kind of random noise consisting of black and white pixels presented randomly in an image, and in a

Figure 1.4. Image after Gaussian blurring: (A) original image and (B) Gaussian filtered.

coloured image instead of black and white dots, random colour dots might appear, as illustrated in figure 1.5 [17, 18].

1.2.7.3 Bilateral blurring
Bilateral blurring is useful where edge preservation is important, as it removes the noise from the image while preserving the edges (see figure 1.6). Based on a weighted average of the pixels around it, the filter determines a value for every pixel. The weight is determined by the intensity difference and how close the neighbouring pixels are to one another [19].

1.2.8 Edge detection

Edge detection refers to the process of identifying boundaries of an object in an image. There are many algorithms available to identify edges, but the base method for detecting edges in an image is to detect certain changes in values of pixel intensities or where a sudden change in brightness occurs. It can be used to detect

Figure 1.5. Image after median blurring: (A) original image and (B) median filtered.

different objects in an image, as seen in figure 1.7, to separate background and foreground, or to identify shapes of objects [14].

1.2.9 Object detection

Identifying items in a digital image, such as automobiles, people, or a workpiece, is the idea behind object detection. It combines localisation with classification with the goal of identifying attributes, such as an object's shape and colour [20].

To detect an object, an image is passed through several steps. For example, if our aim is to detect a square in an image, then our first step is to convert that image to a greyscale image, then convert the greyscale image to a binarized form, since in a

Figure 1.6. Image after bilateral blurring: (a) original image and (b) bilateral filtered.

binary image there are only two colours (black and white) and therefore it would be very easy to detect edges in that image. In the next step the square box can be separated out from the enclosed objects by setting a certain range of criteria [21]. The steps can be visualised in figure 1.8.

1.2.10 Region of interest (ROI)

During image processing in computer vision applications a region of interest (ROI) is to be defined. Selecting an ROI in an image means selecting an area on which image processing is to be applied. Areas other than the ROI are of no use and can be excluded, and only relevant parts can be considered for further processing. This process can be very helpful, saving a lot of computational resources and a lot of time, as the processing is to be done on a selected region only [22].

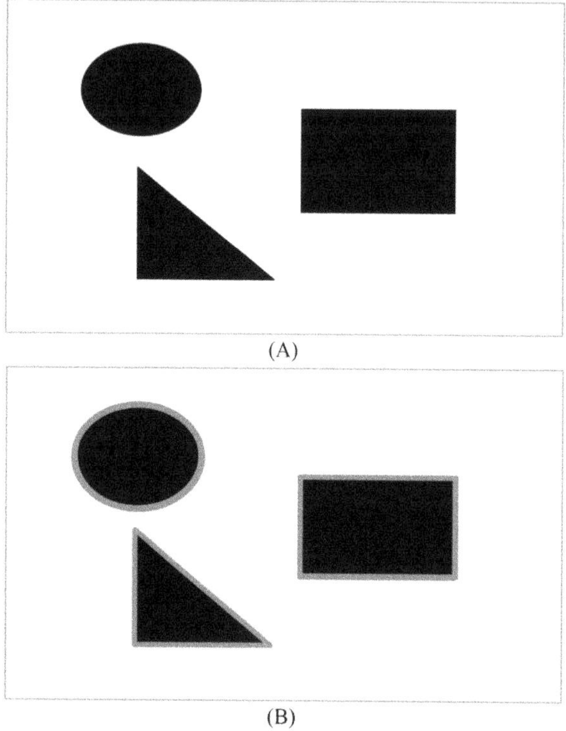

Figure 1.7. Edge detection: (A) original image and (B) edges detected with red colour in original image.

1.3 Applications of vision system

1.3.1 Vision controlled robotic arm

The development of technology is also causing industries to get more high-tech and modern, ND with the development of modern industries, the applications of modern machineries and robotic technologies are increasing rapidly. In particular, robotic arms have gained a lot of applications in industries. These arms are capable of doing heavy and repetitive tasks which a human cannot do easily [23]. Currently in most of the automobile manufacturing companies robotic arms are replacing humans because a lot of power and precision is needed to manufacture the parts and components of a vehicle [24, 25]. A robotic arm consists of links and joints, and can be described by transformation matrices with the help of a Cartesian coordinate system and joint angles provided by the arm. The mathematical modelling of a robotic arm is classified as one of two types: forward kinematics and inverse kinematics. Forward kinematics takes joint angles as an input and calculates the position and orientation of the end point, whereas inverse kinematics takes the orientation and position of the end effector as an input and provides the joint angles as an output. In a vision-based pick-and-place manipulator the robotic arm has a camera in it, which is used to calculate the position of the object which is to be picked [26, 27]. The coordinates or the position of the object is calculated with the

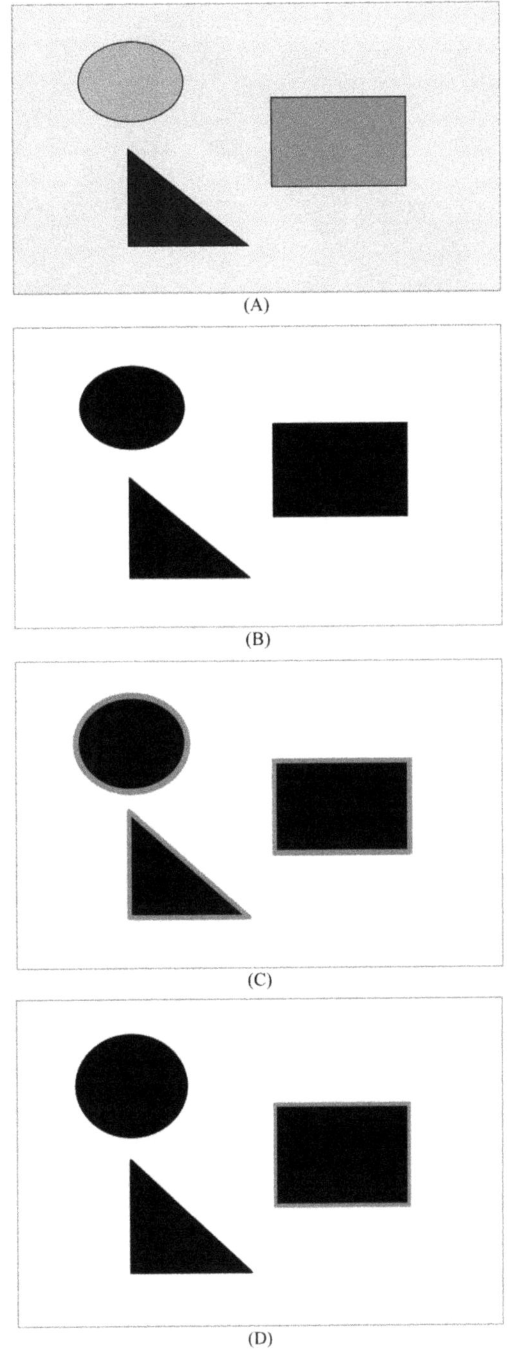

Figure 1.8. Object detection: (A) original image, (B) segmented image, (C) edge detection and (D) object detection.

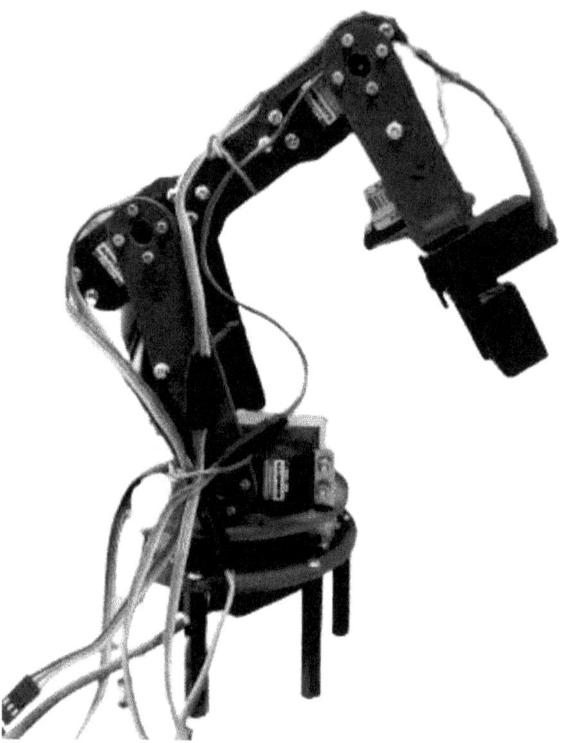

Figure 1.9. Camera mounted robotic arm. Image Courtesy-Mechatronics Lab PDPM IIITDMJ.

help of image processing algorithms and then with the help of inverse kinematics the joint angles are calculated and provided to the arm to reach the object [28, 29]. A vision controlled robotic arm is shown in figure 1.9.

1.3.2 In manufacturing and mining

1.3.2.1 Quality control

Quality control is a very important part of every industry, as it regulates and maintain the quality of the product. It is the process of ensuring that manufactured items fulfil the manufacturer's quality standards. The conventional method of quality control is to personally verify a sample in a batch to confirm the overall batch quality. When it comes to the pharmaceutical industry it is very important to maintain the product quality as it can effect human health. Therefore, high precision computer vision systems are used in their manufacturing line, which assures the quality and grade of the product, and if any kind of error or defect is found in the product it will be rejected [30–32].

1.3.2.2 Intelligent monitoring systems

Companies can undertake remote inspections of their sites and assets with the help of drone-assisted technology. This use of computer vision is crucial in the mining industry, which is dangerous for employees and requires operators to gather visual data in challenging locations. Operators have been able to cut back on routine site visits by half thanks to visual inspections of well sites utilising the Osprey Reach system [33].

1.3.3 In industry application

1.3.3.1 Warehouse management system

A warehouse management system's primary role is to track inventory arrivals and departures. From there, capabilities such as documenting the specific placement of merchandise inside the warehouse, maximising the use of available space, and coordinating tasks for optimal efficiency are implemented. It is a collection of policies and practises designed to organise the work of a warehouse or distribution centre and guarantee that such a facility can run efficiently and accomplish its goals. Utilising computer vision technology in inventory management can reduce a lot of effort and can cut time from hours to minutes, resulting in significant operational cost reductions. One of these technologies is the Gather AI platform, which links to IoT devices and employs drones to scan and count goods. Another example is Amazon, which claims that the Pegasus robot technology it is implementing at its sorting facilities would increase sorting accuracy by 50% [24, 34].

1.3.3.2 Safety-monitoring solutions

Safety monitoring is a subset of error detection and recovery where the malfunction poses a safety risk. When the automated system sensors indicate that a safety situation has evolved that might be dangerous to the equipment or individuals in the area of the equipment, decisions must be made. The goal of the safety-monitoring system is to recognise hazards and take the right action to eliminate or decrease them. This may just involve pausing the operation and notifying maintenance staff of the problem, or it may entail a more sophisticated series of procedures to resolve the safety issue. Computer vision assisted in maintaining public space safety during the pandemic by identifying sick staff or pupils and keeping track of social distance or exposure times. Computer vision systems like IRIS, which monitor behaviour-based safety on the production floor and at construction sites, can safeguard workers by warning machine operators of impending danger [35].

1.4 Proposed work

1.4.1 Proposed model

A colour-based sorting system is implemented in the proposed model, which identifies the colour of the object using a basic sorting algorithm and sorts it accordingly. In the proposed work two models are implemented; the first model separates out red and green coloured objects, while the second model separates red, green, and blue coloured objects.

1.4.2 Model design

In order to implement a computer vision-based colour sorting machine, it is important to understand its basic functionality and working. There are four main components in this sorting machine. These components are the conveyor belt, camera module, raspberry pi as an image processor, and servo motor for separating out the objects.

1.4.2.1 Conveyor belt

A conveyor belt is used to carry an object from one location to another, and is a part of the belt conveyor system. It consists of two or more pulleys, with a closed loop of carrying medium which rotates about the conveyor belt [36]. Refer to figure 1.10.

1.4.2.2 Camera module

A camera module is an image capturing device, which is used to take images of the objects whose colour is to be detected. In this model a Logitech C270 webcam camera system has been used to capture images. The reason behind using this camera is that the image quality of the pictures obtained from this camera is of very high quality with a resolution of 720p. Refer to figure 1.11.

1.4.2.3 Raspberry pi

A raspberry pi is a low-cost computer system, which is used to create hardware projects, projects related to image processing, and home and industry automation. It works on the Linux operating system, and it provides a set of general input/output (GPIO) pins which allow a user to take input from a sensor or can be used to connect a motor as an output. Refer to figure 1.12.

1.4.2.4 Servo motor

A servo motor is a custom DC motor which uses DC power for its operation and a pulse width modulating (PWM) signal to change its position. The advantage of using a servo motor is that the position or angle of the tool attached to the servo

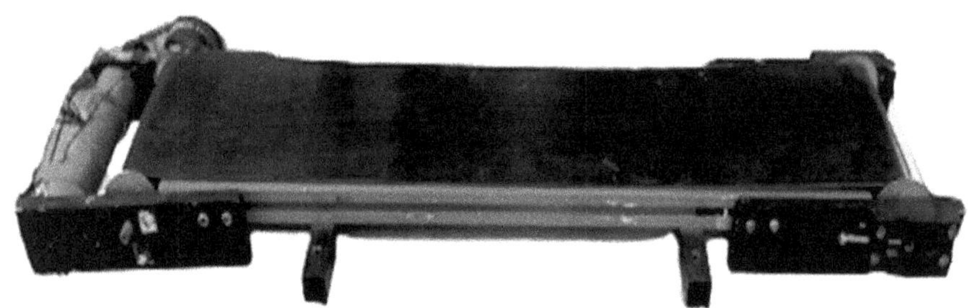

Figure 1.10. Conveyor belt. Image Courtesy-Mechatronics Lab PDPM IIITDMJ.

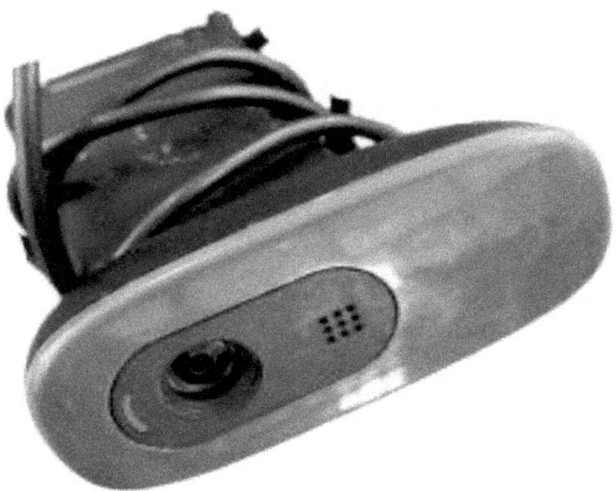

Figure 1.11. Camera module. Image Courtesy-Mechatronics Lab PDPM IIITDMJ.

Figure 1.12. Raspberry pi. Image Courtesy-Mechatronics Lab PDPM IIITDMJ.

motor can be changed or can be set easily by changing the width of the PWM signal. Generally the frequency of a PWM signal is 50 Hz. Refer to figure 1.13.

1.4.3 Working (model-I)

The process starts with placing an object on to the conveyor belt. The placed object passes through a camera module system which continuously captures images of the object. Then the captured images are processed by the image processor. First, the RGB colour format image is converted into a HSV colour format image and then a sample from the centre of the image is taken out to extract its colour properties. Values of hue, saturation, and value are taken out from the pixel's colour properties,

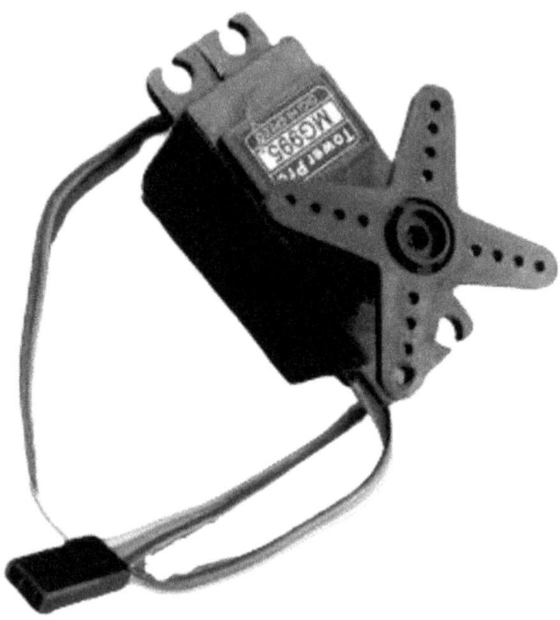

Figure 1.13. Servo motor. Image Courtesy-Mechatronics Lab PDPM IIITDMJ.

then these HSV values are compared to a range of standard HSV colour values obtained from the HSV colour format. After comparison a known colour is obtained and then accordingly the angle of the servo motor is set such that the object reaches its predefined location.

1.4.3.1 Using servo motors

The total length of the conveyor belt is 88 cm and the distance between the camera and the servo motor is 30 cm. It takes around 2 s for the object to travel from camera to servo motor. When a red object passes through the camera, the raspberry pi sends instructions to the servo motor and the servo motor actuates after 2 s, separating the red object from the conveyor. When a blue object is detected it allows the object to pass through and blue objects are collected at the end of the conveyor belt, and hence red and blue objects get separated. Refer to figure 1.14.

1.4.3.2 Using robotic arm

A robotic arm is a sort of mechanical arm that is generally programmable and performs duties comparable to a human arm; the arm may be the sum total of the mechanism or it may be part of a larger robot. The manipulator's linkages are joined by joints that allow either rotating motion (as in an articulated robot) or translational (linear) displacement. The manipulator's linkages may be thought of as a kinematic chain. The end effector is the terminal of the manipulator's kinematic chain, and it is equivalent to the human hand. A 5 degrees of freedom (DOF) robotic arm is used to separate red and blue objects. The distance between camera and robotic arm is 20 cm,

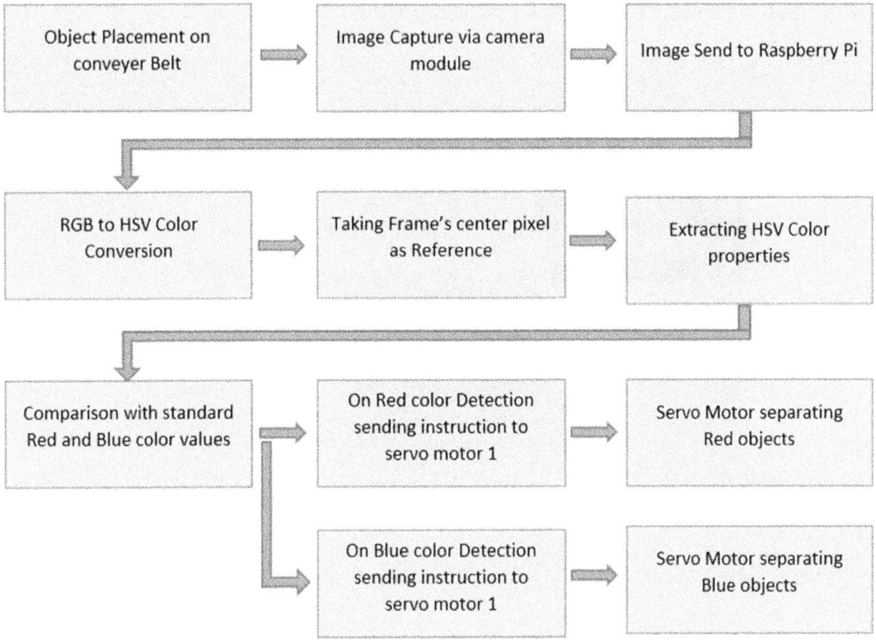

Figure 1.14. Working block diagram-I.

and it takes around 1.2 s for an object to reach near the robotic arm. If the detected object is blue, the arm displaces that object on the left-hand side of it while if the object is red it displaces it on the right-hand side of the arm. Refer to figure 1.14.

1.4.4 Evolution (model-I)

1.4.4.1 Using servo motors

In this section, red and blue objects are used to determine whether the model is working properly or not. For this the objects are placed on a conveyor that passes through a camera sensor, and the image data will be calculated and sent to the raspberry pi for further processing. The algorithm inside it will check whether the object's colour is red or blue and accordingly it commands the servo to separate out the object. Refer to figures 1.15 and 1.16.

1.4.4.2 Using robotic arm

A 5 DOF robotic arm is used to separate red and blue objects. The distance between camera and robotic arm is 20 cm, and it takes around 1.2 s for an object to reach near the robotic arm. If the detected object is blue, the arm displaces that object on the left-hand side of it while if the object is red it displaces it on the right-hand side of the arm. Refer to figure 1.17.

- **Red and blue object detection and separation**

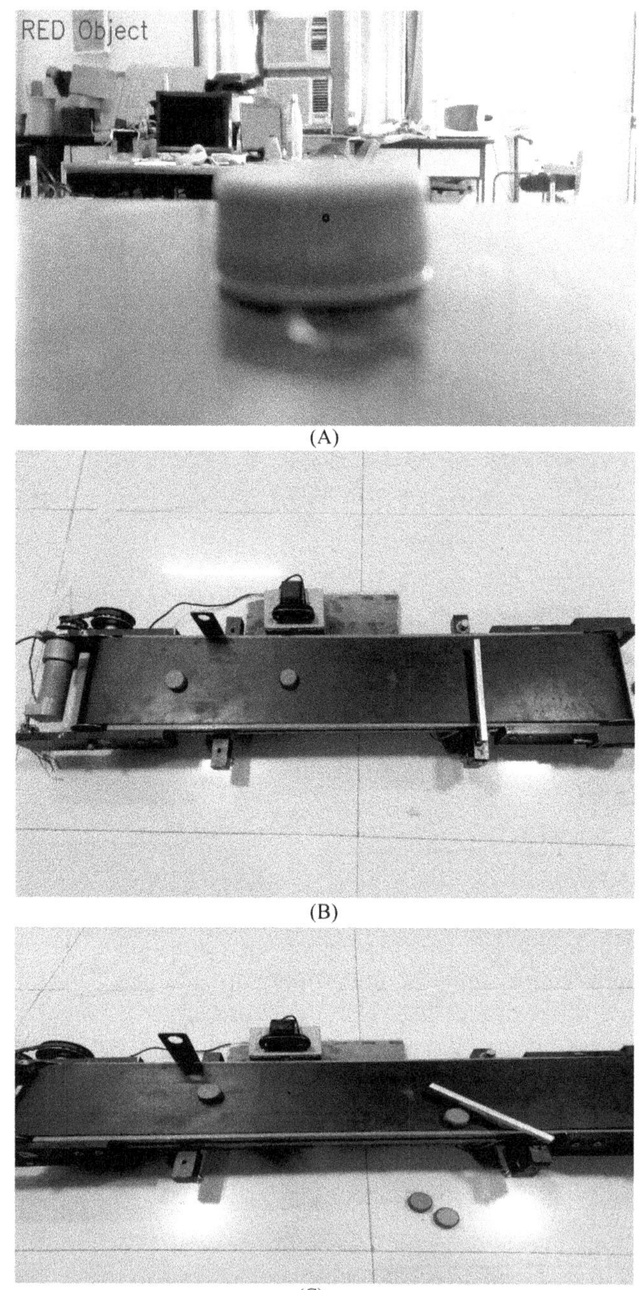

(A)

(B)

(C)

Figure 1.15. Red object detection and separation: (A) object, (B) belt, (C) separation mechanism. (Courtesy: Mechatronics Lab IIITDMJ.)

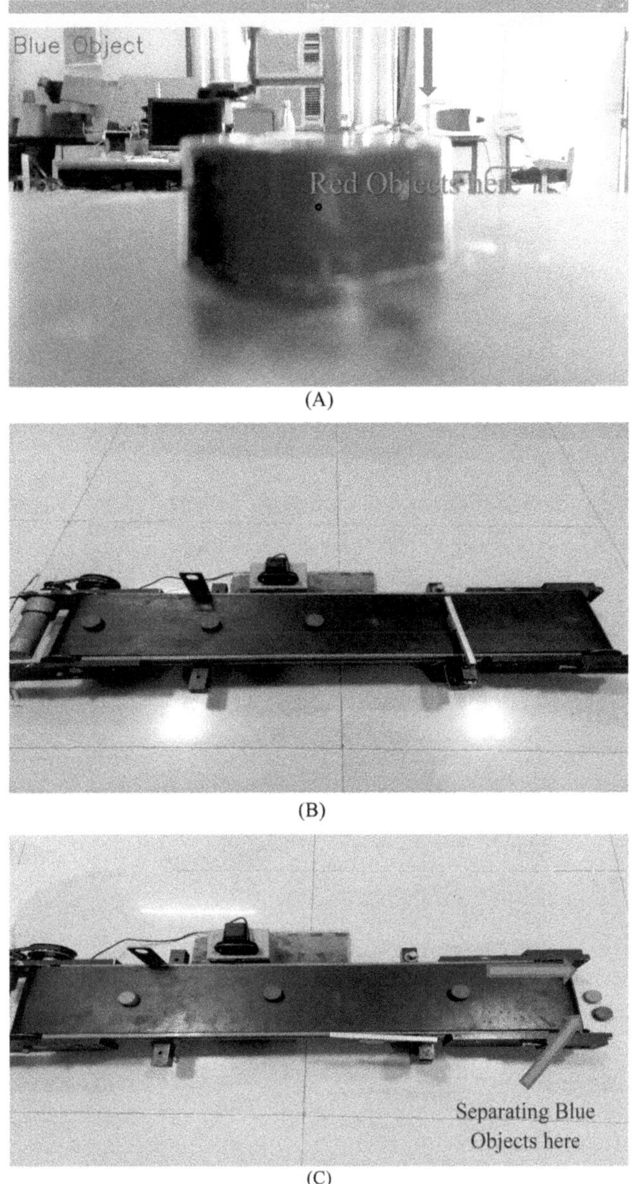

Figure 1.16. Blue object detection and separation: (A) object, (B) belt, (C) separation mechanism.

1.4.5 Working (model-II)

The process starts with placing an object on to the conveyor belt. The placed object passes through a camera module system which continuously captures the images of the object. Then the captured images are processed by the image

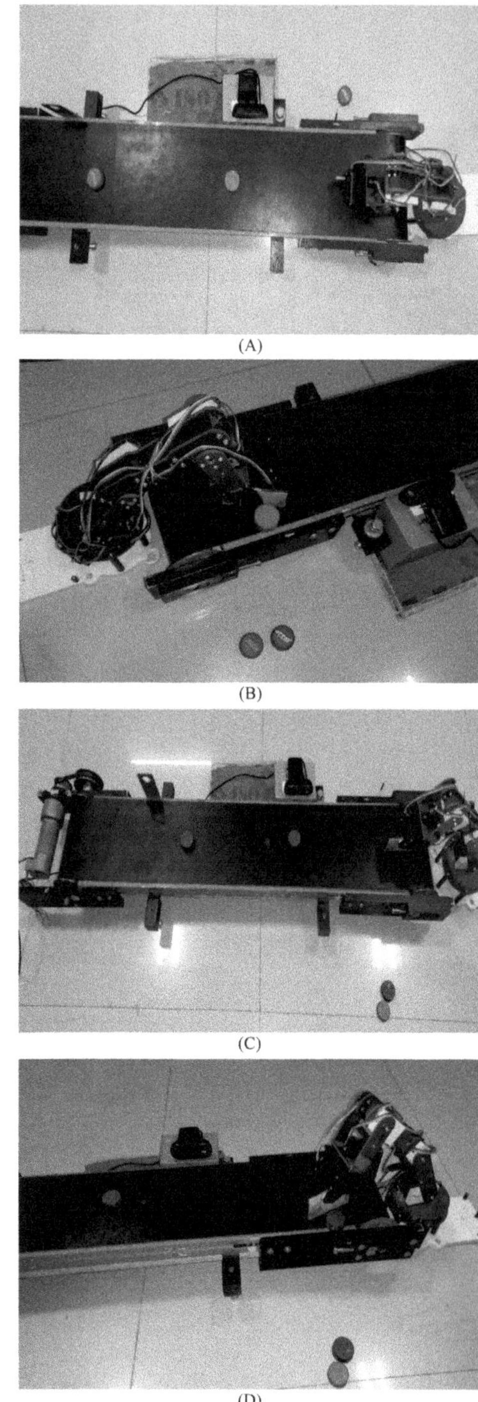

Figure 1.17. Red and blue object detection and separation: (A) red object detection, (b) red object separation, (c) blue object detection and (D) blue object separation. (Courtesy: Mechatronics Lab IIITDMJ.)

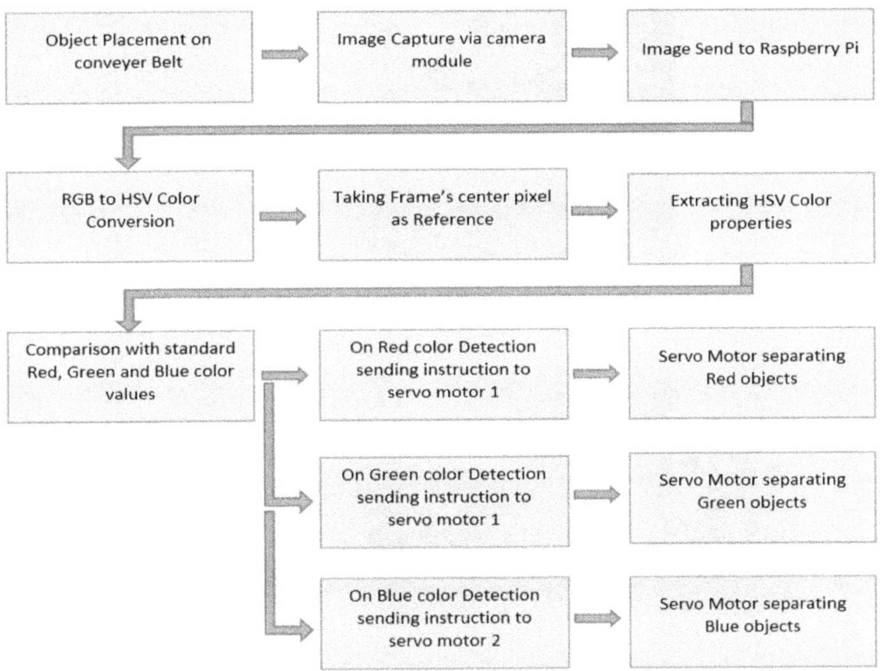

Figure 1.18. Working block diagram-II.

processor. First, the RGB colour format image is converted into a HSV colour format image and then a sample from the centre of the image is taken out to extract its colour properties. Values of hue, saturation, and value are taken out from the pixel's colour properties, then these HSV values are compared to a range of standard HSV colour values obtained from the HSV colour format. After comparison a known colour is obtained and then accordingly the angle of the servo motor is set such that the object reaches its predefined location. Refer to figure 1.18.

1.4.5.1 Using servo motors

When a red object passes through the camera, the raspberry pi commands both servo motors to pass the red object from the conveyor, while when a blue object gets detected it commands the servo1 to separate the object from the conveyor and when a green objects is detected servo2 separates the object from the conveyor, hence red, green, and blue objects gets separated. Refer to figures 1.19–1.21.

1.4.6 Evaluation (model-II)

In this section, red, green, and blue objects are used to determine whether the model is working properly or not. For this, the objects are placed on the conveyor and

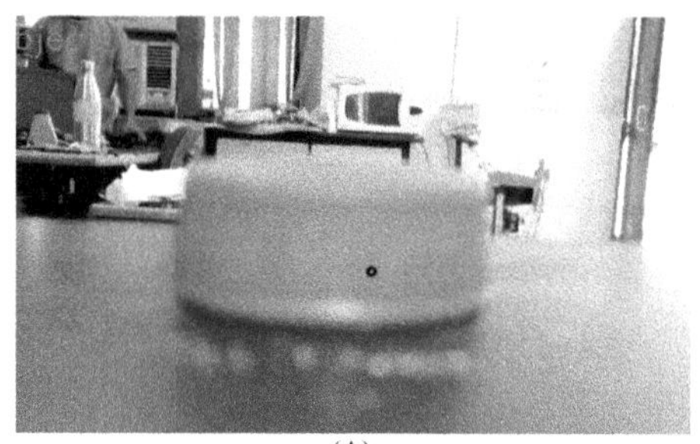

(A)

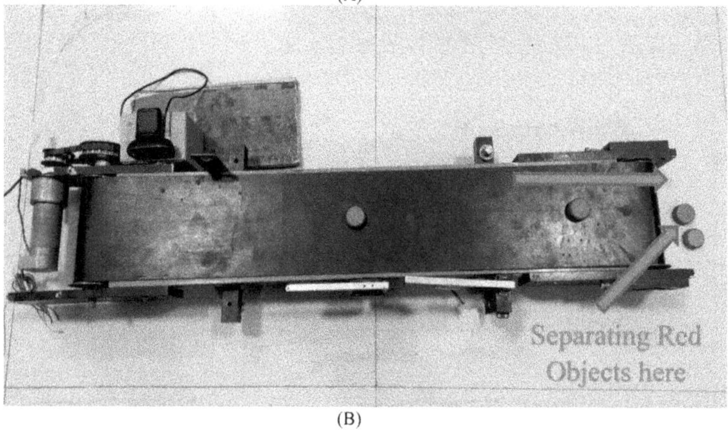

(B)

Figure 1.19. Red object detection and separation: (A) object and (B) belt. (Courtesy: Mechatronics Lab IIITDMJ.)

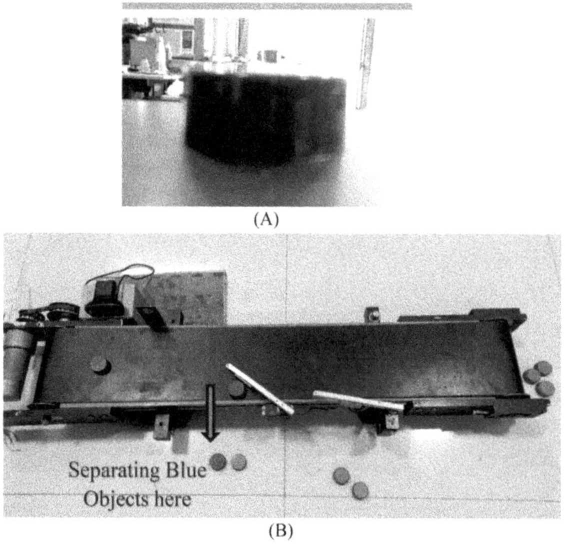

(A)

(B)

Figure 1.20. Blue object detection and separation: (A) object and (B) belt. (Courtesy: Mechatronics Lab IIITDMJ.)

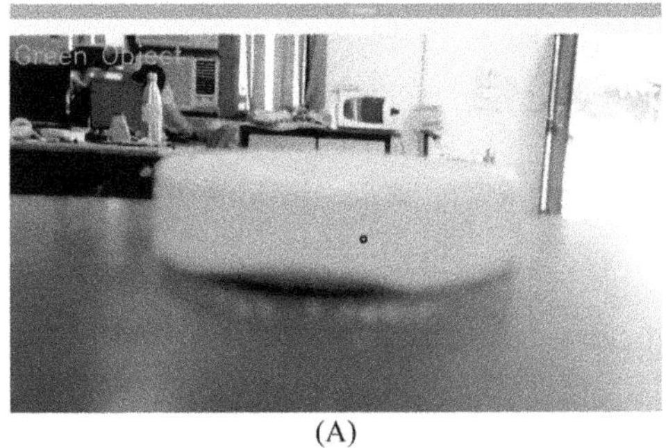

(A)

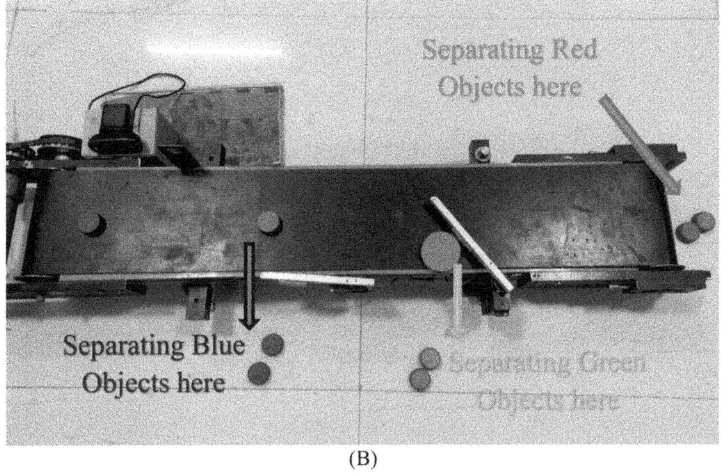

(B)

Figure 1.21. Green object detection and separation: (A) object and (B) belt. (Courtesy: Mechatronics Lab IIITDMJ.)

passed through camera sensor. The image data is calculated and sent to the raspberry pi for further processing. The algorithm inside it checks whether the object's colour is red, green, or blue and accordingly it commands the servo to separate out the object.

1.5 Industrial demo (apple sorting)

A demo example of colour-based apple sorting is implemented based on sorting done in the food industry. The process implemented is very much similar to the process implemented in examples above.

In this process, apples of two colours, red/dark red and yellowish/greenish colours, are passed via a camera system on a conveyor belt. The images of apples are continuously captured by the camera and monitored with a raspberry pi system. If the colour of apple is found to be red, then the red apple is collected at the end of the conveyor belt, and if the colour of the apple is found to be yellow/green then the apple is separated by a servo motor to the side of the conveyor belt. Refer to figures 1.22 and 1.23.

1.5.1 Yellow apple sorting

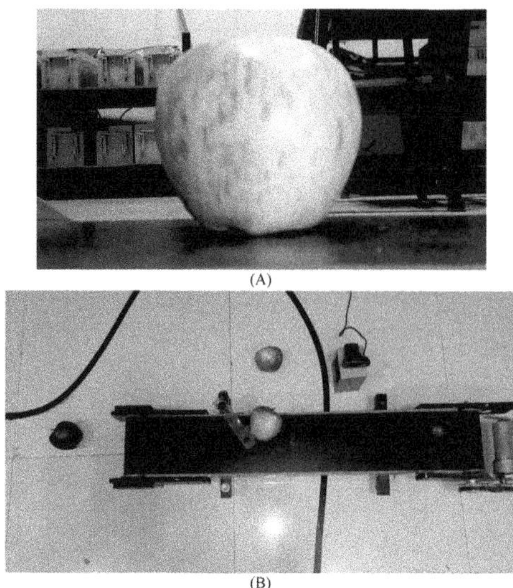

Figure 1.22. Yellow apple detection and separation: (A) object and (B) belt. (Courtesy: Mechatronics Lab IIITDMJ.)

1.5.2 Red apple sorting

Figure 1.23. Red apple detection and separation: (A) object and (B) belt. (Courtesy: Mechatronics Lab IIITDMJ.)

1.6 Conclusion

This chapter gives a brief overview about the role of computer vision in industries, and has implemented an autonomous colour-based sorting system using a servo motor for object separation. A camera was installed on the conveyor system for object image capturing and a raspberry pi module was used for image processing and motor control. The objectives of the project were (see also figure 1.24):

- To determine the object's specification using image processing techniques.
- To extract colour properties of the object from images.
- To separate the object based on colour.

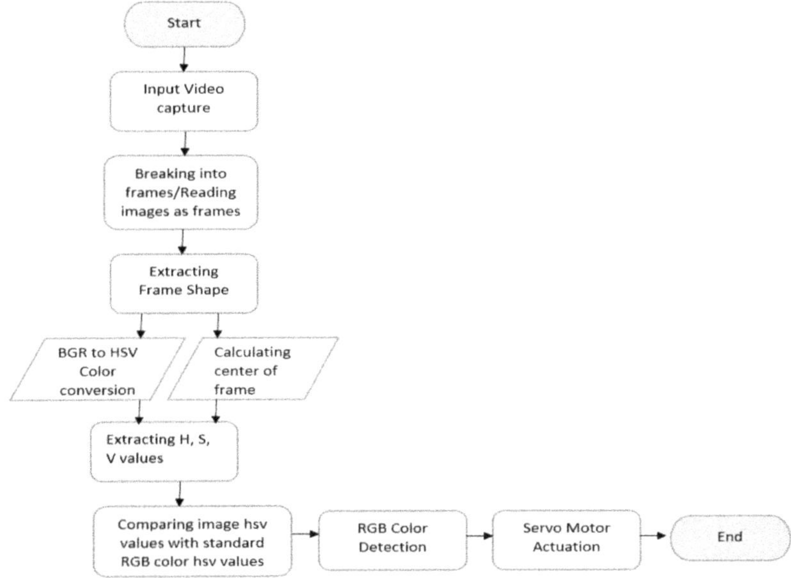

Figure 1.24. Working algorithm flow-diagram.

References

[1] Berglund J and Cedergren J. 2021 *Real-time Computer Vision in Industrial Automation, Master's thesis* Lund University https://lup.lub.lu.se/student-papers/

[2] Fernandez-Carames T M and Fraga-Lamas P 2018 A review on human-centered IoT-connected smart labels for the Industry 4.0 *IEEE Access* **6** 25939–57

[3] Tiwari S 2017 An introduction to QR code technology *Proc.—2016 15th Int. Conf. Inf. Technol. (ICIT 2016)* **1** 39–44

[4] Pavlidis T, Swartz J and Wang Y P 1990 Fundamentals of bar code information theory *Computer* **23** 74–86

[5] Ali M H, Aizat K, Yerkhan K, Zhandos T and Anuar O 2018 Vision-based robot manipulator for industrial applications *Procedia Comput. Sci.* **133** 205–12

[6] LaValle S M 2006 *Planning Algorithms* (Cambridge: Cambridge University Press)

[7] Henry T, Laurence I and Jie F 2017 Design and construction of colour sensor based optical sorting machine *Proc. 2017 5th Int. Conf. Instrum. Control. Autom. (ICA 2017)* 36–40

[8] Mehta S S and Burks T F 2014 Vision-based control of robotic manipulator for citrus harvesting *Comput. Electron. Agric.* **102** 146–58

[9] Andi A F, Nuha H H and Abdurohman M 2021 Fruit ripeness sorting machine using colour sensors *2021 Int. Conf. Intell. Cybern. Technol. Appl. (ICICyTA 2021)* 149–52

[10] Yu L S, Chou S Y, Wu H Y, Chen Y C and Chen Y H 2021 Rapid and semi-quantitative colourimetric loop-mediated isothermal amplification detection of ASFV via HSV colour model transformation *J. Microbiol. Immunol. Infection* **54** 963–70

[11] Lies B T, Cai Y, Spahr E, Lin K and Qin H 2018 Machine vision assisted micro-filament detection for real-time monitoring of electrohydrodynamic inkjet printing *Proc. Manuf.* **26** 29–39

[12] Li S and Guo G 2010 The application of improved HSV colour space model in image processing *Proc. 2010 2nd Int. Conf. Futur. Comput. Commun. (ICFCC 2010)* **2** 10–3

[13] Saravanan C 2010 Colour image to greyscale image conversion *2010 2nd Int. Conf. Comput. Eng. Appl. (ICCEA 2010)* **2** 196–9

[14] Hagara M, Stojanović R, Bagala T, Kubinec P and Ondráček O 2020 greyscale image formats for edge detection and for its FPGA implementation *Microprocess. Microsyst.* **75** 103056

[15] Thresholding (image processing)—Wikipedia https://en.wikipedia.org/wiki/Thresholding_(image_processing)

[16] Hu Q, Hou Z and Nowinski W L 2006 Supervised range-constrained thresholding *IEEE Trans. Image Process.* **15** 228–40

[17] Smoothing and blurring | Computer vision https://cvexplained.wordpress.com/2020/05/26/smoothing-and-blurring/

[18] Brownrigg D R K 1984 The weighted median filter *Commun. ACM* **27** 807–18

[19] Liu C and Pang M 2021 One-dimensional image surface blur algorithm based on wavelet transform and bilateral filtering *Multimedia Tools Appl.* **80** 28697–8711

[20] Solovyev R, Wang W and Gabruseva T 2021 Weighted boxes fusion: ensembling boxes from different object detection models *Image Vis. Comput.* **107** 104117

[21] Hoang N D 2020 Image processing-based spall object detection using gabor filter, texture analysis, and adaptive moment estimation (ADAM) optimized logistic regression models *Adv. Civil Eng.* **2020** 8829715

[22] Yeum C M, Choi J and Dyke S J 2018 Automated region-of-interest localization and classification for vision-based visual assessment of civil infrastructure *Struct. Health Monit.* **18** 675–89

[23] Sangeetha G R, Kumar N, Hari P R and Sasikumar S 2018 Implementation of a Stereo vision based system for visual feedback control of Robotic Arm for space manipulations *Proc. Comput. Sci.* **133** 1066–73

[24] Zaera M, Esteve M, Palau C E, Guerri J C, Martínez F J and Fernández de Códoba P 2001 Real-time scheduling and guidance of mobile robots on factory floors using Monte Carlo methods under windows NT *IEEE Int. Conf. Emerg. Technol. Fact. Autom. ETFA* **1** 67–74

[25] Farag M, Ghafar A N A and Alsibai M H 2019 Real-time robotic grasping and localization using deep learning-based object detection technique *2019 IEEE Int. Conf. Autom. Control Intell. Syst. I2CACIS 2019—Proc. (June)* pp 139–44

[26] Liu Y *et al* 2015 Geometric approach for inverse kinematics analysis of 6-DOF serial robot *2015 IEEE Int. Conf. Inf. Autom.* pp 852–5

[27] Wu X and Xie Z 2019 Forward kinematics analysis of a novel 3-DOF parallel manipulator *Sci. Iran.* **26** 346–57

[28] Fang B, Ma X, Wang J and Sun F 2020 Vision-based posture-consistent teleoperation of robotic arm using multi-stage deep neural network *Robot. Auton. Syst.* **131** 103592

[29] Neto P, Pires J N and Moreira A P 2009 Accelerometer-based control of an industrial robotic arm *Proc. — IEEE Int. Work. Robot Hum. Interact. Commun.* 1192–7

[30] Bahaghighat M, Mirfattahi M, Akbari L and Babaie M 2018 Designing quality control system based on vision inspection in pharmaceutical product lines, *Int. Conf. on Comput., Math. and Eng. Technol. (iCoMET)* (Piscataway, N.J., IEEE) 1–4

[31] Kakani V, Nguyen V H, Kumar B P, Kim H and Pasupuleti V R 2020 A critical review on computer vision and artificial intelligence in food industry *J. Agric. Food Res.* **2** 100033

[32] Applications of computer vision in the pharmaceutical industry—viso.ai https://viso.ai/applications/computer-vision-in-the-pharmaceutical-industry/

[33] Top computer vision applications in different industries in 2021—ITRex https://itrexgroup.com/blog/computer-vision-applications-in-different-industries/#

[34] Dzodzo B, Han L, Chen X, Qian H and Xu Y 2013 Realtime 2D code based localization for indoor robot navigation *2013 IEEE Int. Conf. Robot. Biomimetics (ROBIO 2013, December)* 486–92

[35] Shan S E, Faisal M F, Rezaul Haque S and Saha P 2018 IoT and computer vision based driver safety monitoring system with risk prediction *Int. Conf. Comput. Commun. Chem. Mater. Electron. Eng. (IC4ME2 2018)* pp 28–31

[36] Jakubovičová L, Kopas P, Vaško M and Handrik M 2021 Technical solution of the modern conveyor system *IOP Conf. Ser.: Mater. Sci. Eng.* **1199** 012031

Advanced Signal Processing for Industry 4.0, Volume 1
Evolution, communication protocols, and applications in manufacturing systems
Irshad Ahmad Ansari and Varun Bajaj

Chapter 2

Capnography signal processing in trend with Industry 4.0 advancement

Santheraleka Ramanathan and M B Malarvili

Capnography has been implemented throughout history for detecting airway and pulmonary obstructions in critical care settings. The basic principle of capnography relies on the measurement of carbon dioxide (CO_2) and provides details in terms of ventilations and metabolisms in a spontaneously breathing patient. The analysis of capnogram waveforms reveals multiple features and instantly determines the state of breathing. This chapter details the crux of capnography systems for delivering precise capnogram signals in the advancement of Industry 4.0. The chapter outlines the systematic analyses and prospective tools utilized in literature for incorporating capnography in medical facilities. The use of capnography is extensively reviewed to evaluate the current evidence of capnography efficiency. The vital role of capnography for the remote assessment of respiratory condition is discussed. The evolution of Industry 4.0 has broadened the technology for vibrant capnography signal processing with the aim of developing wireless networking in health facilities.

2.1 Capnography 4.0

Industry 4.0 is a wholesome package which includes the Internet of Things, artificial intelligence, machine learning, big data analytics, cloud computing, etc., and solely upgrades smart manufacturing and application for the latest industrial revolution [1]. Industry 4.0 aims for the flexibility of operation, effective flow of information, real-time outcomes, and upgraded productivity [2]. It is a package of high-performance sensors and advanced signal processing for self-operating and intelligent decision-making technology. The evolution of Industry 4.0 is in great demand for healthcare facilities, where the field of services is enclosed with technology and telecommunication advancement. The contribution of Industry 4.0 in healthcare facilities is associated with traditional medical health, transformation of biological systems, and the virtual world to develop a high efficiency healthcare system. The transformation of biological knowledge into readable and usable evidence improves the quality of medical diagnosis and therapeutics [3]. A remarkable growth in Industry 4.0 is grounded with the development of high-performance sensors, which are applicable in diverse healthcare sectors. The Big Data terminology classified under Industry 4.0 is closely related to sensor technology [4]. The developing sensors are expected to support Industry 4.0 by having massive capacity for data processing and interpretation. Industry 4.0 promises to be a key component of the future medical sector and smart manufacturing, with software, technologies, and processes requiring far less time and money in terms of efficiency and performance [5, 6]. In this chapter, the technology of capnography signal processing in line with the Industry 4.0 is discussed. Among many healthcare applications, capnography has a significant role in ensuring patient safety with prompt diagnostics of airway illness. Additionally, capnography outside the healthcare environment is denoted and the potential for further development with the smart technology is discussed.

2.2 Capnography measurement and physiology

Capnography has been well received since the late 1980s, where it was highly approached for anesthetic practices. Anesthesiologists have embraced the role of capnography as the implementing tool for standard monitoring during anesthesia. Capnography is used in medical facilities for monitoring and enhancing patient safety. In current medical practices, capnography is compulsory in the medication of laparoscopic surgeries. The developed countries around the world have affirmed the application of capnography in laparoscopic surgeries, considering it an integral part of anesthesia [7, 8]. However, capnography was limited in its use in medical healthcare and not appreciated for its significant contribution. For instance, capnography was used in the operation room for patient monitoring, but it was not used in intensive care units (ICUs). Such an act has questioned the use of capnography for continuous intubation or ventilation of patients in various domains. The primary reason of the exceptions for using capnography is the safety of patient monitoring in and outside of healthcare facilities. Nonetheless, the role of capnography has been well recognized inside and outside of medical facilities. It is well implemented for monitoring the physiology and clinical condition of a patient [9, 10]. Along with the development of Industry 4.0, the technology of capnography has been extended to various fields with a specific functional condition for the advancement of future directions. A broad range of sensing technology has been developed for sensing the gas samples for capnography analysis. Infrared sensing technology is the most recognized and applied technology for capnography measurement in medical facilities. Capnography technology is upgraded from time to time and researchers are keen in enhancing capnography performance by increasing the sensitivity and accuracy of the tool and reducing the response time of the capnography signal processing. The sensitivity of a capnography tool is ensured to be superior when it can measure the CO_2 volumes and respiratory rate in premature babies [11].

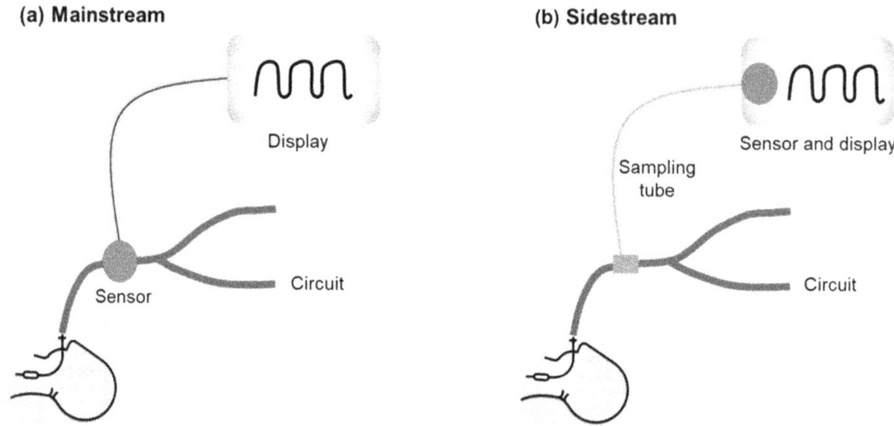

Figure 2.1. Types of capnographs: (a) the mainstream method incorporates a processor and monitoring unit (b) sidestream method incorporating a sample tube, sensor, and monitoring unit.

There are two types of capnography measurement system, which are classified based on the location of the CO_2 measuring sensor (figure 2.1). The first type is mainstream capnography, where the CO_2 measurement system is placed across the tracheal tube and the breathing circuit. The second type is sidestream capnography, where the CO_2 measuring sensor is connected to the breathing circuit. The sampling takes place through an air passage, which is then passed to the sensor through a sampling port. The system delays the CO_2 gas measurement and displaying the capnogram waveform due to the lagging encountered in the airway passage [12, 13]. However, both mainstream and sidestream capnography tools are well implemented in medical and industrial applications, prior to the functional need of a system. Capnogram waveforms for healthy individuals are identical with very little variation in the shape of the waveform. A huge variation in the capnogram waveform is evaluated for the physiological changes and its causes. Capnogram waveforms are classified as time capnograms or volume capnograms. The time capnogram is the commonly applied waveform in multiple facilities, where the end-tidal CO_2 is plotted against time. The end-tidal volumes are plotted against the expired volume in the volume capnogram. The time capnogram is segmented into inspiration and expiration [13, 14]. The expiration of the capnogram is divided into three phases (figure 2.2). The phases indicate the passage of expired CO_2 from the lungs. Phase I is known as dead space gas. There is no expired CO_2 present in the phase. Phase II is the expiratory phase, where the volume of expired CO_2 rapidly increases. It indicates the high expiration rate from the alveoli of human lungs. A perfectly steep alveolar phase is attained if the partial pressure of CO_2 gas (P_{CO_2}) is constant throughout the expiration. However, it is rarely encountered due to the spatial and temporal differences in the lungs. The differences result in the variation of ventilation to perfusion (V/Q) ratio. When the V/Q is small, a longer time constant is shown and results in the later part of phase III. At this point, a slight upward slope is shown in

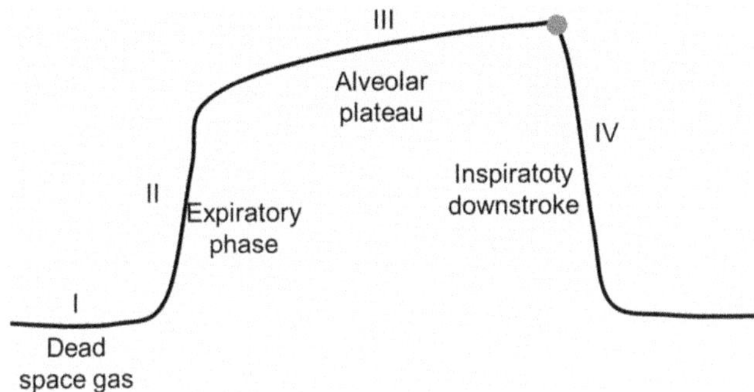

Figure 2.2. Illustration of volumetric capnogram consisting of three phases: phase I, II, III, and the last phase of capnogram waveform, phase IV.

the alveolar phase, which is classified as an alveolar plateau. The alveolar phase of the capnogram waveform indirectly indicates the ventilation and perfusion rate in the lungs. The angle (α angle) between the expiratory and alveolar plateau phase is generally used to determine the physiological state of a patient. Phase I, II, III, and the α angle signify the diagnostic and therapeutic effects of the bronchodilator drugs. The alveolar plateau also shows the ratio of cardiac out to ventilation. For a constant ventilation, the alveolar plateau varies in height, which results in the abstraction of cardiac output. The maximum of the alveolar plateau, which represents the maximum P_{CO_2}, is called the end tidal CO_2 ($EtCO_2$). After the $EtCO_2$ peak, the P_{CO_2} drops gradually, indicating the inspiratory downstroke (phase IV). The angle between $EtCO_2$ and inspiratory is known as the β angle, which is usually 90°. The β angle signifies the occurrence of rebreathing [15]. As per the volume capnogram, the waveform indicates the $EtCO_2$ at various components. Volume capnograms do not indicate the inspiratory phase. Both time and volume capnograms are evaluated based on the differences between P_{CO_2} and $EtCO_2$. The gradient between the P_{CO_2} and $EtCO_2$ represent the interchange between the alveolar phase and the dead phase with zero CO_2 gas. The understanding of waveforms is important for capnography interpretation [16].

2.3 Capnography signal interpretation

The features associated with the time and volume capnograms at the respective phases deliver specific information, which determines the physiological characteristics of respiratory illness, mainly asthma. Time-based capnograms measure the respiratory rate, expiratory CO_2 level, and end-tidal CO_2 ($EtCO_2$). The features are analyzed to understand the time-to-time transition in each breath cycle. In clinical conditions, the possibility of attaining accurate breath cycles and the associated variables is tedious as it is influenced by external factors such as environmental conditions and patient consciousness. At such conditions, volume-based capnogram interpretation delivers quantitative details on the amount of CO_2 gas and the transition between the gas volumes at each phase. Moreover, the volume-based features, such as slopes, angles, peak values, normalized slopes, and many others

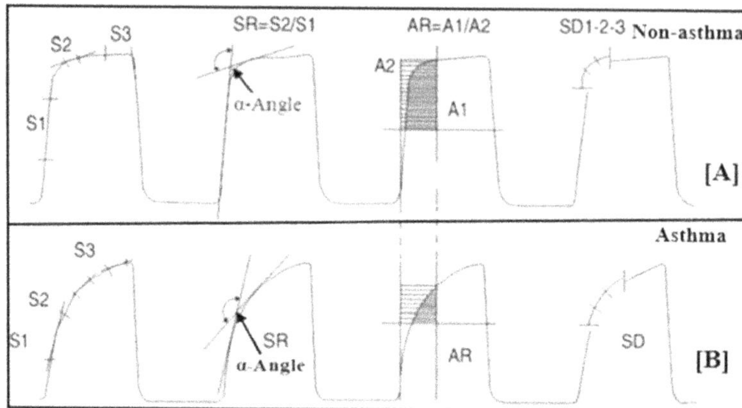

Figure 2.3. Illustration of eight indices extracted from capnograms of non-asthmatic subject [A] and asthmatic patient [B], respectively. S1, S2, and S3 represent the slopes; SR is the slope ratio; A1 and A2 are areas and AR is area ratio; SD indicates double derivatives. (Reproduced with permission from [19]. Copyright 2021 Elsevier B.V.)

provide clear information on the physiological state of asthma of a patient. For instance, figure 2.3 shows the capnogram indexes extracted from asthma and non-asthma patients [17]. Numerous studies have denoted that time- and volume-based capnogram assessments deliver similar terminology in the capnogram waveform, yet the implementations are different, which precisely determines the asthmatic condition of a patient. In present capnography studies, time-based capnograms are referred to as flow waveforms, whereas volume-based capnograms are referred to as CO_2 waveforms. Software is utilized to segmentize breath cycles into flow and CO_2 waveforms, based on the saturation periods and signal disconnections in the capnogram. Real-time asthma monitoring based on capnograms considers factors such as breath initiation, completion interval, inflection point, and threshold amplitudes in the CO_2 waveform. Flow waveforms are tuned with mathematical functions and algorithmic evaluation to exclude invalid findings and deliver appropriate peak values [18]. Time- and volume-based capnogram segmentation plays an important role in the capnography system to interpret the accurate real-time asthmatic conditions.

Capnography is a non-linear quasiperiodic waveform, which reflects the flow of CO_2 varying. Capnography with multiple applications in respiratory and pulmonary assessments requires an appropriate set of criteria and aspects to provide accurate results, regardless of the predominant noises appearing in the system. Capnography waveforms encounter noises arising from bandwidth, and high-frequency components. The effect of the noises will be seen in the capnogram waveform, which results in uneven and blunt end curves. Hence, signal filtering is a critical point in the capnography system, to deliver a sharp and accurate waveform. Filtering is a denoising process, which estimates a clean capnogram signal from its noisy state. The denoiser aims to deliver a clean signal using a denoising method, known as signal filtering techniques. There are several signal filtering techniques applied in denoising the capnogram signals [20]. Low-pass filtering is a common type of denoiser used for high-frequency operating systems. Several systems demand slow dynamics, where the low-pass filters are implemented. The filter then reduces the signal amplifying trend and delivers smooth and palatable signals. Conversely, high-pass filtering is a common denoising technique used in systems that requires shift from a slow-drift to high-drift signal of interest. The filter removes slow noises prior to the requirement of data analysis. It is a standard tool for removing DC components to enable satisfying implementation of limited range digital systems. Besides that, band-pass filtering is a traditional method of data analysis that isolates a particular band and applies the filter. Band-pass filtering interprets capnogram signals based on the frequency bands. Capnogram signal processing usually applies more than one band-pass filter for delivering smooth capnogram waveform [21]. A fixed-coefficient filter is traditionally applied in capnography to suppress the artifacts caused by lung and chest compression. The filter suppresses the spectral content of the capnogram waveform above 1 Hz and develops a digital impulse response. Adaptive filtering is another type of denoiser, which is implemented in capnograms when other types of filters are not able to withstand the artifacts. This filter adapts to the artifact and alters the parameter of the adaptive filters. Filters are ubiquitous in spectrogram studies and an important step to eliminate the noise and reveal the targeted signal. Implementing one or more filters in a capnography system is obligatory, yet the potential effect on the system is critical to be considered for developing novel capnography for asthma monitoring [22].

2.4 Capnography—annotation and classification

Breath annotation is a critical point in capnograms as there are hundreds of breathing patterns that are a few seconds long. Although breath annotation is time-consuming, it demands a significant method to be used by health professionals and implemented in capnography for simple and quick breath analysis. There is no standard method to differentiate normal and abnormal breath. Cohen's Kappa study was one preliminary statistical method applied in differentiating breath types, based on several respiratory conditions such as notched breath, rebreathing, and cardiogenic oscillations. Then, software was developed with automatic integration of algorithms to differentiate the desired and undesired breath cycles in capnography [23]. In breath annotation, the software was developed as it needed to target the first phase and the phase transition in the breath cycle. The remaining phases are consequently labeled according to the inspiration and expiration flowrate. For instance, a MATLAB software designed for breath annotation was able to select breath portions by specifying the starting and ending time intervals in the capnogram. It was also able to eliminate annotated breath and removed it before the signal went for further processing. Literature studies have denoted that software is the reliable and user-friendly method for breath annotation [24]. However, because of the absence of a gold-standard method for breath annotation and a technique to identify normal and abnormal breath, there are endless massive efforts and innovations are being taken to present a reliable and valid breath annotation method for standard capnography conditions. Classification defines the utilization of features extracted from capnogram waveforms to classify into several predefined classes. Classifiers precisely identify the waveform features and subsequently differentiate them into different conditions and hence, evaluate the asthmatic condition of a subject. It is also known as a model, which is developed and improved using software and algorithms for accurate asthma detection. A decision tree is a type of classifier that has been commonly used in the past decade. The classifier targets the induction algorithms and aims to divide data according to features and to grow a tree, then obtain separation of the classes. The class is well known for good generalization and interpretability. Besides that, Naive Bayes is a probabilistic classifier based on Bayes' Theorem that maximizes the class posterior probability, while assuming conditional independence between capnogram features for a given class. K-nearest neighbors is simple algorithm that stores all the available breath segments and classifies the new data or case based on a similarity measure. It is mostly used to classify a data point based on how its neighbors are classified. With the growth of artificial intelligence and deep learning, several sophisticated models have been developed to be used as classifiers in capnography [25]. Artificial neutral networks (ANNs) are a general mathematical functional method used in multiple disciplines for model generation and interpretations. The model is built comprised of many neurons, which are organized and interconnected between layers, and manipulates the input and output variables based on capnogram features. The model parameters are considered to ensure the model works well with new inputs and avoid overfitting. Fast Fourier Transform (FFT) is a fast algorithmic model, which computes signals to successively discrete signals. FFT is mainly applied in capnograms, and analyzes the frequency of the continuous time signal. The periodic capnogram waveform may result in longer segments and an infinite time series, where the FFT is required to attend the signal and result in finite-length waveforms [26]. Besides that, autoaggresive (AR) models are applied in capnography, based on power spectral density estimation. The AR model classifies the power distribution over the function of frequency and then provides the quantitative components based on the power/time series. The AR model is well utilized in biomedical signals, including capnography, yet the inappropriate selection of the power mode in conjunction with the system causes the classifier to not perform at its best [24]. Over the period of advancement, numerous software and deep learning models have been developed for signal classifiers. The type of classifier to be used in the capnography signal is not definite. If the signal meets the mathematical statistics, the classifier may deliver the best outcomes.

2.5 Capnography in medical interpretation

Capnography has been interpreted in medical scenarios based on the three main outputs of capnography signal, which are the shape of the capnogram waveform, the numerical value of $EtCO_2$, and the variation between P_{CO_2} and $EtCO_2$.

The $EtCO_2$ values and the waveform shapes are used as the primary tool for the differential diagnosis. Further analysis of the medical diagnosis using capnography is accompanied by respiratory rate value, heart rate, blood pressure, and partial pressure of oxygen [27, 28]. The accuracy of capnography data analysis is supported and enhanced by the other respiratory readings. This is supported by Ishiwata *et al* (2018). The $EtCO_2$ value from capnography is appropriately optimized to monitor the occurrence of hypoxemia during flexible bronchoscopy. Sedation may be appropriate for the patient during bronchoscopy, yet the risk associated with oversedation results in hyperventilation. It is necessary to monitor the risk and to reduce hypoxemia in patients. During anesthesia, the $EtCO_2$ levels gradually increased. It was later discovered that this was linked to an increase in heart rate and body temperature. A thorough examination of the patient's hemodynamic and respiratory variables, as well as the anesthesia machine, revealed no problems with the anesthetic system or airway obstruction. The study has made strategic execution using $EtCO_2$ values from capnography for hyperventilation monitoring [29]. The tests were conducted by monitoring patients with moderate sedation using capnography and pulse oximetry for comparative analysis. The results concluded patients that underwent capnography showed mild hypoxemic conditions with shorter duration, whereas patients tested with pulse oximetry showed a significant effect on their hypoxemic condition. The precedence of capnography in the early detection of apnea reduces the risk of hyperventilation in sedated patients. In addition, the alarm system in the capnography alerts to the apnea condition using the $EtCO_2$ observed through the visual capnogram waveform and the numerical threshold value [30]. In addition to the numerical values, the gradient of P_{CO_2} and $EtCO_2$ slopes is used in assessing the physiological condition of a respiratory illnesses. A significant difference in the gradient indicates a severe complication in the airway ventilation and perfusion. A constant gradient denotes the stable condition of a patient respiratory airway. As such, capnography is well utilized in medical facilities.

2.5.1 Airway integrity

Adverse respiratory sicknesses encountered during endotracheal intubation is the primary cause of introducing capnography in medical applications. Airway intubation enables medical personnel to handle the miscalculated esophagal intubation and secure the airway ventilation of a patient with difficulty breathing. The proper placement of intubation tubes can be verified by the $EtCO_2$ value obtained through the capnography. The $EtCO_2$ value ensures that the intubation tubes are fixed in major airways. Irrelevant $EtCO_2$ values may indicate that the intubation tube is fixed at the esophagus [31]. In the past decades, a colorimetric capnometer is attached to the end of the intubation tube. The semiquantitative evaluation denotes the presence of CO_2 gas in a patient airway, based on the color change of the colorimetric capnometer. Hereafter, continuous measurement of capnogram waveforms is attained through a capnography tool, which provides the precise quantitative and numerical values of $EtCO_2$. With the advancement of capnography signal processing, the detection of CO_2 gas in the airway of an intubated patient is rapid. Apart from determining the location of intubation tube, capnography shows the waveform of CO_2 gas as in one-third of esophagal intubation [32, 33]. Capnography generally detects the seventh breath from a patient breath. Subsequently, the system removes the CO_2 gas after each successive breath. With this, the precise flow of CO_2 gas flow in an intubated patient is determined. The assessment of intubated patients using capnography with the $EtCO_2$ values may not provide significant output to detect the cruciality of endobronchial mainstream intubation. Hence, the latest technology of capnography has been equipped with the ability of measuring lung auscultation, airway pressures and radiographic elements, which aid the further evaluation of airway morbidity [34].

2.5.2 Asthma

The change in the ventilator is one of the prominent symptoms for asthma detection. The asthmatic condition is denoted by increased respiratory rate, deep inspirations, and increased minute of ventilation. Hypocapnic breathing, which occurs during panic and anxiety situations, leads to asthma exacerbations [35, 36]. Respiratory rate and $EtCO_2$ are the key features for monitoring the asthmatic condition of a patient. Decrement of $EtCO_2$ values is a significant symptom for asthma detection when no hyperventilation symptom is seen in a patient. Due to the significance of the $EtCO_2$ value, capnography has been well established for asthma detection [37]. The variation in the $EtCO_2$ value is categorized as it shows the severity of asthmatic condition of a patient. The capnography monitoring in ambulance provides adjunct diagnosis and treatment for asthma. The performance of capnography has been well studied for asthma detection in the literature. Generally, capnography is compared with spirometer readings [38]. However, a detailed capnogram waveform denotes highly precise outcomes of a respiratory condition, in comparison to a spirometer. A research study compared the capnography with spirometry as a cross-sectional study with 100 asthma patients, 50 cystic fibrosis patients and about 40 healthy volunteers. The outcome of the study was novel in determining the promising capnography and spirometry marker in distinguishing between asthma and cystic fibrosis. The results emphasized that the capnographic index is the best parameter in differentiating asthma and cystic fibrosis in both capnography and spirometry. Capnography was better able to identify the abnormalities in pulmonary function than normal spirometry. With the analysis of the receiver operating characteristic (ROC) curve, the area under the ROC curve, and capnographic indexes the research concluded that capnography is more efficient and accurate in early detection of and discrimination between asthma and cystic fibrosis than spirometry. Capnography better evaluates the movement of air through diffusion, whereas spirometry identifies the air flow in the conducting airways. Thus, it is notable to expect variation in detecting chronic airway obstruction using both devices. Yet, the dominance of capnography in detecting the alterations of parameters enables it to be more reliable than spirometry [27].

2.5.3 Procedural sedation

An adverse growth in the use of sedation is observed in the medical field with the advancement of electrophysiology, radiology, and catheterization procedures. Sedation is usually prescribed by doctors and provided by nurses in a surgical or medical treatment situation. It is evident that hypoxia occurs during procedural sedation [39]. Although a variant of drugs has been administered during sedation, the risk of hypoxia is yet to be reduced. Hence, an advanced monitoring system is in demand for reducing the risk of hypoxia and the risk associated with procedural sedation. Monitoring ventilation using capnography is determined as a promising strategy to ensure the safety of the patient during procedural sedation. The risk of procedural sedation is monitored through the capnography index, which is the $EtCO_2$ value and the oxygen saturation. Hypoxia is diagnosed when the $EtCO_2$ values fluctuates at 50 mmHg with 10% of the baseline volume. Moreover, hypoxia is determined when the oxygen saturation decreases up to 90% and below for 15 s [40]. It is justified that application of capnography during procedural sedation has increased the potential of identifying hypoxia by 17 times, compared to the absence of capnography. Pulse oximetry evaluation and patient observation delays the determination of complications arising from procedural sedation. The use of capnography reduced events of procedural sedation risk by more than half, which emphasized the significance of capnography [41]. However, in the effort to further reduce the occurrence of severe risks in procedural sedation, the variation in $EtCO_2$ baseline is studied to evaluate the defined circumstances with a change in the $EtCO_2$ value. With regard to the study, the respiratory depression and airway obstruction of a patient undergoing procedural sedation is closely monitored [42].

2.5.4 Apnea monitoring

Apnea is a sleep-disordered breathing that occurs in children and adults. It is also recognized as obstructive sleep apnea, where the upper airway tract is repetitively closed during sleep. This results in desaturation of the respiratory airway and causes disruption to breathing. Snoring is the primary symptom for the development of apnea [43]. In addition, upper airway resistance syndrome has been identified as the persisting symptom for apnea development. The breathing pattern is completely different when a patient is conscious and in sleep. Hence, the evaluation of breathing obstruction during apnea is complicated. Capnography technology has been highly approached for ventilation monitoring during breathing. During apnea, the $EtCO_2$ signal disappears completely and returns to show the signal when the upper airway tract recovers [44]. Capnography monitoring aids to diagnose such circumstances occurring during apnea. Moreover, the analysis of P_{CO_2} using the waveform shape of capnography helps to indicate the severity of hyperventilation and prolonged upper airway obstruction in sleep. Capnography enables clinicians to determine and manage the ventilatory control mechanism during sleep apnea. The study is significant for developing continuous positive airway pressure therapy during respiratory depression. Besides, patient undergoing capnography during sleep in medical facilities may benefit from the various outcomes of capnography. The altered patient ventilatory responses are closely monitored and the required therapy is provided [45]. In general, apnea is difficult to be diagnosed, unless in the presence of hypoxia. Studied have proved that capnography signals indicate the occurrence of apnea, and hypoventilation in sleep far earlier than the other detecting strategies. It is evident that capnography is the primary detecting tool for identifying apnea and airway obstruction in sleep, and is supported by other medical tools [46].

2.5.5 Cardiac arrest and resuscitation

Cardiopulmonary resuscitation (CPR) is a well practiced method of restoring cardiac output in a patient with cardiac arrest. The perfusion rate to vital organs determines the survival rate of the patient encountering cardiac arrest. Generally, perfusion rate is measured through organ blood flow. However, it is not feasible to do during CPR. Hereby, capnography has shown its greatest potential in determining the effectiveness of CPR with the non-invasive capnography method and the $EtCO_2$ readouts [47]. The $EtCO_2$ values indicated in the capnography device represent the cardiac index during CPR and cardiac arrest. Both the $EtCO_2$ and cardiac index may drop to zero when ventricular fibrillation is induced. During cardiac resuscitation, the $EtCO_2$ value is expected to be at 25% of pre-cardiac arrest value and it is the same as the cardiac index. After a successful CPR and return of spontaneous circulation, the $EtCO_2$ value may increase rapidly. The overshoot of $EtCO_2$ value may not be the same as the cardiac index. The rapid release of CO_2 from body tissues denotes the high $EtCO_2$ value after cardiac resuscitation. However, it does not resemble the heart pumping rate. Besides this, the capnography $EtCO_2$ value correlates well to coronary perfusion rate and cerebral blood flow. Figure 2.4 shows the P_{CO_2} trend with the change of cerebral blood flow and partial pressure of oxygen [48]. The American Heart Association has affirmed the potential

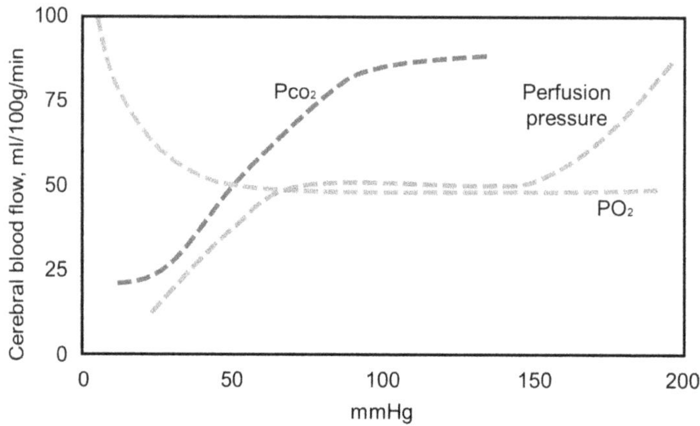

Figure 2.4. Schematic illustration of cerebral blood flow with the change of perfusion pressure, P_{CO_2}, and partial pressure of oxygen.

of $EtCO_2$ as the primary physiological monitor rather than the venous or arterial catheter during CPR. The capnography waveform represents the quality of CPR performed. Elimination of hyperventilation is the primary point to be considered in CPR [16]. Capnography waveforms indicate the ventilation rate, which denotes the occurrence of hyperventilation and the quality of CPR. Since it is anticipated that $EtCO_2$ will reflect organ perfusion during CPR, it may serve as both a target for resuscitation and a predictor of when prolonged CPR is ineffective. The prolonged effectiveness of resuscitation is determined by $EtCO_2$ value [49]. Reports in literature have shown that cardiac resuscitation patients with an $EtCO_2$ value less that 10 mmHg after 20 min of CPR did not survive. Patients with more than 20 mmHG $EtCO_2$ survived the cardiac arrest. The study justified the significance of $EtCO_2$ after cardiac resuscitation [50].

2.5.6 Pulmonary embolism

Pulmonary embolism is a cardiopulmonary dilemma caused by blood clots that flow through the venous and pulmonary circulatory systems. Indirectly, an embolism is caused by respiratory gas obstruction, tumors, bone cement, and cholesterol in the body. Capnography is an alternative diagnostic technique to detect pulmonary embolism [51]. The variations in the capnogram waveform are the features used to differentiate embolisms from the other respiratory illnesses. The slopes and angles of the capnogram waveform are evaluated to determine embolisms. The slope of phase III, which is the alveolar plateau is the focused feature for embolism detection, is generally a horizontal shape. When a pulmonary embolism is detected, the slope of phase III is further reduced towards the horizontal orientation [52]. This is different from other respiratory illness such as chronic obstructive pulmonary disease (COPD), which will increase the slope towards vertical orientation (figure 2.5). Pulmonary embolism limits the blood flow in the pulmonary vasculature to the

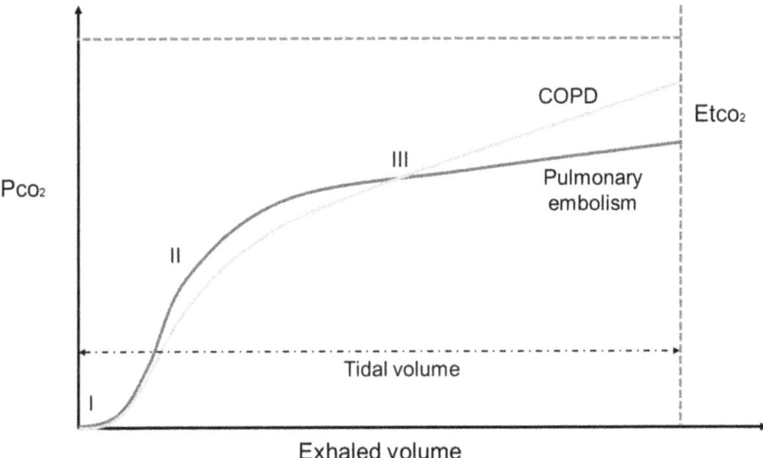

Figure 2.5. Difference between phase III slope of pulmonary embolism and COPD, indicating the reduced and increases gradient, respectively.

respiratory system. Hence, limited or no CO_2 is supplied to the alveoli in the respiratory organ. As the limitation continues, the peripheral alveoli are affected by perfusion. The empty alveoli without gas flow will flatten phase III of the capnogram and result in the reduced horizontal slope [53]. The use of several features derived from capnography to evaluate patients holds promise for early non-invasive embolism detection at the bedside in medical facilities, allowing rapid and effective treatments [54].

2.5.7 Laparoscopy

Laparoscopy is highly preferred in medical surgeries due to its advantages, such as small incisions, reduced postoperative ileus, short-term hospitality, and less pain suffered by patients. Insufflation of gas into the peritoneal cavity is the primary procedure in laparoscopy to enable the endoscope to clearly visualize the anatomic structures. CO_2 gas is the most preferable for the gas insufflation due to the good relative absorption, high lipid soluble, nontoxicity, and inflammable properties. Nevertheless, CO_2 gas may cause sudden and severe hypertension and tachycardia, which may result in myocardial inflammation due to the CO_2 insufflation rate [55]. Such circumstances may cause respiratory acidosis and hypercarbia with the high flow of CO_2 gas in the pulmo-circulatory system. Hence, CO_2 monitoring is mandatory during laparoscopy. $EtCO_2$ is more commonly evaluated than P_{CO_2} for many non-invasive medical testings. However, there is a remarkable difference between $EtCO_2$ and P_{CO_2} monitoring in laparoscopy. The ventilation and perfusion rate cause variation in $EtCO_2$ and P_{CO_2}. During laparoscopy of a preoperative cardiopulmonary patient, CO_2 insufflation may result in increased P_{CO_2} and decreased pH in the blood circulatory system. However, the change will not cause a difference in the $EtCO_2$ value [56]. Therefore, clinicians are aware to perform arterial blood gas analysis to continuously monitor the hypoxemia condition of the patient. Capnography signals and the waveform display have a huge impact in laparoscopy to ensure the patient's safety with prompt diagnostics and therapeutics. A highly sensitive and accurate capnography tool is in great demand for laparoscopy, due to the significant role in displaying the precise $EtCO_2$ and P_{CO_2} numerical values [57].

2.6 Capnography at intensive care units

Although capnography has its praise in medical applications, the technology has been least implemented in the intensive care unit (ICU). A statistical study conducted by the Royal College of Anaesthetists and the Difficult Airway Society of the United Kingdom revealed that about 74% of airway deaths in the ICU could be prevented if capnography technology was implemented with its highest instances [58]. In the current trend of medical advances, capnography has been made compulsory for patients with tracheal tubes in the ICU. A continuous ventilation monitoring is mandatory for ensuring patient safety. There are several factors considered for the routine and intensive application of capnography in the ICU [59]. Capnography suites for the variety of respiratory and circulatory monitoring depend on the expertise of the

ICU setup. Capnography is mandatory to monitor patients with substantial airway or cardiac comorbidity, which enables one to determine the margin of error and consent hypoxia. Such a condition is different from the healthy patients with anesthesia level one and two. Prompt detection of the translocated endotracheal tube is one of the major benefits of continuous monitoring of the patient using capnography in the ICU. Besides the endotracheal tube, the displacement of the nasogastric tube in the tracheobronchial system is detected. In general, capnography helps to detect and monitor the integrity of the airway, cardiac output, and ventilation, while detecting the airway obstruction, bronchospasm, and kinked intubated tubes [60]. Capnography acts as the guide to monitor the metabolic rate and estimate the breathing pattern of a patient in the ICU. Besides the technical advantages in the ICU, capnography eliminates the repetitive blood gas monitoring test and significantly reduces the cost of medical facilities. The high-performance capnography tool reduces the use of multiple devices for specific analyses and improves the spatiality in the ICU. Most importantly, application of capnography technology in the ICU has a high potential of decreasing the death rate to a high extent, with the rapid detection of respiratory and cardiac resuscitation in ICU settings [61].

2.7 Capnogram modeling for respiratory monitoring

A capnography model and simulation is presented for evaluating different states of respiratory condition. Capnogram signal analysis is critical as it determines the variations in capnogram waveforms and consequently differentiates the airway illness. The CO_2 waveform recorded from healthy and asthmatic patients was evaluated using a sidestream capnography system. The capnograph device is developed as it able to generate 100 values of CO_2 in a second with a 0.01 s time interval. The CO_2 data is received from human respiration from a healthy patient and an asthmatic patient. The developed capnography system identifies the valid breath cycles from the human breathing and displays the capnogram waveform. The selection criteria for valid breath cycles eliminates the artifacts of human breath and denotes the precise breath pattern. The sampling was performed using a sampling tube, connected to the subject's nasal cavity. A 100 cm long medical sampling tube was used for obtaining the breathing samples. The breathing pattern of the subject is recorded into the capnograph device. The sampling was performed for a minimum of 2 min and the capnogram waveform patterns were monitored for the healthy and asthmatic subject. Figure 2.6 shows the capnogram waveform recorded using the developed capnograph tool.

The variations were first determined by visual evaluation. Based on the figure above, it is notable that the sinusoidal waveform generated by the healthy subject has consistently shaped breath cycles, whereas the asthmatic patient showed an inconsistent waveform of breath cycles. The plots reveal that the abnormal breath pattern of the asthmatic patient are able to be determined from the capnogram waveform in comparison to the waveform generated by the healthy subject. Further analyses were performed through statistical study from the overall data. The capnogram waveform is divided into several segments by using the capnography

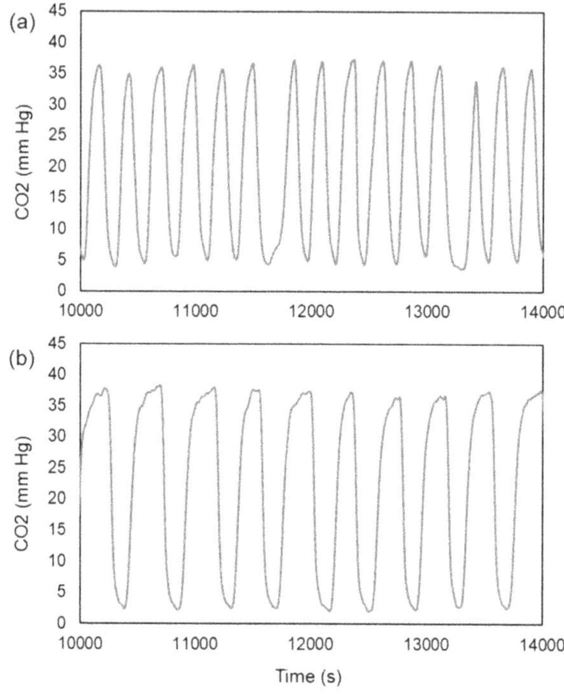

Figure 2.6. Capnogram waveform extracted from (a) asthma patient and (b) healthy subject.

threshold method. The segments were mainly indicating the inhalation and exhalation of CO_2 volume in human breathing. Based on the capnogram segments, the capnography features such as angle, slope, and area were computed. The following equations represent the formula for the computation of slope and area.

$$\text{Area AR}_i = \frac{dt}{6} \sum_{j=0}^{i} (R_{j-1}(t) + 4R_j(t) + R_{j+1}(t)) \tag{2.1}$$

$$\text{Slope}(S_j) = \frac{1}{c} \sum_{j=0}^{c-1} b_j (M_j - S_j)^2 \tag{2.2}$$

A, S, and C represent the area, slope, and the slope length of a capnogram waveform, respectively. M_j and b_j indicate the linear fit and the density of the jth element. The extracted features were evaluated through statistical analysis to determine the significance of each feature in differentiating the respiratory condition. Normality statistical evaluations were performed as the preliminary analysis on the feature distribution. The kurtosis and skewness of the extracted features were determined. In addition, the Kolmogorov–Smirnov and Shapiro–Wilk normality tests were used to verify the normally distributed capnogram features. Thereafter, a paired sample t-test was used to obtain the significance P-values. Table 2.1 shows the significance P-values computed using the paired sample t-test.

Table 2.1. Significance value of features extracted from capnogram waveforms.

Segment	Capnogram segments	Features	P-value
1	Dead space gas	Area	0.01
		Slope	0.002
2	Expiratory phase	Area	0.05
		Slope	0.003
3	Alveolar plateau	Area	0.07
		Slope	0.05
4	EtCO$_2$	Area	0.001
		Slope	0.06
5	Inspiratory phase	Area	0.08
		Slope	0.09
6	Baseline	Area	0.01
		Slope	0.07

The table deduced that the expiratory phase is more significant $(P<0)$ than the inspiratory phase. Additionally, both the slope and area of the capnogram waveform are equally contributing to the significance difference between the healthy group and asthmatic group of patients. The preliminary results deduce that the developed sidestream capnography system is promising in detecting and monitoring the respiratory illnesses [19].

2.8 Capnography signal outside healthcare environment

Apart from the medical interpretation, capnography is applicable to many other fields. Capnography is best used for CO_2 gas monitoring, which enables it to be used in industrial applications. Being the primary greenhouse gas, CO_2 is aimed to be closely monitored for proper venting of the gas to the environment.

2.8.1 Ice core analysis

Anthropogenic is the term used to denote the impact of human activity on climate change. With the current industrial advancements and consumption of natural resources, climate change has dramatically affected the natural environmental cycles. Although the effect is not seen in a day-to-day variation, the impact will be huge in the future. There are several ways to determine the Earth's climate change. The analysis of snow and bubbles from the ice cores is a promising technique to justify climate change. Ice cores obtained from Greenland can provide information for the last 100 000 years, whereas the Antarctic ice cores provides 400 000 years of information. The parameters analyzed are the climatic signals such as ocean pump, biomass, continental shelf hydrates, biogeochemical nitrogen cycles from marine activity, biological systems, and permafrost. Oxygen isotopes are analyzed to approximate the temperature of varying locations. The volcanic activity is examined through dust analyses [62]. With the advancement of capnography, the temperature variation with the level of CO_2 gas in the atmosphere is examined to justify the correlation between CO_2 gas variation and temperature fluctuations. The current interest of industrialists is more concerned with controlling and monitoring the CO_2 gas and other harmful gas emissions to protect the health and safety of people at work. With the support of the Occupational Safety and Health Administration (OSHA) and the National Institute of Occupational Safety and Health (NIOSH), numerous efforts are made to monitor gas levels with capnography to ensure workers' safety and at the same time ensure a safe environment and maintain the natural climate change cycles. Regarding this, capnography-modified gravity models were developed by integrating Social Network Analyses. The model was evaluated through several case studies, mainly to investigate the CO_2 gas emissions in China. The outcome of the model has promoted the accurate CO_2 gas emissions in industry, the spillover effects, and the spatial correlation between different regions of investigation. The developed framework has provided accurate quantitative readings on the average correlation coefficient of CO_2 gas emission in China. Moreover, it evaluated the corresponding network density from 2007 to 2014, which has opened an alternative pathway to determine the social consumption and investment in relation to CO_2 gas emissions. The study is furthered for deeper coordinated gas emissions and expects highly comprehensive outcomes in monitoring CO_2 gas emissions in industry [63].

2.8.2 Mines

Mining is an open industry, where the airflow is constrained and contaminated gases are accumulated in the field. Since the 18th century, deep shaft mining, explosions, and cave-ins have presented significant dangers. Excretion of hazardous gases such as nitrogen dioxide, carbon monoxide, and sulfur dioxide is highly accumulated in the mining industry. Birds were the earliest method of analyzing gas in the mining industry. Birds such as canaries are particularly sensitive to carbon monoxide. The bird stays active with its singing and movement if the gas level is optimum. As the carbon monoxide gas level endangers the bird, it will fall off its perch. Thereafter, the infrared technology replaced the traditional method by using sensors for detecting the toxic gases [64]. In current technology, hand-held infrared devices are implemented in the mining industry to monitor the range of flammable gases and to evaluate the environmental safety of the industry. Capnography is used in mining, specifically to target the range of CO_2. The device acts as an alarm to control the harmful gas emissions to the environment and the event of reduced oxygen saturation.

2.8.3 Space station

A space station is a closed environment, where the atmospheric aspects are well managed. The life support and environmental controls are managed as it is suitable for an estimated two weeks. A water coolant loop system, air revitalization system, atmospheric pressure control system, wastewater system, thermal control, and water supply are closed controlled and monitored in the space station [65]. The gases in the space station are ensured at 20% oxygen and 80% nitrogen with 760 mmHg atmospheric pressure. The partial pressure of oxygen is monitored. Control valves for oxygen and nitrogen gas are tuned to maintain the gas balance in the space shuttle. The humidity of the space station is maintained between 30% and 70% by monitoring the CO_2 and carbon monoxide gases [66]. The cabin fan circulates constantly to maintain air circulation in the space, which also provides the CO_2 removal through an air filter and lithium dioxide canisters. Lithium dioxide canisters convert CO_2 to lithium carbonate, and the odor is eliminated through activated charcoal. The entire space station is controlled by the ventilation and atmospheric monitoring. Capnography application in the space station is introduced for the gas monitoring. Although the technology has not been well established, the experts are keen to upgrade the capnography advancements as it meets the highly complex and unique space station requirements. It paves the way for a significant implementation of capnography in an enclosed environment and exceeding the limit of exploration towards outer space [67].

2.8.4 Submarines

Submarines are designed as enclosed containers to enter the hostile environment of the ocean. They range in size from small, privately owned pleasure boats to large, important naval warships with nearly limitless cruise capabilities. Since their invention up to today's technology, the primary problem encountered in all types of submarines is the challenge of maintaining the internal environment with the secured atmosphere for safety and power. Submarines are maintained at 1 atmospheric pressure [68]. The sources of atmospheric contamination range from exhausts of cooking, smoking, sanitization tanks, weapons, human bodies, air conditioning, refrigeration, batteries, and other unavoidable maintenance activities. Hence, it is a great challenge to control and continuously monitor the safe environment in the submarine. In 1975, the Navy Research Laboratory in the United States developed a central atmosphere monitor for subsequential use in submarines. An infrared CO_2 monitor is installed in the system, which detects the CO_2, carbon monoxide, hydrogen, nitrogen, and oxygen in the submarine's atmosphere [69]. Following the technique, colorimetric measurements are taken as additional readings at certain intervals. With the state-of-art infrared technology, small and portable newer capnography devices are developed for use in submarines. The simple and straightforward technology is implemented to monitor the atmospheric conditions, with the numerical readouts of gas content in submarines [70]. In a recent study, a capnography system has been upgraded into strategic simulation, which precisely measures the CO_2 level in submarines and provides prompt decisions during acute CO_2 exposure. The system successfully reached 5000 ppm of CO_2 gas concentration detection and gives an alarm warning when the CO_2 level reached below 100 ppm. The automatic decision-taking system administered with capnography was found to have several limitations due to the efficiency and safety of the submarine during acute CO_2 levels. Hence, further analysis was performed with qualified submarine sailors. The sailors were exposed to different levels of CO_2 gas concentrations and allowed to handle the strategic simulation at risk. The outcome of the analysis demonstrated the cognitive deficits of the simulated model. With improvements, the system was justified for CO_2 gas level measurement and decision making in submarines [71]. As humanity expands its horizons, new and improved techniques will be necessary to push the capnography boundaries of exploration, whether it is into the depths of the ocean or into space [72].

2.9 Conclusion

In light of the above, this chapter has provided a comprehensive understanding of capnography in medical interpretations and industrial applications. The detailed summary of capnography signals in delivering precise information of airway illness benefits in many circumstances, which is highly preferable for saving the lives of many patients with a variety of illnesses. Along with the Industry 4.0 status, capnography technology requires huge attention in enhancing the technology for healthcare benefits inside and outside the medical facility. The upgraded CO_2 gas ventilation and monitoring system has a high potential in addressing the ever-growing hazardous gases in healthcare sector and heavy industries. More quantitative research is needed to assess the difficulties and enables of capnography adoption in the healthcare and industrial sectors using Industry 4.0 technology.

Acknowledgments

This work was supported by 'Respiratory Assessment Device Based on Internet of Thing (IoT) Technology' under Prototype Development Research Grant Scheme (R.J130000.7851.4L919) from Ministry of Higher Education, Malaysia.

References

[1] Bhavesh M and Joshi H 2022 Categorical framework for implementation of Industry 4.0 techniques in medium-scale bearing manufacturing industries *Mater. Today Proc.* **15** 3531–7

[2] Inuwa H M, Ravi Raja A, Kumar A, Singh B and Singh S 2022 Status of Industry 4.0 applications in healthcare 4.0 and Pharma 4.0 *Mater. Today Proc.* **62** 3593–8

[3] Grybauskas A, Stefanini A and Ghobakhloo M 2022 Social sustainability in the age of digitalization: a systematic literature review on the social implications of Industry 4.0 *Technol. Soc.* **70** 101997

[4] Karatas M, Eriskin L, Deveci M, Pamucar D and Garg H 2022 Big data for healthcare Industry 4.0: applications, challenges and future perspectives *Expert Syst. Appl.* **200** 116912

[5] Iyengar K P, Kariya A D, Botchu R, Jain V K and Vaishya R 2022 Significant capabilities of SMART sensor technology and their applications for Industry 4.0 in trauma and orthopaedics *Sensors Int.* **3** 100163

[6] Ahsan M M and Siddique Z 2022 Industry 4.0 in healthcare: a systematic review *Int. J. Inf. Manage. Data Insights* **2** 100079

[7] Zelikova T J 2020 The future of carbon dioxide removal must be transdisciplinary *Interface Focus* **10** 20200038

[8] Permentier K, Vercammen S, Soetaert S and Schellemans C 2017 Carbon dioxide poisoning: a literature review of an often forgotten cause of intoxication in the emergency department *Int. J. Emerg. Med.* **10** 17–20

[9] El-Betany A M M, Behiry E M, Gumbleton M and Harding K G 2020 Humidified warmed CO_2 treatment therapy strategies can save lives with mitigation and suppression of SARS-CoV-2 infection: an evidence review *Front. Med.* **7** 1–10

[10] Qiao Z, Meng X and Wu L 2021 Forecasting carbon dioxide emissions in APEC member countries by a new cumulative grey model *Ecol. Indic.* **125** 107593

[11] Parker W, Estrich C G, Abt E, Carrasco-Labra A, Waugh J B, Conway A, Lipman R D and Araujo M W B 2018 Benefits and harms of capnography during procedures involving moderate sedation: a rapid review and meta-analysis *J. Am. Dent. Assoc.* **149** 38–50 e2

[12] Gutiérrez J J, Ruiz J M, de Gauna S R, González-Otero D M, Leturiondo M, Russell J K, Corcuera C, Urtusagasti J F and Daya M R 2020 Modeling the impact of ventilations on the capnogram in out-of-hospital cardiac arrest *PLoS One* **15** 1–15

[13] Conway A, Collins P, Chang K, Mafeld S, Sutherland J and Fingleton J 2019 Sequence analysis of capnography waveform abnormalities during nurse-administered procedural sedation and analgesia in the cardiac catheterization laboratory *Sci. Rep.* **9** 1–9

[14] Williams E, Dassios T and Greenough A 2020 Assessment of sidestream end-tidal capnography in ventilated infants on the neonatal unit *Pediatr. Pulmonol.* **55** 1468–73

[15] Balakrishnan M, Kazemi M and Teo A H 2012 A review of capnography in asthma: a new approach on assessment of capnogram *Biotechnol. Bioinform. Bioeng.* **2** 555–66

[16] Manifold C A, Davids N, Villers L C and Wampler D A 2013 Capnography for the nonintubated patient in the emergency setting *J. Emerg. Med.* **45** 626–32

[17] Long B, Koyfman A and Vivirito M A 2017 Capnography in the emergency department: a review of uses, waveforms, and limitations *J. Emerg. Med.* **53** 829–42

[18] Chee R C, Ahmad R, Zakaria M I and Yahya M F 2021 The assessment of end-tidal capnography waveform interpretation and its clinical application for emergency residents in Malaysia: a cross-sectional study *Eurasian J. Emerg. Med.* **20** 161–71

[19] Malarvili M B, Alexie M, Dahari N and Kamarudin A 2021 On analyzing capnogram as a novel method for screening COVID-19: a review on assessment methods for COVID-19 *Life J.* **11** 1–26

[20] Butusov D, Karimov T, Voznesenskiy A, Kaplun D, Andreev V and Ostrovskii V 2018 Filtering techniques for chaotic signal processing *Electron* **7** 450

[21] Chatterjee S, Thakur R S, Yadav R N, Gupta L and Raghuvanshi D K 2020 Review of noise removal techniques in ECG signals *IET Signal Process.* **14** 569–90

[22] de Cheveigné A and Nelken I 2019 Filters: when, why, and how (not) to use them *Neuron* **102** 280–93

[23] Dinis J, Campos G, Rodrigues J and Marques A 2012 Respiratory sound annotation software *Heal. 2012—Proc. Int. Conf. Heal. Informatics* pp 183–8

[24] Chin S T, Romano A, Doran S L F and Hanna G B 2018 Cross-platform mass spectrometry annotation in breathomics of oesophageal-gastric cancer *Sci. Rep.* **8** 1–10

[25] Kazemi M, Krishnan M B and Howe T A 2013 Frequency analysis of capnogram signals to differentiate asthmatic and non-asthmatic conditions using radial basis function neural networks *Iran. J. Allergy Asthma Immunol.* **12** 236–46

[26] Herry C L, Townsend D, Green G C, Bravi A and Seely A J E 2014 Segmentation and classification of capnograms: application in respiratory variability analysis *Physiol. Meas.* **35** 2343–58

[27] Almeida-Junior A, Marson F A L, Almeida C C B, Ribeiro M Â G O, Paschoal I A, Moreira M M and Ribeiro J D 2019 Volumetric capnography versus spirometry for the evaluation of pulmonary function in cystic fibrosis and allergic asthma *J. Pediatr.* **96** 255–64

[28] Verscheure S, Massion P B, Verschuren F, Damas P and Magder S 2016 Volumetric capnography: lessons from the past and current clinical applications *Crit. Care* **20** 1–9

[29] Anderson M R 2006 Capnography: considerations for its use in the emergency department *J. Emerg. Nurs.* **32** 149–53

[30] Ishiwata T, Tsushima K, Terada J, Fujie M, Abe M, Ikari J, Kawata N, Tada Y and Tatsumi K 2018 Efficacy of end-tidal capnography monitoring during flexible broncho-scopy in nonintubated patients under sedation: a randomized controlled study *Respiration* **96** 355–62

[31] Cereceda-Sánchez F J and Molina-Mula J 2019 Capnography as instrument for airway management during basic instrumental CPR manoeuvers *Resuscitation* **138** 132–3

[32] Wang H *et al* 2022 Airway strategy and ventilation rates in the pragmatic airway resuscitation trial *Resuscitation* **176** 80–7

[33] Ritchie J E, Williams A B, Gerard C and Hockey H 2011 Evaluation of a humidified nasal high-flow oxygen system, using oxygraphy, capnography and measurement of upper airway pressures *Anaesth. Intensive Care* **39** 1103–10

[34] Gusti V and Vaghadia H 2021 Hybrid nasopharyngeal and oropharyngeal airway for improving upper airway and capnography in sedated patients *Can. J. Emerg. Med.* **23** 416–7

[35] Chung K F 2017 Clinical management of severe therapy-resistant asthma *Expert Rev. Respir. Med.* **11** 395–402

[36] Nik Hisamuddin N A R, Rashidi A, Chew K S, Kamaruddin J, Idzwan Z and Teo A H 2009 Correlations between capnographic waveforms and peak flow meter measurement in emergency department management of asthma *Int. J. Emerg. Med.* **2** 83–9

[37] Mims J W 2015 Asthma: definitions and pathophysiology *Int. Forum Allergy Rhinol.* **5** S2–6

[38] Jones T L, Melville D M and Chauhan A J 2018 Diagnosis and treatment of severe asthma: a phenotype-based approach *Clin. Med. (Lond)* **18** 36–40

[39] Langhan M L, Shabanova V, Li F Y, Bernstein S L and Shapiro E D 2015 A randomized controlled trial of capnography during sedation in a pediatric emergency setting *Am. J. Emerg. Med.* **33** 25–30

[40] Long M, Green K, Bland E, Jones-Hooker C, Brast S, Duncan K and Wempe E 2016 Capnography monitoring during procedural sedation in radiology and imaging settings: an integrative review *J. Radiol. Nurs.* **35** 191–7

[41] Jach M 2020 Improved sedation capnography and enhanced patient safety for sedation anesthesia *Trends Anaesth. Crit. Care* **30** e52–3

[42] Rose Bovino L, Brainard C, Beaumier K, Concetti V, Lefurge N, Mittelstadt E, Wilson T and Langhan M L 2018 Use of capnography to optimize procedural sedation in the emergency department pediatric population *J. Emerg. Nurs.* **44** 110–6

[43] Lin X, Cheng H, Lu Y, Luo H, Li H, Qian Y, Zhou L, Zhang L and Wang M 2022 Contactless sleep apnea detection in snoring signals using hybrid deep neural networks targeted for embedded hardware platform with real-time applications *Biomed. Signal Process. Control* **77** 103765

[44] Scully K R, Rickerby J and Dunn J 2020 Implementation science: incorporating obstructive sleep apnea screening and capnography into everyday practice *J. Perianesth. Nurs.* **35** 7–16

[45] Taghizadegan Y, Jafarnia Dabanloo N, Maghooli K and Sheikhani A 2021 Prediction of obstructive sleep apnea using ensemble of recurrence plot convolutional neural networks (RPCNNs) from polysomnography signals *Med. Hypotheses* **154** 110659

[46] Scully K, Schermer S, Sever E, Jones A, Shah P, Hodgkins C, Stanley E, Dunn J and Rickerby J 2019 Riding the (end) tidal wave to CO_2 monitoring: using capnography for obstructive sleep apnea following anesthesia *J. Perianesth. Nurs.* **34** e2–3

[47] Chicote B, Aramendi E, Irusta U, Owens P, Daya M and Idris A 2019 Value of capnography to predict defibrillation success in out-of-hospital cardiac arrest *Resuscitation* **138** 74–81

[48] Elola A, Aramendi E, Irusta U, Alonso E, Lu Y, Chang M P, Owens P and Idris A H 2019 Capnography: a support tool for the detection of return of spontaneous circulation in out-of-hospital cardiac arrest *Resuscitation* **142** 153–61

[49] Turle S, Sherren P B, Nicholson S, Callaghan T and Shepherd S J 2015 Availability and use of capnography for in-hospital cardiac arrests in the United Kingdom *Resuscitation* **94** 80–4

[50] Sandroni C, De Santis P and D'Arrigo S 2018 Capnography during cardiac arrest *Resuscitation* **132** 73–7

[51] Manara A, D'Hoore W and Thys F 2013 Capnography as a diagnostic tool for pulmonary embolism: a meta-analysis *Ann. Emerg. Med.* **62** 584–91

[52] Abate L G, Bayable S D and Fetene M B 2022 Evidence-based perioperative diagnosis and management of pulmonary embolism: a systematic review *Ann. Med. Surg.* **77** 103684

[53] Songur Yücel Z, Metin aksu N and Akkaş M 2020 The combined use of end-tidal carbon dioxide and alveolar dead space fraction values in the diagnosis of pulmonary embolism *Pulmonology* **26** 192–7

[54] Sacks M and Mosing M 2017 Volumetric capnography to diagnose venous air embolism in an anaesthetised horse *Vet. Anaesth. Analg.* **44** 189–90

[55] Baraka A, Jabbour-Khoury S, Karam V, Assaf B, Kai C, Nabbout G and Khoury G 1998 Correlation of the end-tidal P_{CO_2} during laparoscopic surgery with the pH of the gastric juice *JSLS* **2** 163–7

[56] Jadhav R S, Puram N N, Ramanand J B, Zende A M, Bhosale R R and Karande V B 2017 End tidal CO_2 level (PETCO2) during laparoscopic surgery: comparison between spinal anaesthesia and general anaesthesia *Int. J. Basic Clin. Pharmacol.* **6** 286

[57] Mamta G P and Swadia V N 2017 Role of EtCO2 (End tidal CO_2) Monitoring (Capnography) During Laparoscopic Surgery under General Anesthesia *Int. J. Res. Med.* **5** 148–54

[58] Riley C M 2017 Continuous capnography in pediatric intensive care *Crit. Care Nurs. Clin. North Am.* **29** 251–8

[59] Kerslake I, Mb H, Frca C, Kelly F, Mbbs M A, Frca M and Dicm F 2017 Uses of capnography in the critical care unit *BJA Education* **17** 178–83

[60] Kreit J W 2019 Volume Capnography in the Intensive Care Unit: Potential Clinical Applications *Ann Am Thorac Soc.* **16** 409–20

[61] Wollner E, Nourian M M, Booth W, Conover S, Law T, Lilaonitkul M, Gelb A W and Lipnick M S 2020 Impact of capnography on patient safety in high- and low-income settings: a scoping review *Br. J. Anaesth.* **125** e88–103

[62] Mgbemene C A, Nnaji C C and Nwozor C 2016 Industrialization and its backlash: focus on climate change and its consequences *J. Environ. Sci. Technol.* **9** 301–16

[63] Liu S and Xiao Q 2021 An empirical analysis on spatial correlation investigation of industrial carbon emissions using SNA-ICE model *Energy* **224** 120183

[64] Duarte J, Rodrigues F and Branco J C 2022 Sensing technology applications in the mining industry—a systematic review *Int. J. Environ. Res. Public Health* **19** 2334

[65] Hakkarainen J, Szeląg M E, Ialongo I, Retscher C, Oda T and Crisp D 2021 Analyzing nitrogen oxides to carbon dioxide emission ratios from space: a case study of Matimba Power Station in South Africa *Atmos. Environ.* **10** 100110

[66] Tapoglou N, Taylor C and Makris C 2020 Milling of aerospace alloys using supercritical CO_2 assisted machining *Proc. CIRP* **101** 370–3

[67] Li S, Tan X, Sun W, Zhang L, Jia H, Li L, Zhao C and Zhang X 2021 Design catalytic space engineering of Ag–Ag bond-based metal organic framework for carbon dioxide fixation reactions *Colloids Surf.* **609** 125529

[68] Sun S C, Liu C L and Ye Y G 2013 Phase equilibrium condition of marine carbon dioxide hydrate *J. Chem. Thermodyn.* **57** 256–60

[69] Caserini S, Dolci G, Azzellino A, Lanfredi C, Rigamonti L, Barreto B and Grosso M 2017 Evaluation of a new technology for carbon dioxide submarine storage in glass capsules *Int. J. Greenh. Gas Control* **60** 140–55

[70] Vajta G, Bartels P, Joubert J, De La Rey M, Treadwell R and Callesen H 2004 Production of a healthy calf by somatic cell nuclear transfer without micromanipulators and carbon dioxide incubators using the Handmade Cloning (HMC) and the Submarine Incubation System (SIS) *Theriogenology* **62** 1465–72

[71] Rodeheffer C D, Chabal S, Clarke J M and Fothergill D M 2018 Acute exposure to low-to-moderate carbon dioxide levels and submariner decision making *Aerosp. Med. Hum. Perform.* **89** 520–5

[72] Toro N, Gálvez E, Saldaña M and Jeldres R I 2022 Submarine mineral resources: a potential solution to political conflicts and global warming *Miner. Eng.* **179** 107441

IOP Publishing

Advanced Signal Processing for Industry 4.0, Volume 1
Evolution, communication protocols, and applications in manufacturing systems
Irshad Ahmad Ansari and Varun Bajaj

Chapter 3

The future of Industry 4.0: private 5G networks

P D Selvam, J Sridhar, V Ganesan and R Ravindraiah

The promise of 5G, which has been widely promoted among consumers, is swiftly finding its way into the corporate world. Large manufacturers have found the benefits of 5G to be attractive, and are installing private 5G networks in new facilities. Modernizing a manufacturing operation to support Industry 4.0 applications can be a difficult task. This is especially true if a facility's machines are connected to a central office through outdated fixed Ethernet. Where a facility has already gone wireless, the transition may be easier, but expanding to next-generation capabilities can highlight the limitations of commercially available Wi-Fi and LTE cellular networks. Industry 4.0 employs highly automated, intelligent, and collaborative cyber-physical systems, which necessitate very robust, low-latency wireless connections. As a result, a growing number of businesses in industries like manufacturing, energy, mining, power distribution, logistics, and others are going private. They are abandoning existing network providers in favor of building their wireless networks. The application of private 5G networks to Industry 4.0 and the associated supply chain is the focus of this chapter. In this chapter the benefits of private networks and Industry 4.0 is examined. A number of use cases are outlined in this chapter that benefit from private 5G networks. Citizens Broadband Radio Service (CBRS) will enable businesses and service providers of all types to develop and operate 5G networks, potentially completely revolutionizing wireless communications in the United States. A new generation of wireless applications, such as private 5G networks, will be driven by this CBRS.

3.1 Introduction

A network infrastructure that is only utilized by the devices in the end-user organization is known as a private mobile network. The end-user organization typically owns or occupies one or more of the physical locations where this infrastructure is located. The devices registered in a public mobile network will not function on the private mobile network, except for situations where it has been specifically authorized. While the phrase 'private mobile network' is more frequently used, across all levels of industry they are officially known as 'non-public networks.'

There is no worry about the effect on public users on the number of devices that can be connected, the throughput obtained, or other network performance indicators, because a private mobile network like this solely serves the devices assigned by the end-user organization. Additionally, coverage can be supplied precisely where it's needed, whether it is to indoor or outdoor places like production lines in factories or warehouses, or outdoor locations like port areas or mines. However, a private network's functionality goes beyond its capacity and coverage to include things like specialized security measures and interaction with other operational or commercial systems that are part of the industrial firm.

Using 5G network slicing over the public mobile network, an industrial firm can also receive a virtualized private mobile network. In this scenario, the industry can gain the majority of the benefits of a private mobile network without having to invest in and manage on-site wireless infrastructure or deal with its associated upfront costs, and it has the option for one or more services to operate across the entire public mobile network operator.

When compared to widely used unlicensed frequency bands, which is used for Wi-Fi networks, mobile networks operate in dedicated frequency bands that provide superior stability. However, it is usually anticipated that mobile network carriers would provide solutions for private mobile networks based on their licensed spectrum. The spectrum is being set aside in various countries for local/industrial use, including private networks. The spectrum granted to mobile network providers is typically larger, supporting higher bandwidth services for the industrial company.

Devices from a private mobile network can also be supported on the public mobile network with the right agreements and interworking. This enables the device to continue operating even when it leaves the private network's geographic boundaries, as is the case in wider-area applications like logistics. This also includes international roaming.

Even though there could be hundreds of private 5G networks built across America, Asia, and Western Europe, 5G deployments are still in their early phases, and there are still far more private 4G networks than there are 5G private mobile networks. Akin to the drivers for 4G networks, the drivers for private 5G networks realize faster throughput, lower latency, and boost reliability to meet the needs of particular applications and secure data. The distinction? The establishment of the 5G standard had businesses in mind.

These private 5G (and 4G) deployments are being driven by manufacturing applications (like Industry 4.0), but a plethora of new initiatives and government and military applications that are empowered by 5G technology are in the development phases and are beginning to take shape.

Compared to private 4G networks, 5G private networks provide Industry 4.0 with a larger range of deployment options. These options range from on-premises 5G private mobile networks, which are coupled with Multi-access Edge Computing (MEC) solutions. The deployment and management of private 5G networks can be done in a variety of ways, including using internal technical expertise with the help of any combination of the following: a communications service provider (CSP), a network equipment manufacturer (NEM), a cloud provider, and a systems integrator combination.

The cloud technologies for 5G RAN and the Core network, several potential deployment permutations, and multi-vendor solutions bring the difficulty of private 5G networks. They bring an enhanced level of complexity for private 5G networks [1]. Real-time analytics are intended to feed the console layer for automation and to realize private 5G networks for Industry 4.0. Independent, comprehensive, end-to-end visibility is essential to guarantee service levels.

Unmeasurable things are impossible to guarantee! To provide private 5G networks for Industry 4.0, it is required to predict the performance perceptibility earlier. Measurements for bandwidth, latency, dependability, and quality of experience (QoE) must be tailored to the requirements of the specific services in question. Particularly with constant latency and bandwidth measurement for applications like remote surgery and smart factories, carriers must ensure Service Level Agreements (SLAs). The Transmission Control Protocol (TCP) with three-way handshake, which is the foundation for conventional network latency measures, is ineffective for 5G. To quickly ascertain the 'what, where, and why' of a latency or throughput problem, continuous latency, and throughput tracking are required. The most reliable source of information for developing such measurement capacity is packet data.

3.1.1 Advantage of 5G: high speed and capacity

We all know that 5G boasts faster speeds and greater capacity than 4G, but understanding why also explains why 5G will be more environmentally friendly than earlier generations.

Data transmission through mobile technologies uses radio waves. Our phones encode the data when we make a call, send a text, or stream a video so that it can be sent across radio waves and decoded at the other end. Imagine radio waves as ocean waves with varying amplitudes and frequencies (or height). Small adjustments to the wave pattern are used to encode the digital signal.

Any section of the spectrum can be used for 5G. Governments can give service access to a wider frequency range or more bandwidth at higher frequencies, which enables the transmission of more data. For instance, although channel blocks typically range from 10 to 20 MHz in lower frequencies, they span from 100 to 200 MHz in higher frequencies.

With the variety of 5G technology available, operators have more alternatives when building their networks. Antenna arrays that support 'Massive MIMO' (multiple-input multiple-output) technology, which enables the simultaneous transmission and reception of multiple data signals over a single radio channel, can enhance capacity without using more airwaves. In addition to the coverage provided by the macro cells, 'Small Cells,' which are deployed both indoors and outdoors, offer coverage in areas of high data traffic.

Governments across the world are allotting substantially more of the radio frequency spectrum for 5G than they did for earlier generations.

The distinctive qualities of each spectrum can be used by operators and businesses. Operators can use a higher frequency spectrum to give the high bandwidth and capacity per radio antenna in regions with significant network traffic, such as urban areas, stadiums, airports, etc.

The low frequency spectrum, however, has its own benefits. For lower density network traffic in rural areas, the signal is appropriate since it naturally propagates farther and penetrates structures better than the high frequency spectrum.

The data transmission capacity of the 5G network is 100 times greater per unit of energy. This is accomplished by improvements in equipment design, such as Massive MIMO, which uses less energy than earlier antenna arrangements, and by utilizing a higher frequency spectrum for more effective data transmission.

Additionally, there is an enabling impact in that 5G has the potential to reduce emissions in other industries by 10 times the amount produced by mobile networks alone, thanks to digitization and automation.

3.1.2 Industry 4.0 and mobile connectivity

We are all aware that mobile broadband will be incredibly fast with 5G. The story's least interesting element is the phone, though. The fourth industrial revolution will totally transform how we live, work, and play by enabling connectivity through 5G. Additionally, 5G will make it a sustainable revolution.

The third industrial revolution, which was marked by computers and the internet, had a significant impact on industries that provide services, like banking and the media. However, this 'digital revolution' was largely ignored by sectors of the economy dependent on large amounts of heavy machinery, production lines, and material handling.

The fourth industrial revolution, or Industry 4.0, will use digitalization and automation to sustainably source resources, move them to markets, manufacture, power, operate, and service every aspect of our new, technological world, including the management of our cities and the wellbeing of people and the planet.

Mobile networks are the driving force behind Industry 4.0, starting with carrier-grade 4G in the form of private wireless networks and progressing to 5G when new business requirements arise. The cabled and Wi-Fi networks that function well in an office environment were not made for mobility, so they are inappropriate for use in an industrial environment. Cable will still have a place, but expanding and linking numerous assets are difficult with it. Wi-Fi and other legacy wireless technologies lack coverage, dependability, consistent performance, security, and mobility despite the fact that they are wireless.

Mobile networks are built to successfully connect objects that must move about and in challenging radio settings. They have larger capacity, wide coverage, fewer blind spots, and improved signal penetration. Most importantly, they offer predictable lower latency, which is necessary for machine-to-machine connectivity,

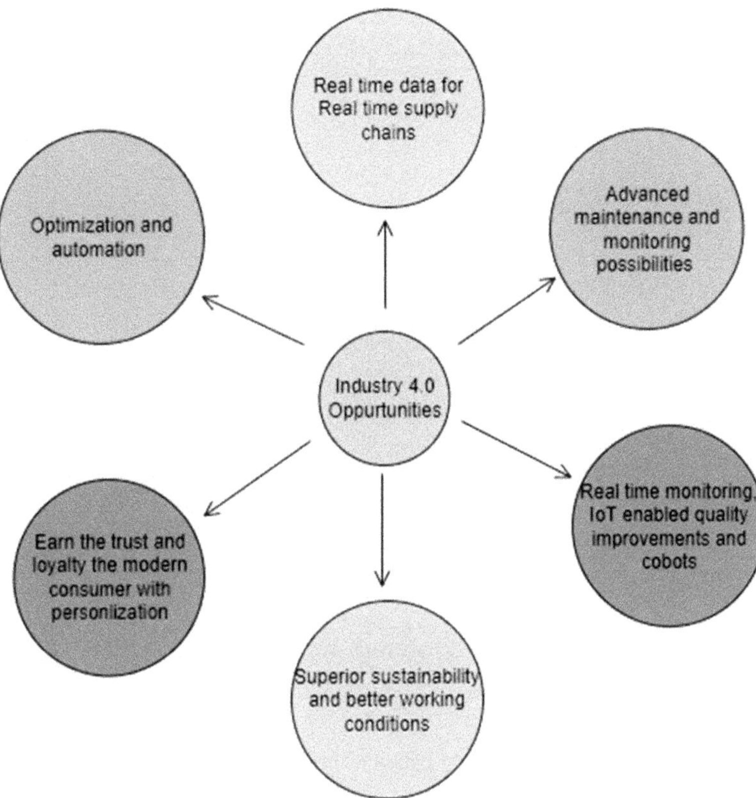

Figure 3.1. Industry 4.0 opportunities.

such as in robotics, where there must be any variance in the time between an instruction and a response.

3.1.3 Industry 4.0 opportunities and challenges

Industry 4.0 opportunities and challenges are shown in figures 3.1 and 3.2.

3.2 Private 5G networks' flexibility in Industry 4.0

Private wireless is a separate network with a focus on users and operational assets in the industrial sector. Similar to a public wireless network, a private wireless network offers wireless broadband connectivity, but it is owned and managed by the business that constructed or bought it. Mining, utilities, factories, warehouses, ports, airports, public safety, and smart cities are just a few examples of industry verticals and public sector domains that are great candidates for private wireless and are already utilising such networks to assist in digitalizing their operations.

Due to its advantages over cabled or Wi-Fi networks, private businesses and corporations are switching to private wireless. The majority of these are already available and will be strengthened even more by 5G.

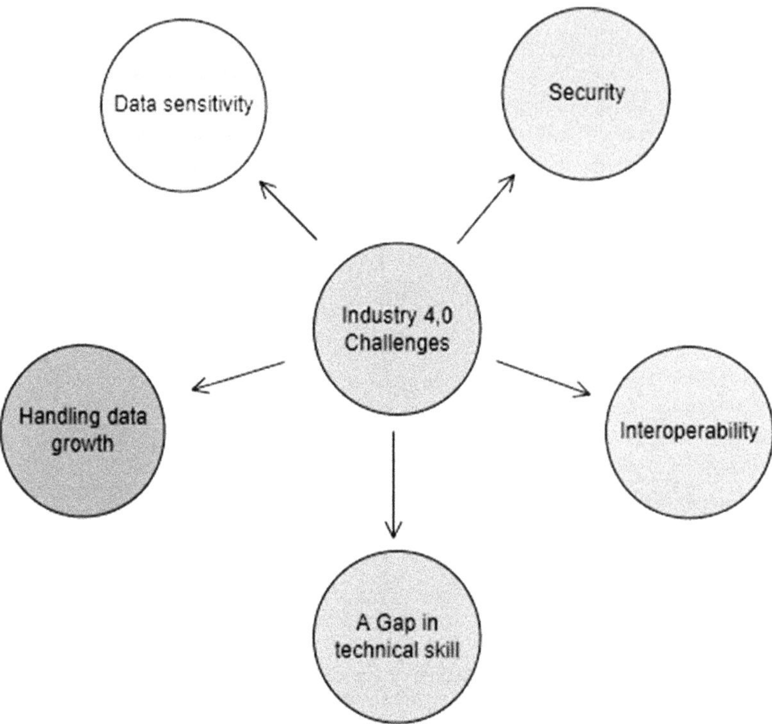

Figure 3.2. Industry 4.0 challenges.

Private 5G deployment options originate in a wide variety of configurations and combinations [11–14], but they all fall into one of two categories: private or public.

For instance, a business may choose a completely private on-site design if it demands guaranteed service quality, superior data privacy, and security. The licensed or unlicensed spectrum that has been leased from a mobile operator is usable by businesses. All three of the MEC, 5G Core, and 5G small cells may be present. If an organization has the necessary internal knowledge and skills, it can opt to create and manage itself.

Alternatives include working with a mobile operator, a provider of network hardware, or a systems integrator. In contrast, Industry 4.0 may choose a hybrid strategy, leveraging the RAN of the mobile operator or the full mobile operator network via 5G network slicing, if they require lower-cost 5G solutions that may be deployed fast while still delivering low latency and strong security. As long as the edge/MEC is on premise, hosted, or in the cloud, businesses can still use tiny cells for coverage.

3.3 Use cases of 5G networks in manufacturing

The use cases and requirements in the industrial sector are extremely diverse, ranging from those that call for the public mobile network's wide area coverage to

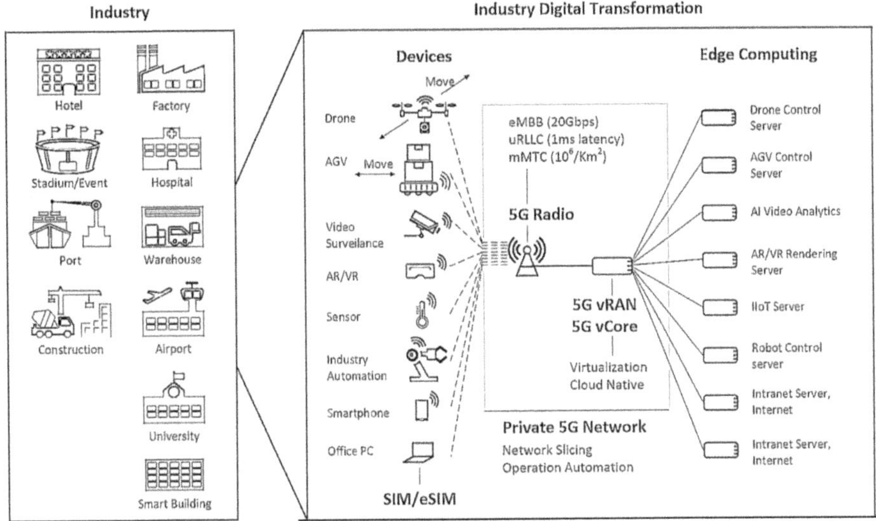

Figure 3.3. Use cases of 5G networks in manufacturing.

those with more specialized requirements that are satisfied by private and dedicated networks (figure 3.3).

Various examples of use cases in Industry 4.0 and the associated supply chain include the following. Some makers may choose to employ a private 5G network to support a larger range of use cases than is available with wired or Wi-Fi industrial networks:

Using 5G private networks, manufacturing line flexibility is made possible, enabling quick production line reconfiguration to provide new goods. Installation and reconfiguration of fixed wire networking are time-consuming, expensive, and unfeasible in some settings [2]. Although in some manufacturing facilities Wi-Fi networks have been installed, they do not scale well for big industries or for use cases with high bandwidth requirements, which are typical of Industry 4.0. Additionally, there also exists a problem in signal penetration. With sensors widely dispersed across the manufacturing facility and enormous gains in available bandwidth thanks to 5G networks, the production line can network completely without wires.

A network of connected machines and sensors may cooperate to carry out production tasks and execute intricate processes to their full potential using machine-to-machine communications. This makes possible, for instance, robotics, 'cobots', automated quality control of production lines, and intelligent warehousing and logistics. Network slicing delivers secure virtual networking between devices while the 5G network offers extremely low latency and excellent dependability.

The private 5G network provides ultra-high bandwidth and reliable mobile handovers across production line areas, warehouses, and dispatch zones, which are advantageous to Automated Guided Vehicles (AGVs). To increase safety and efficiency, real-time sensors are utilized along with image and video processing.

AGVs may be employed wherever they are required across the industrial or warehousing facility, as well as in mines or shipyards, thanks to the extraordinarily high and reliable bandwidth made possible by private 5G networks.

Applications for 'Connected Workers' enable high-bandwidth access for maintenance and facilities workers as well as machine operators, supervisors, and quality inspectors throughout the production site or campus. The initial solicitations include a paperless shop floor, ranging to connected tools like screwdrivers and torque wrenches, and focusing on worker safety with features like location tracking and personal safety monitoring. Advanced use cases include setting up equipment in virtual reality and augmented reality (VR/AR), manually assisting precision assembly, teaching operators, and remotely assisting maintenance engineers.

Production plant, warehousing/storage, input supply chain, and outgoing supply chain all receive support for end-to-end logistics. Massive deployments of inexpensive tracking devices can be made possible by private 5G networks, and these sensors can report on the location, condition, and surroundings of finished goods, parts, assemblies, and supplies. It is also feasible to operate seamlessly between the private 5G network environment and public mobile networks, which is notably advantageous for nationwide and international logistics, especially for pallets, wagons, and shipping containers.

By utilizing 5G network slicing, multi-client serving facilities can provide customers with 'private' sub-networks that are delivered via a shared 5G infrastructure. This can be applied to campus settings, warehouses, or production facilities. Network slicing can also be used to connect a manufacturing or warehousing facility to the national 5G network using seamless private network services over the public network.

Wide-area coverage use cases, like logistics or remote management, depend further on the segregation characteristic of supplied network services than they do on other aspects of network performance, such as bandwidth or latency, which is why they are often less rigorous or demanding. Serving devices over relatively general-purpose and shared network resources (particularly radio access resources) in a wide-area coverage situation is crucial. This can be done by giving the quality-of-service network or best effort services priority.

Contrarily, use cases like AGVs or vital production line management that are focused on small area coverage have a tendency to have more precise and difficult requirements, which may be substantially altered from the traditional network needs of customer mobile services.

A Service Level Agreement (SLA) with high dependability service with a guaranteed service is a promise made by a service provider to a customer regarding supplied network services. Network performance characteristics like reliability, latency, and deterministic communication may be included in the technical requirements of an SLA that is achieved through the use of dedicated private network hardware or network slicing. In addition to performance criteria, an SLA could also include functional and operational requirements, such as real-time monitoring, high-precision positioning, etc.

(i) **New traffic model:** Compared to traditional consumer mobile network services, uplink-to-downlink capacity ratio for such use cases is substantially different, necessitating the potential need for a new uplink-to-downlink capacity ratio.

(ii) **Strict data separation:** Some industry clients choose to keep their data on-site or on their own infrastructure. As a result, stringent data isolation is necessary, if they share the same infrastructure such as the separation of data between customers and data linked to the communication service user plane or control plane. The use cases where stringent data isolation or localization is required are supported by edge computing and network slicing.

(iii) **Privacy and security:** To safeguard industry data, a solid foundation for privacy and security is required (e.g. related to production lines and corresponding management and operation). Only industry clients should have access to the data, which should also be safeguarded from numerous potential assaults.

(iv) **Operation and management separation:** Numerous small and medium-sized industries lack the technical know-how or resources necessary for design, deployment, and operation of the network. The best cost-effective method for these industries to concentrate on their core competencies while offloading the burden of implementing and managing industry connectivity is to purchase these services from mobile network operators.

3.4 Specific technological elements for private networks

The introduction of three new 5G standard services that provide dedicated support for certain industrial applications as given below:

(a) Massive Machine-Type Communications (mMTC)
(b) Ultra-Reliable, Low-Latency Communications (URLLC)
(c) Enhanced Mobile Broadband (eMBB).

A variety of service types with varying characteristics can coexist on the same radio channel thanks to the development of the 5G radio interface [3]. For instance, many devices could be offered the aforementioned 5G standard services instantaneously on the same radio channel.

Additionally, network slicing, which is a recent innovation in 5G standard, allows private 5G networks to have greatly improved characteristics tuned to the necessary industrial applications or use cases, providing a type of 'virtual networking' with the additional advantage of being able to customize particular characteristics, such as uplink/downlink bandwidth ratio.

To reduce the end-to-end transmission latency significantly and to keep data contained within the boundaries of the industries for security and privacy of information, edge computing enables the interface of industrial enterprise systems to the mobile network (private or public) at the most local level possible.

In the future, open networking will be essential for managing and deploying networks, especially because it will allow for the use of cutting-edge AI tools in the network's construction and maintenance. It is anticipated that this new industry initiative would give Industry 4.0 and mobile operators more options and innovation for private 5G network deployments, which is especially pertinent given the rollouts of the 5G era networks.

(a) Massive machine-type communications (mMTC)

Massive machine-type communication (mMTC) is the 5G option for meeting the expectations of the mass market for IoT. According to this estimate, there can be up to 1 million linked IoT devices per square kilometer of space, subject to spectrum availability. This works well for a variety of use cases in the utility industry, commercial campuses, academic campuses, smart homes and cities, agriculture, transportation, and logistics.

Generally speaking, devices connected through mMTC rarely send or receive relatively small data packets. Thus, this method is perfect for periodic sampling or warning of strange events like

- equipment failure in a manufacturing plant,
- pipelines or mining monitoring,
- industrial campuses or across larger geography monitoring with low-frequency environmental sensing,
- detection of shipping containers' condition and periodic location reporting,
- notification of the opening or closing of an emergency access door,
- use of sensors or actuators in smart energy management.

(b) Ultra-reliable low-latency communications (URLLC)

A collection of mobile network specifications created by the mobile technologies standards group, 3GPP, is what the industry refers to when it uses the term '5G.' Data is divided into tiny packets and sent via the Internet. In use-cases like performing remote heart surgery or for self-driving cars, where losing connectivity for a few seconds is not an option, one of 5G standards is for ultra-reliable low latency.

When an instruction is provided and an operation is started, there is a delay in the network called latency. Robots and vehicles can operate remotely or even autonomously thanks to low latency. Consider being able to remotely operate a vehicle hauling a container at a port. It would be necessary for the operator to have as good a control as if they were physically operating the steering and braking systems of the vehicle.

Common 5G use cases include eMBB, mMTC, and URLLC. The most demanding applications, such as motion control, demand communication services that are available for up to 99.99% of the time and have end-to-end latency as low as 500 s. The URLLC challenge extends beyond simply offering a low-latency or extremely reliable link. To provide an accessible, effective, and sustainable service, there are end-to-end ramifications and trade-offs. Use cases are only taken into account if they cannot compromise on either their needs for low latency or high-reliability or both.

It might be advantageous to transmit numerous control signals using 'multi-TRPs' to increase URLLC reliability [7]. Multi antenna array elements using massive MIMO and beamforming for small cells for multi-TRP technology are made possible by high frequencies using mmWave in 5G, which are not yet supported by 4G. The coordinated multi-point (CoMP) concept was introduced with LTE, but 4G CoMP is unable to support realistic situations, such as less-than-ideal backhaul, and cannot, therefore, give deployment flexibility. CoMP necessitates precise coordination between the TRPs and highly detailed feedback (for instance, channel status information). Due to delays or a lack of backhaul capacity, scheduling jointly among numerous TRPs connected by a suboptimal backhaul may not be possible, leading to poor link adaptation or performance loss. Despite the high signal loss and blockage issues associated with 5G in mmWave, if employed in many independent links via numerous TRPs, it offers a robust contrary to blockages and beam failures.

The transmission can be scheduled independently from one TRP to another without the knowledge of channel state information. Relaxing the backhaul and synchronization constraints in the many TRPs will allow non-coherent joint transmission.

The latency in the network is reduced to 1 ms with a 5G network, resulting in delivering much more responsive 'real-time' control applications, particularly when edge computing is used. The typical round-trip latency for radio transmission in 4G networks is about 15 ms with a typical end-to-end system.

A critical benefit of 5G URLLC is that it provides control applications with a network dependability of 99.999% or better, enabling high consistency in real-time control applications. This vastly advances on earlier 4G networks' normal 99% reliability, which was insufficient for applications requiring crucial control.

5G ultra-reliable, low-latency communications: industrial use cases:

- Robot control, including navigation and collision avoidance, for both stationary and mobile robots
- Environmental management of plants
- Monitoring and control of safety
- Services for vehicle-to-vehicle and vehicle-to-infrastructure driver safety
- Autonomous vehicles
- Intelligent traffic management
- Drone-to-drone communication to prevent collisions
- Optimizing the utilization of airspace which includes drones and platooning
- Equipment for precise control power generation
- Access control and monitoring and control of escape routes; fire alarms and suppression

(c) **Enhanced mobile broadband (eMBB)**

The significantly increased accessible uplink and downlink bandwidth made possible by 5G networks eMBB is very advantageous for applications that require a lot of bandwidth.

With downlink data rates of at least 100 megabits-per-second per usable device over a typical dense urban setting, eMBB offers extremely high data rates over the 5G coverage area. As described below and with particular emphasis on the improved uplink and downlink bandwidth in indoor hotspot situations like smart factories or industrial sites, high performance is maintained across various coverage scenarios:

The increased bandwidth made possible by 5G eMBB benefits the following applications:

- CCTV surveillance systems—fixed or portable,
- robotics of production line control for quality,
- real-time sensing such as motor or actuator positioning, voltage or current monitoring with ultra-high rate,
- smart farming and drone-based surveillance,
- ultra-high-resolution mapping data downloads and guidance for autonomous vehicles,
- the ability to see through the vehicle in front in vehicle-to-vehicle applications,
- remote monitoring and maintenance.

In addition to having much lower over-the-air latency than 4G LTE, 5G eMBB provides 'nearer-real-time' data transfer and processing. The 4 ms latency of 5G eMBB is a significant improvement over the normal 50 ms latency of 4G LTE networks.

Applications ranging from industrial machine control to autonomous warehouse robots can make decisions more quickly because of the reduced latency.

3.5 Other benefits of 5G features for private networks

3.5.1 Edge computing

You don't want your car to transmit information to a distant cloud server before making an evasive maneuver if you're being driven by an autonomous vehicle and another vehicle loses control and is headed your way. To reduce the latency associated with this task, you want the computing work to be performed as close to you as possible. This is called edge computing, which provides cloud computing capabilities at the network's 'edge.'

When 5G networks are paired with edge computing capabilities, interactive and delay-sensitive applications like gaming and augmented reality are made possible. Warehouse and factory operators can cut costs by moving the majority of the robots' computing power to the network core in industrial settings where edge computing is already widely used in 4G private wireless networks. This allows for lighter, more maneuverable, energy-efficient, and less expensive robots.

Industries will gain from edge computing techniques being developed for public networks in the form of an edge cloud. In the edge computing technique, the latency is reduced as the cloud advances closer to the edge of the network to transfer, store, and analyze the huge amounts of data created by devices.

Industry 4.0 is made possible in large part by Multi-Access Edge Computing (MEC). The deployment of number of use cases immediately influences and is increased by the proximity of the data processing capability in the industry context. With data processing capabilities on or near the industry, coupled with 5G networks' low latency, more advanced services, like machine vision, can be provided, opening up the possibility of further automating complicated procedures. By isolating data for security, privacy, or confidentiality concerns, the storage and data processing more effectively processed by edge computing.

The crucial part in the context of a private 5G network is whether the edge cloud should be constructed around a private or public cloud and how this relates to the network architecture. The edge cloud's public or private architecture can be thought of in the same way as the private network itself. An exclusive infrastructure and management are needed for a private cloud. The best-of-breed components from public cloud providers will be used by a public or hybrid cloud to offer edge capabilities.

The edge computing capabilities are achieved through any one of the following methods:

Private network and private cloud: There is no external inputs or outputs in an entirely closed system and all the data processing are finished and stored locally.

Public cloud and public network: In this combination either dedicated or shared infrastructure run by the public mobile network operator is utilized to enable edge computing of public edge agents and infrastructure. This utilizes regional, national, or metropolitan data centers in addition to resources from the public mobile network, such as base station sites.

Hybrid network and hybrid cloud: The hybrid model is created to the use of public cloud and public network providers. In this model using isolated edge infrastructure, the majority of data processing is carried out locally, but data can also be accessed and stored from a public network or cloud service for certain use cases, such as logistics.

3.5.2 5G network slicing

The traditional end-to-end network offered all users the same service in earlier generations—2G, 3G, and 4G. It would only have been possible to construct a Virtual Private Network or build a new, physical network if an operator's client, like the police or fire department, needed a guaranteed quality of service for an application. Operators can build virtual, distinct networks within a single physical network via network slicing. The network operator can ensure speed, latency, stability, and security service levels for the slice as if it were a separate network, whether it's linking a handset to mobile broadband or a robot to an automation program.

A sports arena might feature sections for uploading selfies by spectators, security surveillance, and the broadcaster. Functions to enable each slice would be distributed differently, and each would have different needs and priorities. Even if 50 000 supporters attempt to share their recorded footage to celebrate a goal, the traffic in the broadcast and security slices will always receive the specified bandwidth.

The novel feature of 5G networks called network slicing enables certain customers or apps to have the network resources reserved for their sole usage and give a range of performance attributes that are appropriate for their requirements. This is accomplished by configuring the 5G physical infrastructure to create a virtual network that offers a specialized network for customer or application with the same functionality as a physical network [8].

The common mobile network architecture is logically or physically configured through network slicing to provide the necessary competences to the corresponding users and applications. This contrasts with what occurs over 4G and previous networks, where users and devices are fundamentally connected to one network with a shared set of features. Therefore, 5G networks can share many network slices that may give various network properties for various users or applications.

To achieve explicit SLA criteria, such as speed and capacity and ultra-reliable connectivity for a certain application, network slicing enables mobile service providers to provide a virtualized section of their networks for particular customer or industry requirements. Even though network slicing is quiet in its infancy, CSPs are probably beginning with a few fixed network slices for applications that demand URLLC and eMBB, Massive IoT (MIoT) video and gaming. However, network slices will be coordinated in almost real-time down to the specific enterprise level in a few years. If adopted, 5G slices may one day become as widespread as network domain names. An open RAN architecture or distributed small cell, which is less complex than network slicing, can also be used to accomplish many of the same objectives as 5G slicing today.

As it relates to the Industry 4.0, this implies that it is possible to create a virtualized network that tailors performance to the unique requirements of various industrial use cases such as:

- Network communications with low latency, which can be needed for manufacturing applications requiring control in close to real-time.
- High bandwidth availability suitable for sending picture data in AI and Edge applications or low bandwidth availability for industrial applications.
- Network or data dependability or other Quality of Service (QoS) measures, such as ensuring that industrial control applications will get a control command through the network with specific communication features, such as security and privacy.

3.5.3 Open networking

The telecoms sector is abuzz with discussion of open radio access networks (O-RAN) [19–22].

Networks have become more complex, and they have also become more homogeneous in terms of equipment vendors. This is especially true for the RAN, the portion of the network that is readily visible and made up of the base stations and antennas atop towers.

With O-RAN, the protocols and interfaces between the major RAN building elements (radio antenna, hardware, and software) are made 'open' so mobile network operators can employ products from diverse suppliers while still ensuring compatibility.

There is greater innovation and a wider range of alternatives for the operators since more vendors supply the basic components. The RAN Intelligent Controller (RIC), which gives programmability to O-RAN, is another feature. For instance, artificial intelligence can be implemented through the RIC to optimize the network near a football stadium on game days.

Industry 4.0 now has the chance to seek Total Cost of Ownership (TCO) advantages from two architectural changes: open interfaces and virtualization.

Although an organization may adopt one without the other, technology is now emerging that combines virtualization of RAN (centralization) with open interfaces.

- Virtualization—To improve economics through capacity aggregation or cloudification in centralized systems such as C-RAN or V-RAN with an operator.
- Open interfaces—For a network's end product to offer more variety and innovation.

There are two ways that an Industry 4.0 can establish an open network: either by utilizing localized private, public, or hybrid network coverage and capacity (such as edge computing), or by connecting to a wide area network that is provided by a mobile operator.

When an enterprise deploys a dedicated network on its premises, it can connect to an operator's network via a variety of open interfaces (such as the RAN or core network) for wide-area coverage with services (like edge computing), which are supported through a Network Slice, and in the Network Slice's management console.

Mix and Match refers to a situation in which the network coverage is implemented on the premises of the industry and maintained by the operator, with customized local and wide area services (such as computing) enabled through a network slice. Industry 4.0 has the option of choosing and owning the RAN vendor that supports O-RAN interfaces.

3.6 Advantages and applications of private 5G networks

3.6.1 Advantages of private 5G networks

Compared to conventional solutions like wired Ethernet, Wi-Fi, or unlicensed wireless networks, private 5G networks offered via 5G mobile technologies offer a wide range of advantages [5].

- When reconfiguring conventional Ethernet networks, it is expensive and difficult to move cables and equipment around; nevertheless, private 5G networks allow flexibility in device deployment and movement efficiency.
- Wi-Fi and other unlicensed wireless technology networks do not always offer scalability and reliability in areas where cable access is appropriate. 5G's URLLC can provide the coverage and capacity needed with high quality and dependability on mobile private and dedicated networks.
- The sophisticated variety of features offered by 5G network solutions can especially help to meet strict security, privacy, and data isolation needs.
- For efficiency and safety, autonomous guided vehicles (AGVs) must have high bandwidth coverage, highly focused service, probable error and delay characteristics, and device capability. These Wi-Fi networks do not support them well, especially when the network is busy or while switching between access points, or handovers. Industry 4.0 can choose in what way the mobile network is deployed for their particular situation, whether this is for coverage, capacity, or associated aspects such as data privacy and security, or the necessary SLA. 5G specifically delivers the characteristics needed for AGVs operating on campus over a private network, or over widespread areas over the public network. They might also ask the mobile provider to provide essential services including network design, deployment, operations, and maintenance.
- When edge equipment is deployed for the private network, transmission latency can be decreased because there is no longer a need to transport user data across occasionally great distances, as is the case with older public mobile networks, mostly 2G and 3G. This makes it possible to provide real-time services, especially with the advancements made possible by ultra-low latency services provided by 5G when combined with edge computing.
- For 5G, the transmission latency and error rate are extremely predictable, enabling the use of trustworthy software-based compensation. This is crucial for industrial and autonomous applications in contrast to Wi-Fi networks, where device congestion is frequently a problem.
- Bandwidth can be distributed at scale to meet business demands without necessarily relying on the public mobile network operator's larger deployment plans.
- Depending on the industry use cases, uplink/downlink bandwidth ratios can be altered using the new 5G radio interface and time division duplexing (TDD) technology. This means that manufacturers are not constrained by network design decisions that are important for users of public mobile networks. The uplink bandwidth need might be significantly more than the downlink, making it pertinent for use cases like video image processing in industry automation or driverless vehicles.

- With the introduction of on-site edge computing, network slicing can also be used in private networks to separate devices and applications, allowing for significant improvements to data security and containment.

- The characteristics called time synchronization network (TSN) included in 5G networks, which permit the incorporation of legacy Ethernet TSN networks, are able to support time-critical applications.

3.6.2 Automated and robotic deployments in retail

To perform several repetitive operations, retailers of today in the retail sector use remote, automated equipment like inventory robots, guided vehicles, forklifts, and many others to manage inventory and track assets.

To reduce complexity on the user equipment (UE) side, uplink and downlink data transmission is designated as having a high priority in all automated use cases and handled on a local server. The majority of automated devices that demand extremely low latency and great dependability need a high signal-to-interference-plus-noise ratio (SINR), which is only attainable in deployments of private networks built on 5G technology.

High-resolution video cameras and cutting-edge detection technology are required in other areas, such as industry and government, in order to spot and record any incidents that pose a threat to human life as well as errors and flaws in industrial applications. Through the air interface, the camera continuously monitors and transmits its UHD video stream. High bandwidth, sensitivity to latency, secure transmission, and processing on a secure private server are required for this data connection. These sectors (private and public) choose not to share the data with the public cellular carrier for such processing. By using spectrum-efficient 5G wave-forms, multi-spectrum support, flexible deployments, and the placement of inter-mediate user plane function (UPF)-like nodes or by branching in UPF adjacent to edge server, 5G private networks can handle such designs [6]. Since a more trustworthy network can be controlled and updated as necessary, these use cases serve as a great illustration for Industry 4.0 that demand high performance.

3.6.3 Smart cities, smart offices, smart factories, and the gaming industry

In order to provide a richer experience, 5G is frequently combined with Wi-Fi and used in complimentary modes. For the continuation of sessions, several industrial sectors require high bandwidth, low latency in the range of 1 ms, multiple connections for reliability, and mobility supported by micro and macro on private-private and private-public networks. Due to control user separation, support for multiple connectivity via multiple radio access technologies through a single core, support for multiple device profiles, and the capability to integrate RAN and Core with non-3GPP (Third Generation Partnership Project) technology through cloud-native interfaces, 5G-based private network topology facilitates these use cases. The following are additional crucial qualities that support 5G private networks in businesses like the gaming industry and the smart factory, office, and city [9].

- Reliable bandwidth and service both inside and outside of the office
- Workforce locations with greater density (for example in real-time game)

- Rationalizing devices and ensuring seamless employee mobility
- Improved cooperation through high-definition multi-media and AR/VR that is both reliable and latency-free
- Enhanced connectivity across Wi-Fi and 5G
- Using intent-based networking with consistent policy and security
- Critical manufacturing quality and efficiency with 5G high assurance wireless
- Closed-loop connectivity platform for AI/ML processes

3.6.4 Applications in healthcare

The healthcare system must rely on connection in order to continue vital and routine patient procedures while ensuring the security and safety of front-line employees, such as doctors and nurses. Additionally, a lot of healthcare facilities, including hospitals and dental offices, are utilising the potential of virtual reality (VR) to lessen patient discomfort and help with pain management. Patients going through normally stressful or even painful treatments and recoveries are finding much-needed relief thanks to VR headgear. The growing library of streaming virtual experiences and the popularity of this technology means that ordinary network architecture cannot handle VR devices without consuming too much bandwidth or degrading the stream for your patients.

In order to choose a virtual experience, the patient using the VR headset browses a library that is saved on a nearby server. This information is then sent to the local network by the headset. Patients' chosen virtual experiences can be streamed through the VR headset with little to no lag since it receives and responds to locally processed, low-latency data.

3.6.5 Fixed wireless access

Industrial IoT (IIoT), one of the outside use cases for a private network using 5G as an access and core topology is Fixed Wireless Access (FWA). It makes it possible to offer ultra-high-speed mobile broadband in rural and suburban regions for the service provider. It is also a possible replacement for fiber broadband for home and school broadband. When fiber is impractical to maintain over the long term for residential and commercial applications, FWA can be employed. As a result of free spectra like CBRS and the adaptability of heterogeneous networks (Het-Net), there is less reliance on a public network these days for such use cases.

A further benefit is network sharing and neutral host integration with the public network, particularly when used in conjunction with the implementation of O-RAN and the 5G converged core with an open stack, and virtualized platforms that offer flexibility and openness. The RAN Baseband Unit (BBU), which was a combined hardware/software solution, was also very reliant on proprietary technology and did not allow open designs through its interfaces. Decoupling and virtualization are currently steps toward more individualized and flexible solutions that meet various needs. Above all, open architectures were not supported by interfaces.

3.7 Citizens broadband radio service (CBRS) for private 5G network

3.7.1 CBRS overview

The 3550–3700 MHz band is used by CBRS [15–18], a shared wireless broadband technology also referred to as the 3.5 GHz band. In order to enable shared usage of this band by the federal government and non-federal entities, CBRS, which is known as the innovation band, is designed to allow a three-tier shared spectrum architecture.

The important characteristics of CBRS are as follows:

(i) **Greatest use of the spectrum:** General Authorized Access (GAA) is used for private networks and Priority Access License (PAL) is used on public ones.

(ii) **Wide-area networks for macro:** business models that are sustainable, scalable, and do not just focus on indoor solutions for location-specific connection. There is no reliance on Tier 1 mobile network operators (MNOs). The distinction with CBRS is in the spectrum assignment; both CBRS and LTE share the same RF interface whether operating in the licensed frequency or the unlicensed 5 GHz band.

3.7.2 Requirements for CBRS

The requirements for Citizens Broadband Radio Service Device (CBSD), End User Device (EUD), Priority Access License (PAL), and General Authorized Access (GAA) are detailed in the section below. These requirements are intended to specify the operation and standards interfaces required to affect a properly functioning environment for spectrum sharing in the 3550–3700 MHz band.

3.7.3 Components of the CBRS network architecture

Dynamic spectrum access in a tier system is the core idea behind CBRS. A mechanism has been developed to make the sharing of the spectrum in real-time. A distributed system serves as the foundation for the CBRS spectrum coordinating architecture. The database maintained by the FCC, which centralizes spectrum allotment, is at the top of the hierarchy. The Spectrum Access System (SAS) is the next level. The sensor network known as the Environmental Sensing Capability (ESC) is the following tier. To provide shared spectrum access, the ESC system recognizes and notifies an SAS of the presence of a signal from an incumbent user. The SAS user network, which interacts with the SAS for PAL and GAA usage, is the next tier. Information from the FCC database, other SASs, Environmental Sensing Capability (ESC), and CBRS Broadband Service Devices (CBSD) are collected by it as the gatekeeper. After that, each CBSD is given a specific amount of frequency and power according to FCC regulations.

3.7.4 Network structure for CBRS

The fixed base stations (BSs), or networks of BSs, that make up the CBSDs can only be controlled and managed by a single SAS. Users of the PAL and GAA are expected to use only CBSDs that have been certified by the FCC and approved, and these CBSDs are required to register with the SAS with the necessary information, such as operator ID, device identity and parameters, and position data. The controlled CBSD network in a typical MNO deployment scenario consists of the domain proxy (DP).

The CBSDs are similar to LTE and new radio (NR) base stations, however they differ in that they can only function with the SAS's permission. Together PAL and GAA users are expected to adhere to the technical criteria and must pass a test in a lab. Such a CBSD receives an FCC-approved ID and serial, which are likewise kept in the FCC database. These authorized CBSDs will also register with SAS with the data necessary to comply with the standards, such as operator ID, device identity and parameters, position information, and more. It is advised that in a large commercial deployment, the new network element should handle all CBSD devices in addition to the element administration.

In such a network, the DP may be a bidirectional information routing engine or a more sophisticated mediation function that enables flexible self-control and inter-ference optimizations. Additionally, DP offers a translational competency to inter-face old radio equipment with an SAS, for instance, integrating the small cells of a mall or sports arena into a virtual BS entity. CBSDs are provisioned and configured, similar to traditional LTE systems, an element management system (EMS) is a necessary component. A network management system is discretionary but recom-mended method for big CBSD deployments (like MNO), as it offloads and simplifies the individual CBSDs while centralizing communication to the SAS network.

The primary responsibility of SAS is to manage the interference environment, impose exclusion zones and protection measures to safeguard higher priority users, and dynamically establish and enforce CBSD maximum power levels in both space and time. All SASs must have identical models for calculating interference, according to the FCC. In addition to the aforementioned, SAS also handles user information identification, registration, authentication, and SAS-SAS message exchange.

The FCC passed regulations requiring ESC in and near the CBRS band to identify mandatory radar activity in coastal areas and close to inland military sites to meet the mission-critical criteria of the DoD Incumbent Access. Once incumbent access activity is discovered, the ESC notifies an SAS of it for processing. If necessary, the SAS issues a directive to a commercial user to leave an interfering channel within 300 s of any change in time, frequency, or location.

3.8 Modeling of 5G private networks' economic impact

3.8.1 Overview

Private networks offer a number of specialized possibilities to assess the economic merit of various business model options. The flexibility of ownership within the value chain is the main differential. Businesses, service providers, infrastructure vendors, and anybody else in the value chain may choose to use private 5G networks for their own gain or the benefit of others [4].

The choice to participate for their benefit is probably not influenced by outside variables, but rather by an innate awareness of various value drivers and how they stack up against current or potential alternatives. The main value drivers were privacy, security, control, and performance listed in the use case section.

3.8.2 Business perspective

The problems that modern businesses confront as they embark on their digital transformation journey have risen. A successful digital transformation drive requires the merging of the operational technology (OT) and information technology (IT) worlds, and network infrastructure serves as the thread that binds everything together, making it a key component of this convergence. The flexibility, future-proofing, and control mechanics of the network fabric should, nevertheless, take into account the fact that the necessities for OT and IT are extremely dynamic and constantly changing. Private networks provide this level of flexibility and control in addition to performance, security, and privacy—all essential characteristics.

Examining present capabilities—or lack thereof—as well as the harm caused by new problems and the potential benefits of private networks in addressing some of these problems—are some ways to measure each of these traits. By implementing private 5G networks, such as using remotely connected robotic arms with real-time capability in production facilities, a particularly harmful influence will be completely eliminated. The accuracy of the increased number of wirelessly connected sensors may help to reduce incidents and decrease downtime associated with malfunctions during retooling. Wireless connectivity may also help to improve downtime during predictive maintenance to reduce downtime associated with malfunctions.

3.8.3 Perspective of a service provider

Despite the saturated consumer arcade, service providers are finding new ways to make money in the business world. However, there are other prospects such as new IoT business opportunities, macro network offload, premium SLA services, and managed services. Connectivity revenue from industry clients appears to be the main expectation of installing private network solutions. But it's important to carefully consider how much each of these extra services will cost. The goal is to understand the economics of when things are running properly rather than relying on typical investment plans, which may offer realistic predictions of working conditions. High expectations make it difficult to foresee worst-case scenarios before implementing the solution, and the decentralized structure of these networks makes this task even more difficult. Some of the difficulties may be reduced by network automation, such as more precise incident management, predictive maintenance, or self-healing networks. There will be a certain number of occurrences that necessitate traditional managed service solutions or break-fix, both of which are not always affordable to deliver.

3.8.4 View from an infrastructure vendor

Because there is additional shared and unlicensed spectra available for private 5G networks, there is not only the typical vendor and provider relationship opportunity, but also a direct-to-market opportunity. Because there will be fewer entry barriers, vendors will be able to expand their traditional revenue models and add recurring contracts to them, just like a service provider. But in order to provide such solutions and services, it will be necessary to build up a new operational personnel and set up new operational capabilities, which would require a sizable investment and continuing capital commitment. Some of these difficulties may be reduced, although not entirely, by network management automation. So, before making such a commitment, it is crucial to conduct a thorough risk/benefit analysis.

3.8.5 New player perspective

With increased shared and unlicensed spectrum possibilities, entrance hurdles are marginally lessened, and new participants like tower businesses or real estate corporations may view private networks as a niche option, among others, to tap into new markets. The assets these entities presently own or plan to hold in the future may result in considerable operational cost synergies. The core competencies and capital structure are better suited to control facility-related expenses and provide an operational foundation for these networks.

3.8.6 Various models of funding for private networks

One of the key factors in the enterprise's acceptance is ease of use. It is critical to take into account how simple it is for the business to deploy and manage the solution. IT affects adoption and, in the end, total cost of ownership (TCO). TCO is constantly a priority for IT departments. The operational models must take into account old infrastructure or provide a mechanism to do so. Consolidating the workloads will make it easier to maintain, view the big picture, and eliminate a variety of servers [10]. Finally, how businesses are using the cloud in terms of cost, data control, and general operations must be considered.

Because the ecosystem has not yet been unified around a single dominating business model structure, the ownership structure and roles and duties are also changeable. Due to the wide range of needs and capabilities, both solo and hybrid funding opportunities, as shown in figure 3.4, may be available.

3.8.7 Finding synergies in common

It is expected that every participant in the value chain will assess the competitiveness of a 'one-size-fits-all' strategy. However, the absence of an ecosystem capable of providing end-to-end solutions today suggests large investment requirements, which could be difficult given the state of the industry. On the other side, in the early stages of private networks, there may be an opportunity to find possible collaborations and overlaps between different players in order to explore accessible business models.

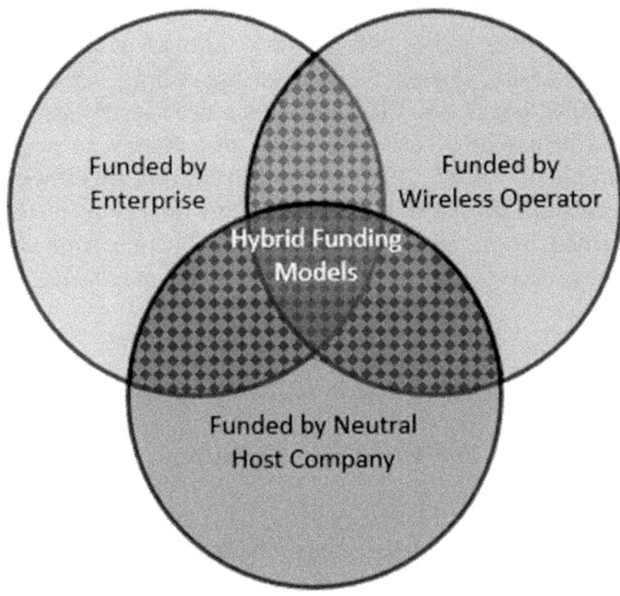

Figure 3.4. Various models of funding for private networks.

It might also be conceivable to containerize the physical components of each potential offering to select and combine different ranges of solution stacks to varied company demands in an increasingly digital environment where virtual solutions are more containerized and stackable.

The one-size-fits-all type of solution might be too challenging to adopt for any enterprise, which means it may be beneficial to offer that personalized strategy. Instead, this degree of customization will call for a large amount of testing, standardization, and coordination of each solution stack to deliver an end-to-end capability. Without significant commitment and alliances among industry participants, achieving such mature cooperation will be very difficult.

3.9 Conclusion

The ability to build a factory without wires and cables is perhaps the most obvious benefit of 5G for manufacturers. While expensive investments like these provide network bandwidth, they also create barriers to factory mobility and the ability to reconfigure robots and assembly lines as product lines change. By doing away with wires, 5G lowers costs, improves adaptability, and provides extremely high bandwidth and low latency. According to estimates, 5G networks will be 100 times faster than 4G LTE. There are organizations that support it and invest funds in it. You can utilize 5G instead of a wired solution because the performance is so much better. For a plant, it is much better. The greatest locations for private 5G include large manufacturing, ports, and storage companies. Network slicing, which enables multiple independent logical networks to operate on the same physical network while sharing physical capacity among them, is an appealing feature of 5G networks. Each slice is an end-to-end network tailored to a certain application's requirements. In a factory, one slice might service robots, which require minimal latency, while another serves AR/VR apps, which need faster speeds. Additionally, network slices can be assigned based on workload demands to remove bottlenecks and increase throughput. CBRS, a 3.5 GHz spectrum that was reserved for the U.S. Navy but has seen little use, especially in locations remote from military stations, may be implemented by organizations making the switch from 4G LTE to 5G. CBRS has demonstrated its appeal for 4G LTE deployments thus far.

References

[1] Curwen P and Whalley J 2023 5G literature review *Understanding 5G Mobile Networks* (Emerald: Bingley)

[2] Rao S K and Prasad R 2018 Impact of 5G technologies on Industry 4.0 *Wirel. Pers. Commun.* **100** 145–59

[3] Wollschlaeger M, Sauter T and Jasperneite J 2017 The future of industrial communication: automation networks in the era of the internet of things and Industry 4.0 *IEEE Ind. Electron. Mag.* **11** 17–27

[4] Rüßmann M, Lorenz M, Gerbert P, Waldner M, Justus J, Engel P and Harnisch M 2015 *Industry 4.0: The future of Productivity and Growth in Manufacturing Industries* (Munich: Boston Consulting Group) https://bcg.com/publications/2015/engineered_products_project_-business_industry_4_future_productivity_growth_manufacturing_industries

[5] Soós G, Ficzere D, Seres T, Veress S and Németh I 2020 Business opportunities and evaluation of non-public 5G cellular networks—a survey *Infocommun. J.* **12** 31–8

[6] Sachs J and Landernäs K 2021 Review of 5G capabilities for smart manufacturing. *Proc. 2021 17th Int. Symp. on Wireless Communication Systems (ISWCS)* (Berlin, Germany) 6–9 September 2021, pp 1–6

[7] 5G URLLC-Achieving Industrial Automation. Available at: https://ericsson.com/en/cases/2020/accelerate-factoryautomation-with-5g-urllc Access date: Apr 27, 2023

[8] GSMA 2023 *5G for Smart Manufacturing Insights on How 5G and IoT Can Transform Industry* Available at: https://gsma.com/iot/wp-content/uploads/2020/04/202004_GSMA_SmartManufacturing_Insights_On_How_5G_IoT_Can_Transform_Industry.pdf Access date: Apr 27, 2023

[9] Cheng J, Chen W, Tao F and Lin C L 2018 Industrial IoT in 5G environment towards smart manufacturing *J. Ind. Inf. Integr.* **10** 10–9

[10] Smart Manufacturing Powered by 5G. Available at: https://pwc.com/gx/en/industries/tmt/5g-in-manufacturing.html Access date: Apr 27, 2023

[11] Role of Private 5G Networks in Industry 4.0 and Beyond. Available at: https://stl.tech/blog/role-of-private-5g-networks-in-industry-4-0-beyond/ Access date: Apr 27, 2023

[12] 5G Network Slicing Use Cases-Tracing Untold Possibilities. Available at: https://stl.tech/blog/5g-network-slicing-use-cases-tracing-untold-possibilities/ Access date: Apr 27, 2023

[13] Private 5G: Industry 4.0 Use Cases. Available at: https://stlpartners.com/articles/private-cellular/private-5g-industry-4-0-use-cases/ Access date: Apr 27, 2023

[14] Private 5G Takes Industry 4.0 to the Next Level. Available at: https://ibm.com/blogs/digital-transformation/in-en/blog/private-5g-takes-industry-4-0-to-the-next-level/ Access date: Apr 27, 2023

[15] 5G, Private Spectrum and CBRS. Available at: https://celona.io/cbrs/cbrs-5g#:~:text=So%20What%20Is%20CBRS%20and,LTE%20and%205G%20cellular%20technologies Access date: Apr 27, 2023

[16] 5G Deployment: CBRS or C-Band? Available at: https://solid.com/us/5g-deployment-cbrs-or-c-band/ Access date: Apr 27, 2023

[17] Citizens Broadband Radio Service (CBRS). Available at: https://commscope.com/solutions/5g-mobile/citizens-broadband-radio-service-cbrs/ Access date: Apr 27, 2023

[18] What Is the Citizens Broadband Radio Service (CBRS)? Available at: https://metaswitch.com/knowledge-center/reference/what-is-the-citizens-broadband-radio-service-cbrs Access date: Apr 27, 2023

[19] Why Should MNOs Care about Open RAN? Available at: https://moniem-tech.com/questions/why-should-mnos-care-about-open-ran/ Access date: Apr 27, 2023

[20] The 5G Open RAN Revolution: Freedom for MNOs, A Better Deal for Customers. Available at: https://forbes.com/sites/forbestechcouncil/2021/02/02/the-5g-open-ran-revolution-freedom-for-mnos-a-better-deal-for-customers/?sh=26223ef165a3 Access date: Apr 27, 2023

[21] 5G Open RAN: What You Need to Know. Available at: https://stl.tech/blog/5g-open-ran-what-you-need-to-know/ Access date: Apr 27, 2023

[22] OpenRAN – 7 Vital Benefits for MNOs. Available at: https://parallelwireless.com/blog/openran-7-vital-benefits-for-mnos/ Access date: Apr 27, 2023

IOP Publishing

Advanced Signal Processing for Industry 4.0, Volume 1
Evolution, communication protocols, and applications in manufacturing systems
Irshad Ahmad Ansari and Varun Bajaj

Chapter 4

Applications of infrared imaging for non-destructive testing and evaluation of industrial components

Navpreet Kaur, Geetika Dua, Ravibabu Mulaveesala and Vanita Arora

Numerous research groups have focused on the idea of non-destructive evaluation 4.0, which addresses new technologies such as infrared thermal wave imaging technologies for non-destructive testing and evaluation and particularly new experimental thermal methods along with appropriate processing and visualisation approaches. A completely new class of intelligent and effective inspection procedures and digital data ecosystems are made possible by this concept. It is a multidisciplinary approach with the goal of implementing Industry 4.0 technology as well as new process structures to design and enhance inspection methodology. Due to its whole-field, non-coupling, safe, quick, and quantitative inspection capabilities to evaluate various structures regardless of their electrical and magnetic properties, thermal wave imaging (TWI) has gained tremendous popularity in the recent past. Due to its high testing resolution and thermal sensitivity for detecting subsurface flaws, the recently introduced aperiodic pulse-compression assisted frequency-modulated thermal wave imaging (FMTWI) is a good choice among the many possible TWI techniques. Compared to traditional TWI methods, FMTWI provides improved testing resolution and thermal sensitivity even with thermal sources that have relatively low peak power. In this chapter, we propose an algorithm for automatic flaw detection and defect diameter estimation in fiber-reinforced polymer materials using signal- and image-processing approaches. The results demonstrate the potential of the suggested approach for automated inspection of fiber-reinforced polymer materials with the identification, inspection, and characterization of subsurface flaws.

4.1 Introduction

A testing technique called non-destructive testing and evaluation (NDT&E) is used to identify inconsistencies and anomalies in a component without affecting how well it works. NDT&E is crucial in many industries to guarantee the quality of the manufactured component. It contributes to dependable system performance and evaluates the component's health monitoring, preventing catastrophic failure. Typically, production or in-service manufacturing flaws will impair the component's operation. A thorough testing and evaluation procedure, preferably using an NDT&E process, is required to maintain its integrity. NDT&E is a process that allows a test structure to be inspected, assessed, or examined without compromising its continued serviceability. It is important in a wide range of industries, including the electrical, mechanical, civil, aerospace, automotive, and building sectors. An NDT&E method is selected based on the test specimen material, the type of flaw, and its geometry [1–20]. The current work's primary goal is to identify and characterize any voids, anomalies, or hidden flaws in glass fiber-reinforced polymer (GFRP) material.

GFRP composites have grown in popularity as a building material for aircraft structures due to their adequate strength-to-weight ratio, light weight, corrosion resistance, longevity, and ease of maintenance. Defects can change the structure's toughness, so NDT plays an integral role in monitoring and determining the condition of material components [2–15]. Nowadays, several techniques are employed for anomaly/void detection in composite structures, such as shearography, x-ray, and ultrasound. Compound methods are used in a variety of situations to detect and quantify various defects. Timely monitoring and inspection of the components are required using a rapid and easy inspection technique to limit costs. Therefore, it is essential to construct automatic, fast, remote practices and procedures to analyze the recorded data to obtain the quantitative characterization of defects. For the purpose of ensuring the quality of these materials, numerous strategies have been researched and proven to work. Each technique has advantages and disadvantages when used for composites inspection. Infrared thermography (IRT) provides attributes suitable for investigating almost all types of materials irrespective of their physical properties. It is a remote method that is easily automated, and the testing time is shorter than that of other well-established traditional NDT techniques. IRT is extensively used to inspect composite materials [20–30].

Both passive and active thermographies using infrared technology are possible. The thermal response over the test structure is observed with a thermal camera under ambient conditions in passive IRT without any induced thermal stimulus. Although it is inexpensive and simple, extrinsic influences such as surface emissivity variations and thermal reflections from the environment have an impact on the measured heat distribution over the test sample. The passive IRT is unreliable in terms of its usefulness for NDT&E due to these variables. The voids with smaller sizes and those found deeper into the test object are not mentioned. To find minor and more serious flaws, active thermography is preferred. In active IRT, temperature distribution over the test component is mapped while being influenced by a predetermined, controlled external thermal source [30–45]. It aids in turning active IRT into a quantitative method of locating the test specimen's lesser and deeper flaws. The relevant dynamic IRT technique can be classified as either pulse-based pulse thermography (PT) and pulse phase thermography (PPT) or modulated thermography (lock-in thermography (LT), linear frequency modulated thermal wave imaging (LFMTWI), coded thermal wave imaging (CTWI), and Golay coded thermal wave imaging (GCTWI)) depending on the shape of the external thermal signal [35–65].

All these techniques use a volumetric heat source to stimulate the surface of the testing material and analyze the thermal data response over the component's surface. The common feature, standard with all the IRT techniques, is that the defect in the material inducing different thermophysical properties will also yield an anomaly during thermal diffusion of the thermal waves and temperature gradients on the component's surface. These techniques identify areas of varying thermal responses related to an anomaly and are typically based on thermal–physical properties involved in thermal wave diffusions, such as the density of the material, the specific heat at constant pressure, and thermal conductivity.

In PT and PPT, short-duration flashes are imposed on the test structure. The subsequent thermal gradients at the sample's surface are measured with an infrared camera to visualize the defects. This approach is straightforward, swift, and easier to implement. However, the surface features and other environmental perturbations influence the results obtained. PPT and PT are different in that PPT involves post-processing (spectral analysis) of the resultant temperature distribution data to get the magnitude and phase thermal images. Phase thermal images are of more interest to many researchers as they provide a qualitative representation of defects and help obtain quantitative information. In addition, PPT results are less susceptible to surface features of the sample, such as emissivity variations and uneven heating. Furthermore, it uses a short-duration pulse and demands high peak power. So, as far as the high cost is involved, their application is limited and may also deteriorate the surface of the test specimen to resolve the defects with enough resolution and sensitivity.

The sample is heated periodically in lock-in thermography, typically using sinusoidally modulated thermal waves. Thermograms of the test structure are recorded throughout the active period, which lasts for at least one complete excitation cycle. The amplitude of the sinusoidal thermal flux change at the surface and its phase for the applied modulated heating is then computed by frequency domain analysis of the recorded thermal data. LT overcomes the difficulties posed by the high peak power requirements of PT or PPT, but its mono-frequency thermal excitation constrains the test resolution. Therefore, LT must be conducted with several excitation frequencies in order to identify the flaws located at varying depths of distinct spatial dimensions. It turns LT into a laborious method [66–73]. The current work presents an advantageous LFMTWI with high depth-resolved pulse compression for GFRP material fault detection. Using relatively low peak power heat sources with adequate experimentation time, LFMTWI confirms improved detection resolution and sensitivity.

LFMTWI employs a frequency modulated evoking signal applied to the test structure, with frequencies changing within a pre-defined band of nearly equivalent energies. The resulting thermal waves diffuse into the sample structure, resulting in a time-varying thermal distribution. Defects within the sample disrupt heat flow, resulting in thermal gradients across the surface. This thermal response over the sample surface is recorded and then processed with pulse-compression analysis to create pulse-compressed thermograms. Pulse compression concentrates all available energy into a single instant, improving depth resolution [25–35]. Additionally, this work proposes an algorithm for the automated detection of subsurface flaws in GFRP composite material using pulse-compressed thermograms. The algorithm also determines the defect's lateral dimensions (diameter), as well as marks the defect centers. The suggested algorithm is firstly validated on a modeled GFRP sample and then subsequently confirmed for an experimental GFRP test sample in this work. A GFRP sample is subjected to a three-dimensional (3D) finite element analysis (FEA) using COMSOL Multiphysics. The obtained findings unmistakably demonstrate the potential of the suggested scheme for the quantitative estimate of the diameter of hidden flaws as well as the automatic detection of subsurface faults.

4.2 Theory

This section describes the theoretical analysis of the LFMTWI and the different signal and image processing methods adopted in this work for automated detection and estimation of the defects present in the GFRP sample structure.

4.2.1 Linear frequency modulated thermal wave imaging

Theoretically, the thermal gradient profile over the test structure's region is determined by solving the one-dimensional (1D) heat equation for the homogeneous and semi-infinite medium with no heat sink or source present. Equation (4.1) represents the 1D thermal equation [25–45]:

$$\frac{\partial^2 Th(z,\, t)}{\partial^2 z} = \frac{1}{D}\frac{\partial Th(z,\, t)}{\partial t} \tag{4.1}$$

where D stands for thermal diffusivity, $(D = \lambda/\rho\, cp)$; ρ, λ, cp are the test structure's density, thermal conductivity and specific heat, respectively, and $Th(z,t)$ denotes the surface's thermal distribution profile. z stands for diffusion direction, and t stands for time.

Equation (4.2) represents the linear frequency modulated (LFM) signal S with a duration of τ seconds and a bandwidth BW:

$$S = S_0 e^{2\pi j\left(f_i t + \frac{BWt^2}{2\tau}\right)} \tag{4.2}$$

where S_0 is the envelope of the LFM thermal stimulus, f_i is the base frequency and $2\pi j\left(f_i t + \dfrac{BWt^2}{2\tau}\right)$ gives the phase of the signal.

The solution to equation (4.1) for incident heat flux (equation (4.2)) is derived by taking precise boundary and initial conditions into account as [35–45]:

$$Th_{\mathrm{LFM}}(z,\,t) = Th_0 e^{2\pi j\left(f_i t + \frac{BWt^2}{2\tau}\right)} e^{-z\sqrt{\frac{\pi}{\alpha}\left(f_i + \frac{BWt}{\tau}\right)}} e^{-jz\sqrt{\frac{\pi}{\alpha}\left(f_i + \frac{BWt}{\tau}\right)}} \tag{4.3}$$

where τ is the duration of excitation, BW/τ is the modulation factor, and Th_0 is the initial temperature.

The equation for thermal diffusion length μ_{L} is obtained as:

$$\mu_{\mathrm{L}} = \sqrt{\frac{\acute{\alpha}}{\pi\left(f_i + \dfrac{BWt}{\tau}\right)}} \tag{4.4}$$

4.2.2 Pre-processing using polynomial fit

An appropriate fitting polynomial is used to pre-process the thermal response data over the surface of the test structure to feature the zero-mean response. Figure 4.1 depicts a flowchart illustrating the computation of zero mean data. $Th(x_i, y_j, t)$ denotes the captured thermal distribution profile. Here, i and j refer to pixel sequence along the x and y dimension, respectively, and t signifies the frame number. From this captured response, a single pixel temporal temperature profile (i.e., for $i = 1, j = 1$, for all the frames) is extracted. On this extracted profile, a linear fit function is applied to obtain the zero mean single pixel profile, which is then stored. In a similar manner, all the pixel profiles are processed one by one by incrementing i and j until $i = m$ and $j = n$, where m and n are the dimensions of each frame. Further, the zero mean temperature data ($Th_{\mathrm{zeromean}}(x_i, y_j, t)$) is reconstructed from the stored profiles. Zero mean thermal data is further processed using a pulse-compression post-processing scheme as described next.

4.2.3 Signal processing using pulse compression

Signal processing, specifically zero mean thermal distribution data, is done with pulse compression. Pulse-compression data ($T_{\mathrm{PC}}(\tau)$) is obtained by estimating the cross-correlation of the obtained zero-mean thermal distribution profile data ($Th_{\mathrm{zeromean}}(x_i, y_j, t)$) for each pixel with the thermal profile of any chosen reference pixel ($Th_{\mathrm{Ref}}(x_{i,j}, t)$) as explained with the help of the flowchart given in figure 4.2. The equation is shown as:

$$T_{\mathrm{PC}}(\tau) = \int_{+\infty}^{-\infty} Th_{\mathrm{zeromean}}\left(x_i,\, y_j,\, t\right) Th_{\mathrm{Ref}}(x_{i,j},\, t + \tau)\mathrm{d}t. \tag{4.5}$$

Here, (i,j) denotes the reference pixel location. Pulse compression concentrates all of the energy imparted into a confined time instant, improving depth resolution and signal-to-noise ratio. The flowchart is as described in figures 4.1 and 4.2.

$Th_{\mathrm{zeromean}}(x_i, y_j, t)$ here denotes the computed zero mean thermal distribution profile. From this data, extract one pixel temporal profile (mainly a pixel at any

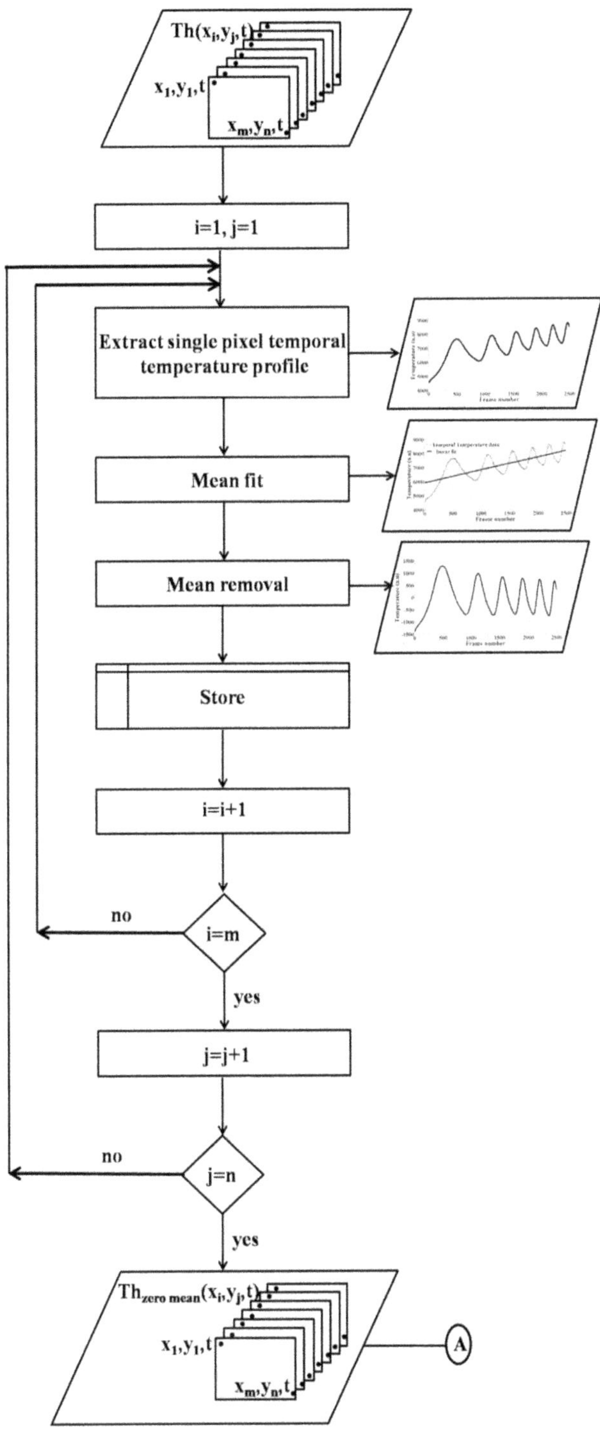

Figure 4.1. Flowchart illustrating the calculation of zero-mean thermal response.

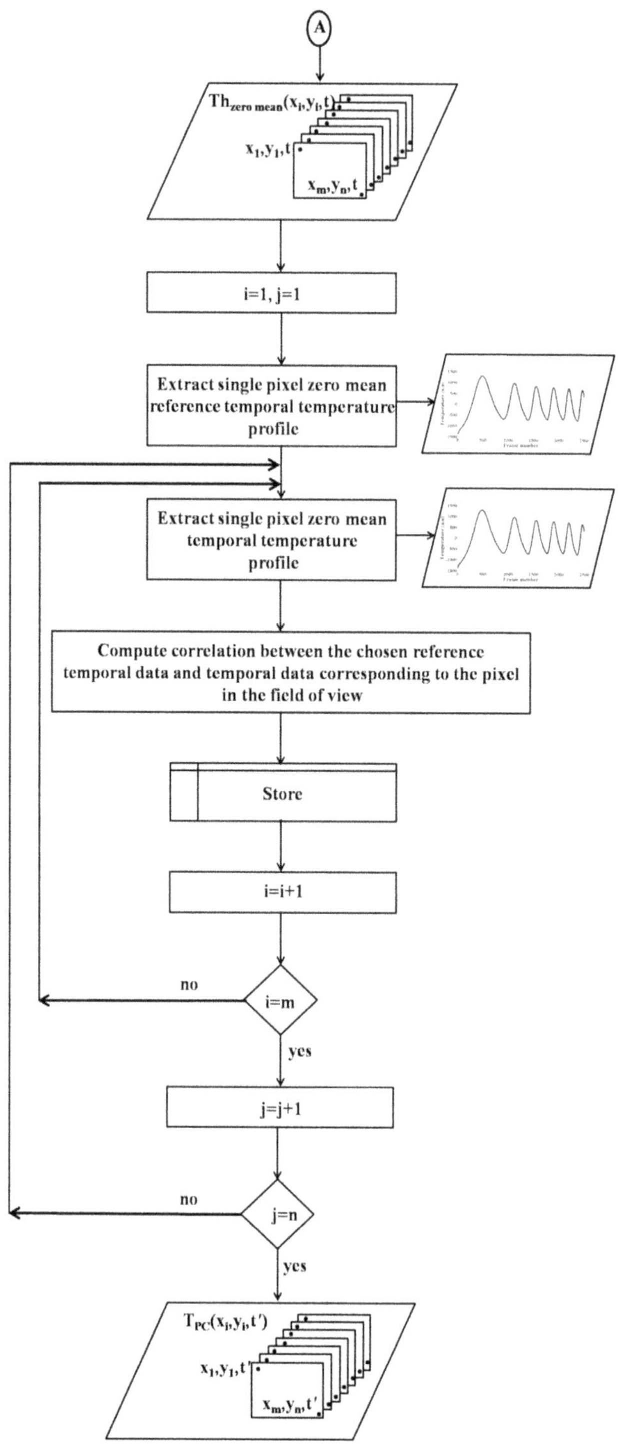

Figure 4.2. Flowchart illustrating the computation of a pulse-compressed thermogram.

non-defective location) and take it as a reference profile. On the other side, extract a pixel profile from the zero mean temperature profile starting with $i = 1, j = 1$, and compute cross-correlation of this extracted pixel with the chosen reference pixel. This will be a pulse-compressed single pixel profile.

In similar manner, extract each pixel from the zero mean data by incrementing 'i' and 'j', till $i = m$ and $j = n$, and compute cross-correlation with the reference. At last, from the stored pulse-compressed single pixel profiles, complete pulsed compressed data $(T_{PC}(x_i, y_j, t))$ is reconstructed for further processing.

4.2.4 Image processing: segmentation, edge detection

Defect identification is the first and most crucial stage in automated void/defect investigation. In digital image processing, semantic segmentation most accurately simulates human perception of fault features. The best pulse-compressed thermogram (one that picks up the most flaws) is chosen to start the process. Additionally, it is utilized to identify the border of flaws, identify centroids, and calculate the lateral dimensions of the defect. The procedures followed follow those outlined in the flowchart in figure 4.3.

Segmentation is partitioning an image (here, a pulse-compressed thermogram) into analogous pixel groups. Segmentation is applied on the pixel level employing a global threshold. In this work, the shape is identified not simply as a collection of parts and adopts a recognition method based on a comprehensive shape boundary-based recognition. For the application of a boundary-based model in scrambled thermograms, precise shape-used segmentation is necessary to isolate defective regions from the non-defective areas. However, without familiarity with the target shape, accurate automatic segmentation of the object from realistic clutter can be difficult. Human perception evidence suggests that familiarity plays a prominent role in figure assignment [5–20]. The proposed method operates on assuming that most defects have a round shaped high-intensity spot (or a void) and, thus, can be estimated applying a template matching-like algorithm. To improve defect resolution, segmentation isolates defective locations from sound (non-defective) regions. The applied method involves the following three steps:

Step I: For this application, not much pre-processing is required as any severe noise removal method would also affect the defects. Thus, we use a Gaussian filter with mean zero and variance 3.5. The result is stored in matrix F.

Step II: Convolution of the obtained image F (to get the filtered image $F2$) with a system t (as given below). The purpose of using this step is to emphasize the parts of the image that have the same variation as t. This filter is selected from recordings of the typical pattern among the defects.

The maxima are selected to have the round shape with a radius of about four pixels.

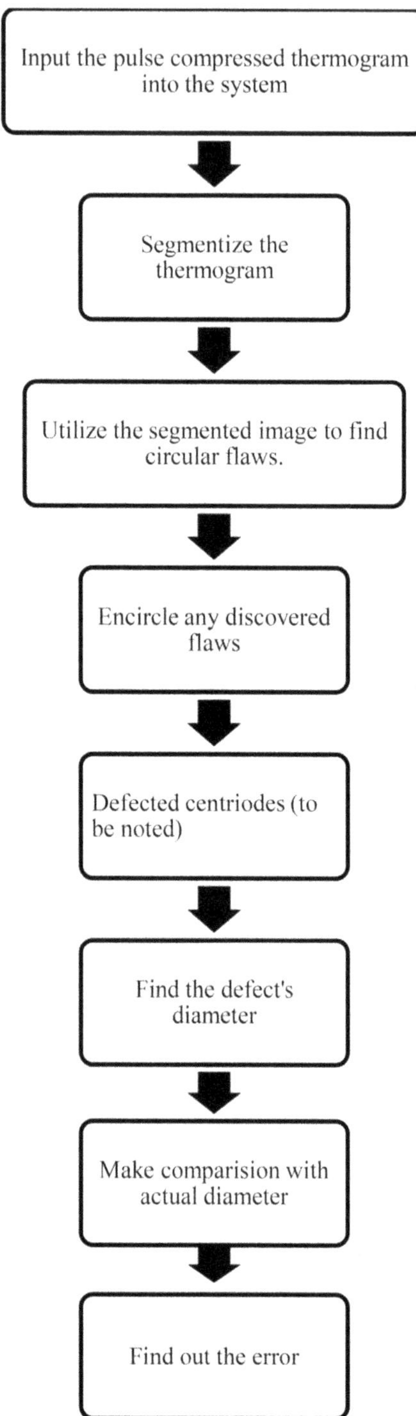

Figure 4.3. Flowchart showing the procedures for calculating the parameters of the fault.

$$t = \begin{matrix}
0 & 0 & -1 & -1 & -1 & -1 & -1 & -1 & 0 & 0 \\
0 & -1 & -1 & 0 & 0 & 0 & 0 & -1 & -1 & 0 \\
-1 & -1 & 0 & 1 & 1 & 1 & 1 & 0 & -1 & -1 \\
-1 & 0 & 1 & 1 & 1 & 1 & 1 & 1 & 0 & -1 \\
-1 & 0 & 1 & 1 & 2 & 2 & 1 & 1 & 0 & -1 \\
-1 & 0 & 1 & 1 & 2 & 2 & 1 & 1 & 0 & -1 \\
-1 & 0 & 1 & 1 & 1 & 1 & 1 & 1 & 0 & -1 \\
-1 & -1 & 0 & 1 & 1 & 1 & 1 & 0 & -1 & -1 \\
0 & -1 & -1 & 0 & 0 & 0 & 0 & -1 & -1 & 0 \\
0 & 0 & -1 & -1 & -1 & -1 & -1 & -1 & 0 & 0
\end{matrix}$$

Step III: We then use a threshold F2 to get a binary image B, representing the mismatches. According to the experimental results, a fixed threshold seems to be better for this purpose.

When segmenting a thermogram based on a void/discontinuity, often edge detection is crucial. If the thermogram contains more than one defect, undoubtedly, there shall be a boundary between the defective and sound regions, leading to the formation of edges. Based on these clearly defined edges, it is possible to distinguish between sound (non-defective) and defective areas, i.e., an edge comes from a change in the intensity distribution of the image at a local level in these thermograms. However, edge detection is an important step in image processing because edges in images are created by the boundary between two objects or between the object of interest and the backdrop. It is possible to separate the defective places from the sound (non-defective) regions by identifying these edges in order to improve fault resolution. In this work, a Canny edge detection technique is used. It performs edge point correlation coefficient comparison and distance estimation within the range of 0–255. By this, boundaries of the circular defects are located, and centroids are marked. The centroid is determined using:

$$c_{n_x} = \frac{\sum_{k=1}^{N} c_{n_k} I_k}{\sum_{k=1}^{N} I_k} \tag{4.6}$$

where c_{n_k} is the nth co-ordinate of pixel k. The number of pixels in the region of interest is denoted by N, k is the pixel index, and I_k is kth pixel intensity. By analyzing the thermogram at the single pixel level, the defect size (diameter) is estimated. Area measurement functions provide excellent defect size within the detected boundaries, i.e., a region of interest (ROI) and defining the pixel count inside the ROI. The size of a void can then be computed by translating the pixels to their corresponding equivalent size and area [42, 43].

4.3 Modeling and simulation

A modeled GFRP sample is subjected to a three-dimensional (3D) finite element analysis (FEA) using COMSOL Multiphysics. This section focuses on how the suggested technique might be used to find flaws in a GFRP sample that has been modeled and is 160 mm by 160 mm in size and 5 mm thick. The sample consists of

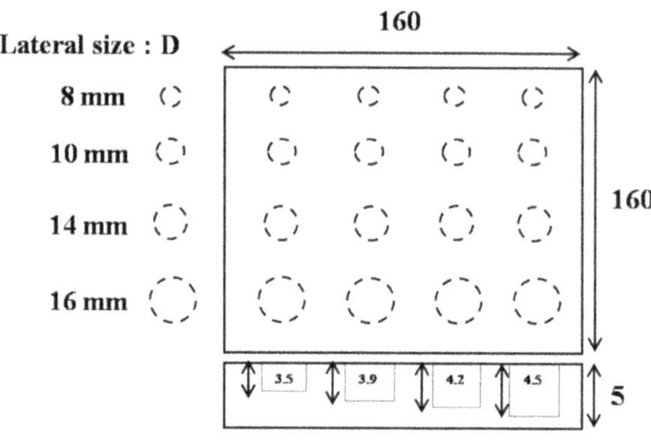

Figure 4.4. Schematic representation of the modeled GFRP specimen (all dimensions are in mm).

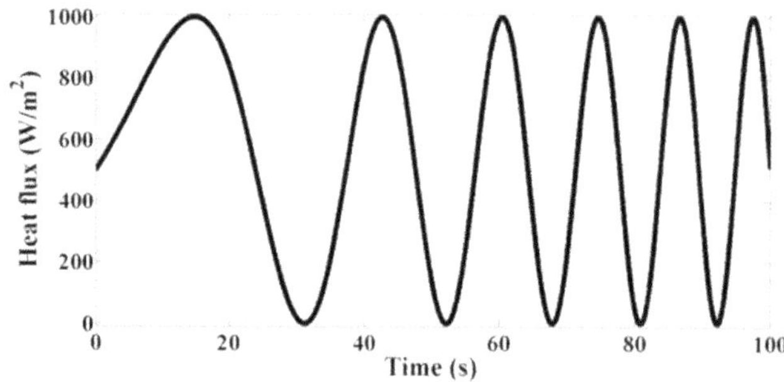

Figure 4.5. LFM signal lasting 100 s with frequencies varying from 0.01 to 0.1 Hz.

16 flat-bottom holes, each with a different radius, arranged in four rows (8 mm, 10 mm, 14 mm, and 16 mm). Each row has four faults spaced at different depths from the front surface of the construction, as shown in figure 4.4.

By stimulating the sample surface with a 1000 W m^{-2} linear frequency modulated (LFM) thermal flux with a frequency varying from 0.01 to 0.1 Hz for a time span of 100 s, the 3D finite element-based analysis is conducted out. In figure 4.5, the imposed LFM heat flux is depicted.

At a frame rate of 25 frames per second, the resulting heat distribution on the surface of the test structure is captured and recorded. Each simulation's recorded thermal response is pre-processed to produce a zero-mean thermal response using an appropriate fitting polynomial, and then it is treated further using the pulse-compression method. The results of further signal and image processing are presented in the results and discussion (section 4.5).

4.4 Experimentation

This chapter focuses on applying the suggested method to find flaws in an experimental GFRP sample with dimensions of 28 × 24 × 6.99 mm (length × width

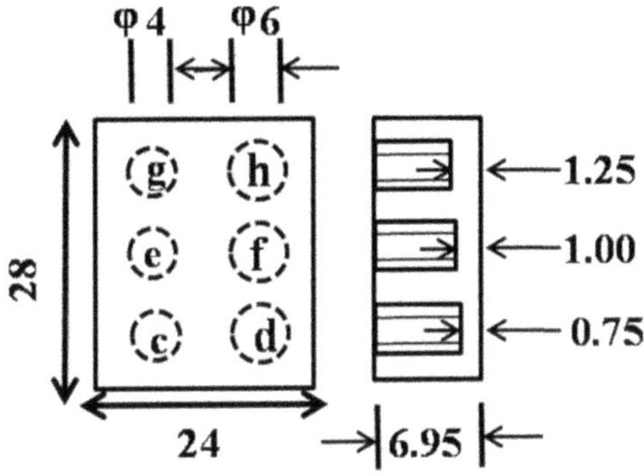

Figure 4.6. Schematic representation of the experimental GFRP specimen (all dimensions are in mm).

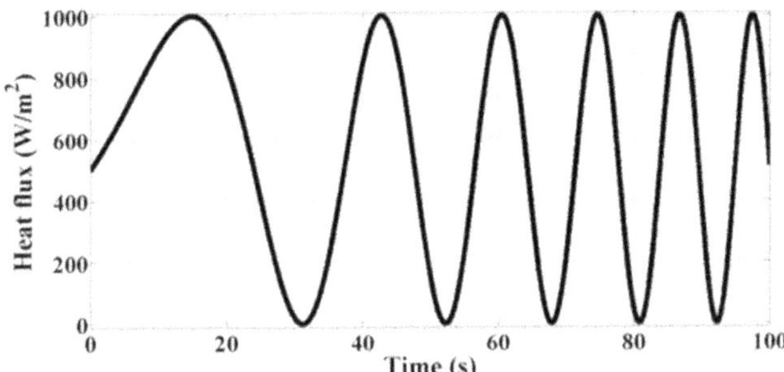

Figure 4.7. Linear frequency modulated signal with frequencies ranging from 0.01 Hz to 0.1 Hz for a time duration 100 s.

× thickness). Six flat-bottomed holes with two columns each measuring 4 mm and 6 mm in diameter make up the test specimen. As seen in figure 4.6, each column has three void defects that are spaced differently from the specimen's front surface.

In order to perform the experimentation, a linear frequency modulated thermal flux of 1000 W m^{-2} is excited over the test specimen surface for 100 s duration with frequencies varying from 0.01 to 0.1 Hz. In figure 4.7, the excitation signal is displayed. Using an infrared camera, the resulting thermal distribution profile over the structure surface is recorded at a frame rate of 25 Hz. Figure 4.8 displays the recorded temperature profile at a specific (average of 3 × 3 region) non-defective location.

The collected response is further processed using the signal and image processing methods described above and results are presented in the next section.

4.5 Results and discussions

This section presents the results using different signal and image processing techniques for automatic flaw detection and estimation in modeled and experimental

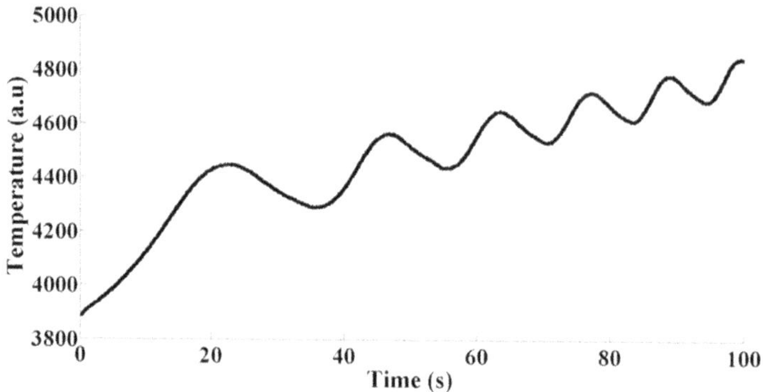

Figure 4.8. Captured thermal profile at a (average of 3 × 3 region) non-defective location.

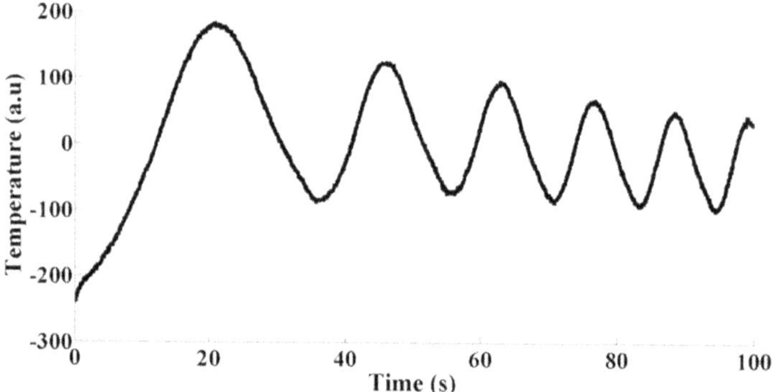

Figure 4.9. Zero-mean single pixel thermal profile at a particular location.

GFRP sample structures. The average temperature rise during active heating is corrected by applying an appropriate fit to the captured thermal response data described in the theory section. The resultant zero-mean thermal distribution profile over the exact non-defective location (taken as an average of the 3 × 3 region) on the surface is shown in figure 4.9. The zero-mean data is further modified using the pulse-compression approach to reconstruct the pulse-compressed thermograms. Figure 4.10 shows the corresponding pulse-compressed single-pixel thermal profile. The reconstructed pulse-compressed thermograms (identifying the maximum number of defects) obtained at a time instant of 8 s for the modeled sample and 10.5 s for the experimental sample are shown in figures 4.11 and 4.12.

Additionally, the flaws' boundaries, centroids, and estimated lateral dimensions are located, marked, and calculated using the chosen pulse-compressed thermogram. Additionally, the flaws' boundaries, centroids, and estimated lateral dimensions are located, marked, and calculated using the chosen pulse-compressed thermogram.

Segmentation and several other edge detection refining methods are combined to locate the boundaries of the defects, as shown in figures 4.13 and 4.14, for modeled and experimental samples, respectively. In order to improve the resolution of defects

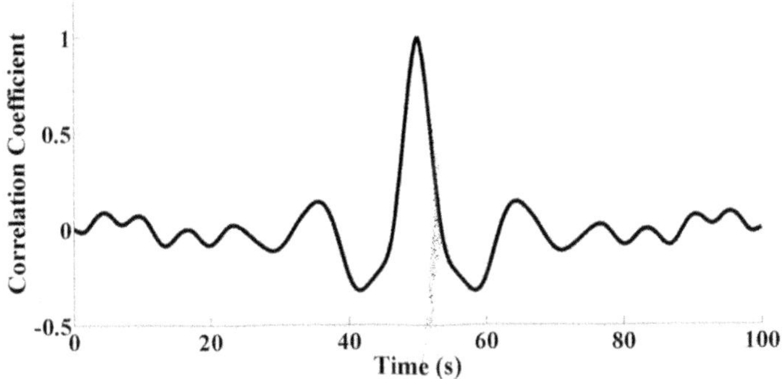

Figure 4.10. Pulse compressed single pixel thermal profile at a chosen location.

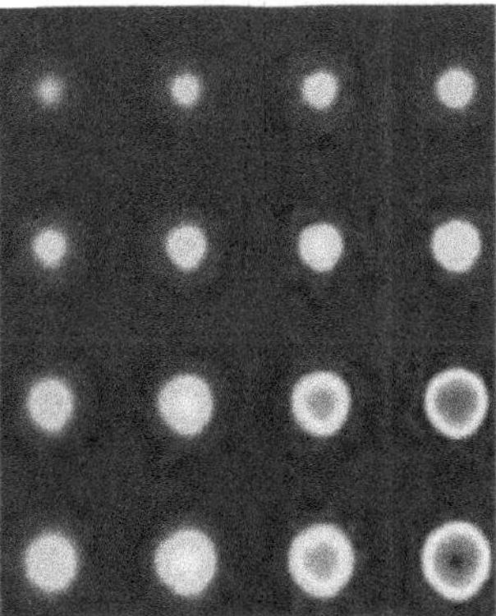

Figure 4.11. Pulse-compressed thermogram for modeled GFRP sample obtained at 8 s.

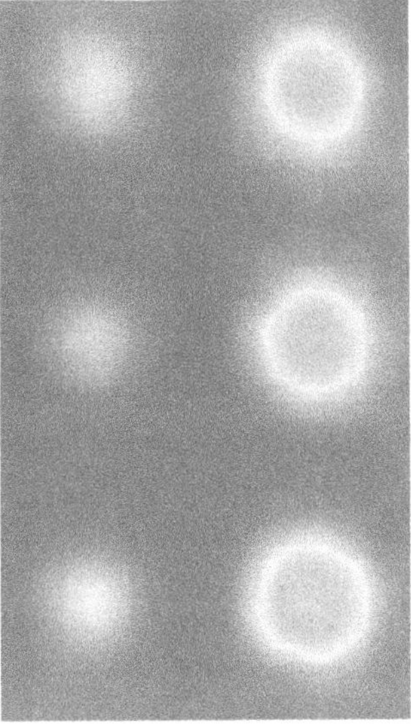

Figure 4.12. Pulse-compressed thermogram for experimental GFRP sample obtained at 10.5 s.

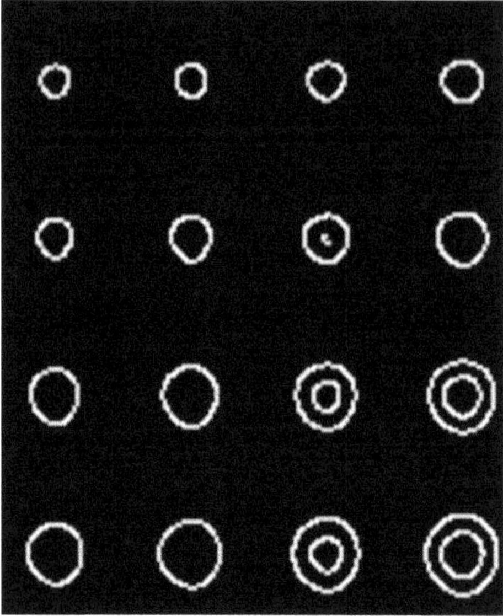

Figure 4.13. Thermogram for the modeled sample that displays the boundaries of the flaws.

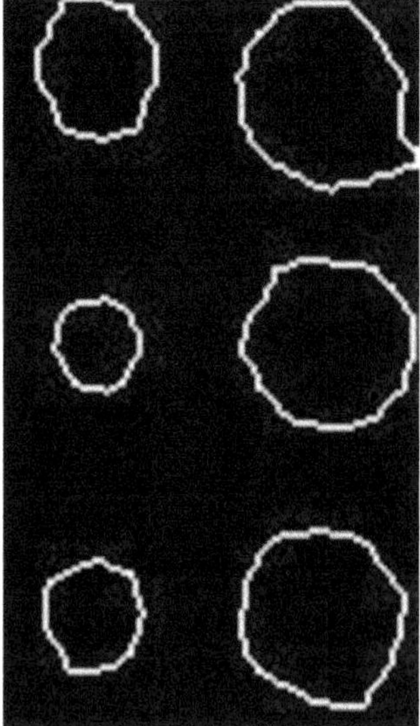

Figure 4.14. Thermogram for experimental sample displaying the borders of flaws.

for the modeled sample and the experimental sample, respectively, figures 4.15 and 4.16 show the isolation of defective sites from the sound (non-defective) regions.

Additionally, the maximum and minimum radius of segmented objects are discovered and tested to see if they follow the circle equation in order to detect circular faults using the segmented image. The one that meets the circular requirements is designated as a circle. Figures 4.17 and 4.18 show the thermograms

Figure 4.15. Isolation of defective locations from the sound (non-defective) regions for modeled sample.

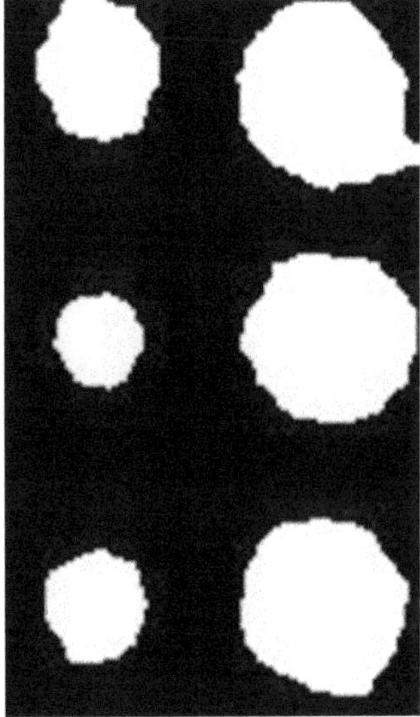

Figure 4.16. Isolation of defective locations from the sound (non-defective) regions for experimental sample.

identifying all the circular defects of the modeled and experimental samples, respectively.

Additionally, the centroids of these discovered circular faults are indicated to determine their diameter. In figures 4.19 and 4.20, thermograms with the specified centroids are shown. Using the Matlab® command regionprops, these usual properties of a void with a specific form have been estimated. Additionally, the ROI's major and minor axes, which contain the form of the ROI, are described, along with the centroid for each void.

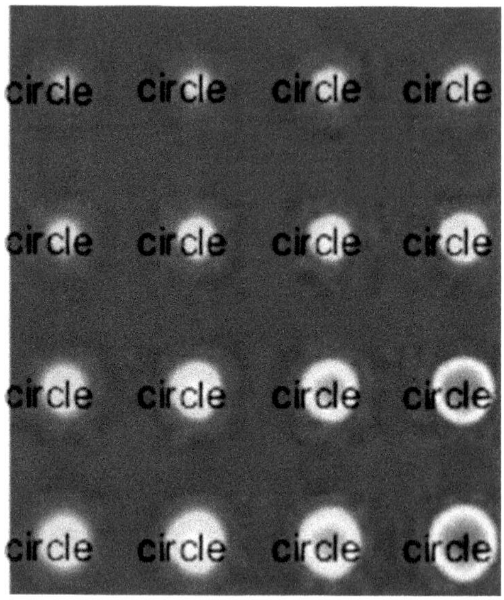

Figure 4.17. Thermogram identifying the circular defects in modeled sample.

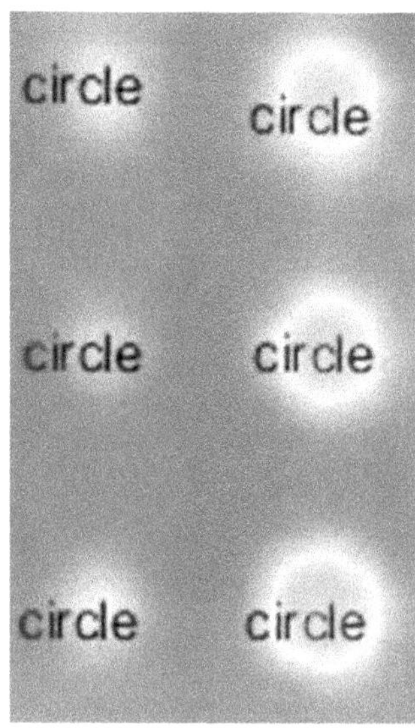

Figure 4.18. Thermogram identifying the circular defects in experimental sample.

Table 4.1 shows the actual and computed diameters along with % error in modeled sample.

Table 4.2 shows the actual and computed diameters along with % error in experimental sample.

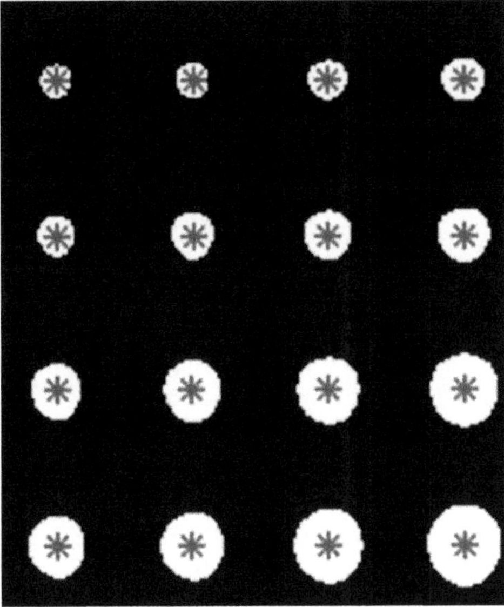

Figure 4.19. Thermogram with marked centroids at the defect's location in modeled sample.

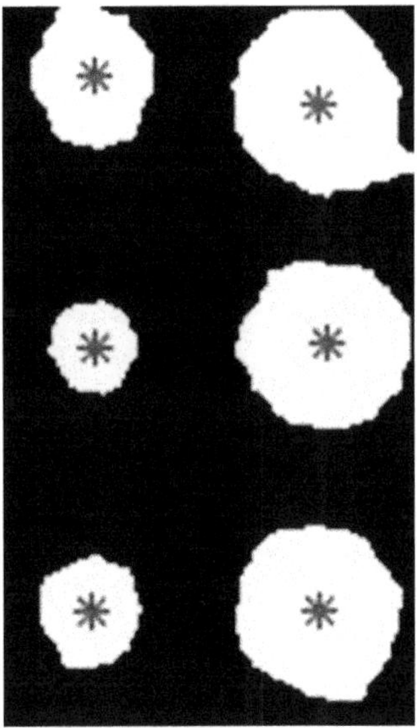

Figure 4.20. Thermogram with marked centroids at the defect's location in experimental sample.

The values in the table indicate that as depth increases, keeping the diameter the same, the estimated error % also increases. Also, the error % rises with an increase in diameter, maintaining depth the same. The variation of error % with depth and diameter for the modeled sample is represented as graphs in figures 4.21 and 4.22, respectively.

Table 4.1. Computed and actual diameters along with % error (modeled sample).

Serial No.	Sample's depth from the front surface	A = Actual Radius (mm)	C = Computed Radius (mm)	Percentage Error (%) = ((A~C)/A)
1	0.5	8	8.02	0.25
2	0.8	8	8.05	0.625
3	1.1	8	8.1	1.25
4	1.5	8	8.37	4.625
5	0.5	10	10.07	0.7
6	0.8	10	10.16	1.6
7	1.1	10	10.25	2.5
8	1.5	10	10.6	6
9	0.5	14	14.1	0.7
10	0.8	14	14.24	1.7
11	1.1	14	14.44	3.1
12	1.5	14	14.88	6.2
13	0.5	16	16.15	0.9
14	0.8	16	16.3	1.8
15	1.1	16	16.55	3.4
16	1.5	16	17	6.2

Table 4.2. Computed and actual diameters along with % error (experimental sample).

Serial No.	Sample's depth from the front surface	A = Actual Radius (mm)	C = Computed Radius (mm)	Percentage Error (%) = ((A~C)/A)
1	0.75	4	4.02	0.5
2	1	4	4.1	2.5
3	1.25	4	4.25	6.25
4	0.75	6	6.05	0.8
5	1	6	6.24	4
6	1.25	6	6.5	8.3

Also, the graphs in figures 4.23 and 4.24 show the variation of error % with depth and with diameter for the experimental sample, respectively.

4.6 Conclusion

This chapter demonstrates the potential of the recommended technique for automatic flaw detection and defect diameter estimation. The experiment is performed on a GFRP structure sample made up of flat-bottomed, circular voids using linear frequency modulated thermal wave imaging. The algorithm used to identify the defect's lateral dimensions (diameter) and designate the defect centers uses the

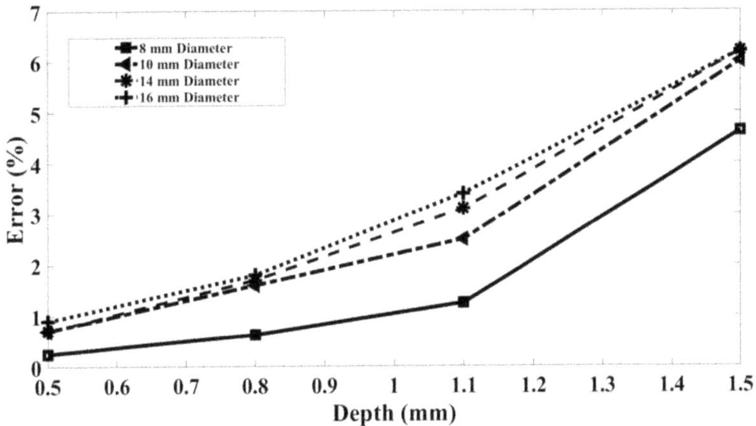

Figure 4.21. Error percentage vs depth for different diameters for modeled sample.

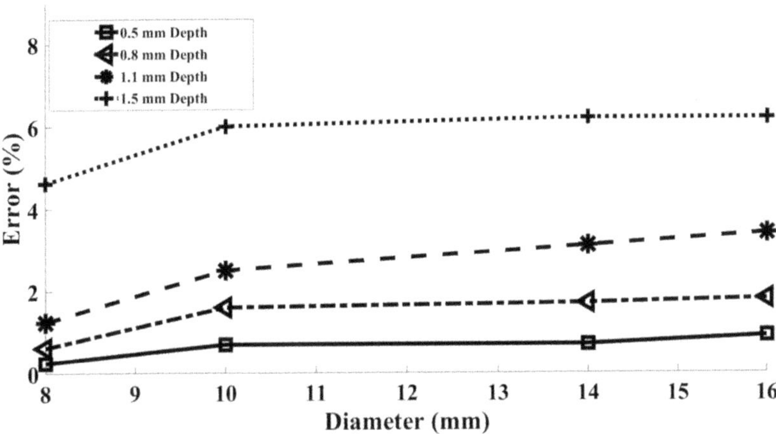

Figure 4.22. Error percentage vs diameter for varying depths for modeled sample.

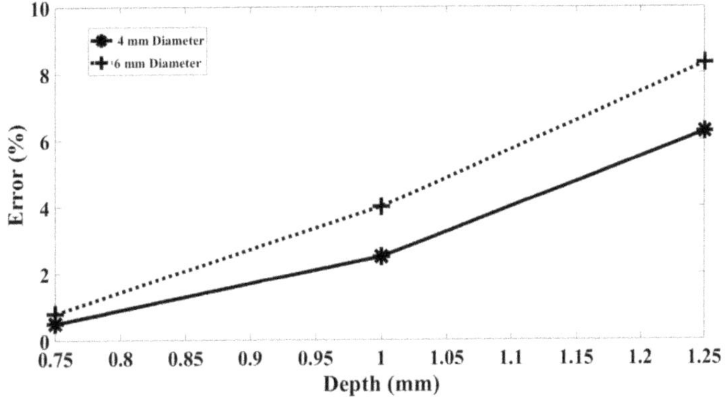

Figure 4.23. Error percentage vs depth for different diameters for experimental sample.

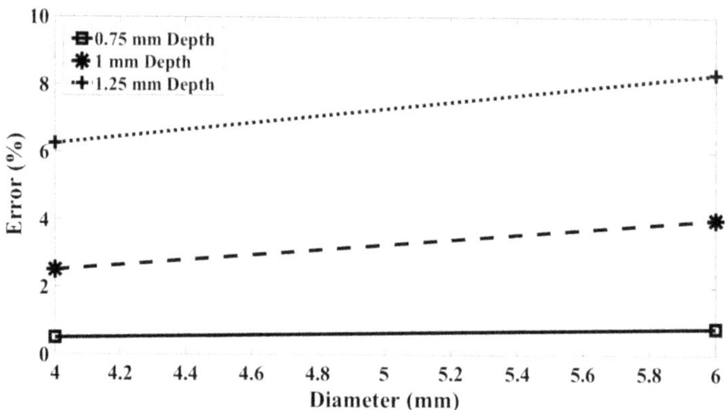

Figure 4.24. Error percentage vs diameter for varying depths for experimental sample.

computed pulse-compressed thermogram as input. The outcomes clearly demonstrate the effectiveness of the suggested method for automatically identifying subsurface flaws, figuring out their shape, and quantitatively estimating the defect's lateral dimensions. The results further indicate that as depth increases, keeping the diameter the same, the estimated error % increases. Also, the error % rises with an increase in diameter, maintaining depth the same.

Acknowledgments

Authors wishing to acknowledge the funding agencies like DRDO, SERB and GITA-DST for their financial support to establish the Infrared Imaging Laboratory (IRIL) to carry out the experimentation for the presented work in this chapter.

References

[1] Al H N S 2014 A novel approach to gradient edge detection *J. Glob. Res. Comput. Sci.* **5** 1–4

[2] Balageas D L, Krapez J-C and Cielo P 1986 Pulsed photothermal modeling of layered materials *J. Appl. Phys.* **59** 348–57

[3] Busse G, Wu D and Karpen W 1992 Thermal wave imaging with phase sensitive modulated thermography *J. Appl. Phys.* **71** 3962–5

[4] Castellano A, Fraddosio A, Marzano S and Piccioni M D 2017 Some advancements in the ultrasonic evaluation of initial stress states by the analysis of the acoustoelastic effect *Procedia Eng.* **199** 1519–26

[5] De Angelis G, Meo M, Almond D P, Pickering S G and Angioni S L 2012 A new technique to detect defect size and depth in composite structures using digital shearography and unconstrained optimization *NDT E Int.* **45** 91–6

[6] Duan Y, Liu S, Hu C, Hu J, Zhang H, Yan Y, Tao N, Zhang C, Maldague X and Fang Q 2019 Automated defect classification in infrared thermography based on a neural network *NDT E Int.* **107** 102147

[7] Fang Q, Nguyen B D, Castanedo C I, Duan Y and Maldague X II 2020 Automatic defect detection in infrared thermography by deep learning algorithm *Thermosense: Thermal Infrared Applications XLII* (Bellingham, WA: SPIE), pp 180–95

[8] Huang Q K and Zhou Y C 2010 Edge detection algorithm of core image based on the improved Canny operator *Proc. –2010 3rd IEEE Int. Conf. on Computer Science and Information Technology, ICCSIT 2010* pp 411–3

[9] Jie G and Ning L 2012 An improved adaptive threshold canny edge detection algorithm *2012 Int. Conf. on Computer Science and Electronics Engineering* (Piscataway, NJ: IEEE) pp 164–8

[10] Madruga F J, Ibarra-Castanedo C, Conde O M, López-Higuera J M and Maldague X 2010 Infrared thermography processing based on higher-order statistics *NDT E Int.* **43** 661–6

[11] Maldague X 2001 *Theory and Practice of Infrared Technology for Nondestructive Testing* (New York: Wiley)

[12] Maldague X and Marinetti S 1996 Pulse phase infrared thermography *J. Appl. Phys.* **79** 2694–8

[13] Mery D, da Silva R R, Calôba L P and Rebello J M 2003 Pattern recognition in the automatic inspection of aluminium castings *Insight, Non-Destr. Test. Cond. Monit.* **45** 475–83

[14] Nikishkov Y, Airoldi L and Makeev A 2013 Measurement of voids in composites by X-ray computed tomography *Compos. Sci. Technol.* **89** 89–97

[15] Vavilov V, Maldague X, Dufort B, Robitaille F and Picard J 1993 Thermal nondestructive testing of carbon epoxy composites: detailed analysis and data processing *NDT E Int.* **26** 85–95

[16] Xiao W and Hui X 2010 An improved canny edge detection algorithm based on predisposal method for image corrupted by Gaussian noise *2010 World Automation Congress* (Piscataway, NJ: IEEE), pp 113–6

[17] Yao Y, Sfarra S, Lagüela S, Ibarra-Castanedo C, Wu J-Y, Maldague X P and Ambrosini D 2018 Active thermography testing and data analysis for the state of conservation of panel paintings *Int. J. Therm. Sci.* **126** 143–51

[18] Yousefi B, Sfarra S, Sarasini F, Castanedo C I and Maldague X P 2019 Low-rank sparse principal component thermography (sparse-PCT): comparative assessment on detection of subsurface defects *Infrared Phys. Technol.* **98** 278–84

[19] Zhu H, Meng F, Cai J and Lu S 2016 Beyond pixels: a comprehensive survey from bottom-up to semantic image segmentation and cosegmentation *J. Visual Commun. Image Represent.* **34** 12–27

[20] Dudek G and Dudzik S 2018 Classification tree for material defect detection using active thermography *Adv. Intell. Syst. Comput.* **655** 118–27

[21] Ahmad J, Akula A, Mulaveesala R and Sardana H K 2021 Probability of detecting the deep defects in steel sample using frequency modulated independent component thermography *IEEE Sens. J.* **21** 11244–52

[22] Ahmad J, Akula A, Mulaveesala R and Sardana H K 2020 Probability of detection of deep defects in steel samples using barker coded independent component thermography *Electron. Lett.* **56** 1005–7

[23] Ahmad J, Akula A, Mulaveesala R and Sardana H K 2020 Defect detection capabilities of independent component analysis for Barker coded thermal wave imaging *Infrared Phys. Technol.* **104** 103118

[24] Ahmad J, Akula A, Mulaveesala R and Sardana H K 2019 An independent component analysis based approach for frequency modulated thermal wave imaging for subsurface defect detection in steel sample *Infrared Phys. Technol.* **98** 45–54

[25] Ahmad J, Akula A, Mulaveesala R and Sardana H K 2019 Barker-coded thermal wave imaging for non-destructive testing and evaluation of steel material *IEEE Sens. J.* **19** 735–42

[26] Arora V and Mulaveesala R 2017 Application of golay complementary coded excitation schemes for non-destructive testing of sandwich structures *Opt. Lasers Eng.* **93** 36–9

[27] Arora V and Mulaveesala R 2014 Pulse compression with Gaussian weighted chirp modulated excitation for infrared thermal wave imaging *Prog. Electromagn. Res. Lett.* **44** 133–7

[28] Arora V, Mulaveesala R and Bison P 2016 Effect of spectral reshaping on frequency modulated thermal wave imaging for non-destructive testing and evaluation of steel material *J. Nondestr. Eval.* **35** 1–7

[29] Arora V, Mulaveesala R, Dua G and Sharma A 2020 Thermal non-destructive testing and evaluation for subsurface slag detection: numerical modeling *Insight: Non-Destr. Test. Cond. Monit.* **62** 264–8

[30] Arora V, Mulaveesala R, Kumar S and Wuriti S 2021 Non-destructive evaluation of carbon fiber reinforced polymer using Golay coded thermal wave imaging *Infrared Phys. Technol.* **118** 103908

[31] Arora V, Mulaveesala R, Rani A, Kumar S, Kher V, Mishra P, Kaur J, Dua G and Jha R K 2021 Infrared image correlation for non-destructive testing and evaluation of materials *J. Nondestr. Eval.* **40** 1–7

[32] Arora V, Mulaveesala R, Rani A and Sharma A 2019 Digitised frequency modulated thermal wave imaging for non-destructive testing and evaluation of glass fibre reinforced polymers *Nondestruct. Test. Eval.* **34** 23–32

[33] Arora V, Mulaveesala R, Siddiqui J A and Muniyappa A 2014 Hilbert transform-based pulse compression approach to infrared thermal wave imaging for sub-surface defect detection in steel material *Insight: Non-Destr. Test. Cond. Monit.* **56** 550–2

[34] Arora V, Mulaveesala R and Subbarao G V 2015 Non-destructive testing of steel sample by non-stationary thermal wave imaging *Int. Conf. on Signal Processing and Communication Engineering Systems – Proc. SPACES 2015, in Association with IEEE* pp 527–30

[35] Arora V, Siddiqui J A, Mulaveesala R and Muniyappa A 2015 Pulse compression approach to nonstationary infrared thermal wave imaging for nondestructive testing of carbon fiber reinforced polymers *IEEE Sens. J.* **15** 663–4

[36] Awasthi S, Mulaveesala R and Tuli S 2007 Thermal non-destructive evaluation of scaling in boiler tubes *Proc. SPIE – Int. Soc. Opt. Eng.* **6541** 304–11

[37] Bhowmik A, Repaka R, Mishra S C and Mulaveesala R 2013 Detection of subsurface skin lesion using frequency modulated thermal wave imaging: a numerical study *ASME Int. Mechanical Engineering Congress and Exposition, Proc. (IMECE)* **56215** p V03AT03A064

[38] Dua G, Arora V and Mulaveesala R 2021 Defect detection capabilities of pulse compression based infrared non-destructive testing and evaluation *IEEE Sens. J.* **21** 7940–7

[39] Dua G, Kumar N and Mulaveesala R 2015 Applications of digitized frequency modulated thermal wave imaging for bone diagnostics *Int. Conf. on Signal Processing and Communication Engineering Systems – Proc. SPACES 2015, in Association with IEEE* pp 518–21

[40] Dua G and Mulaveesala R 2018 Thermal wave imaging for non-destructive testing and evaluation of reinforced concrete structures *Insight: Non-Destr. Test. Cond. Monit.* **60** 252–6

[41] Dua G and Mulaveesala R 2018 Applicability of active infrared thermography for screening of human breast: a numerical study *J. Biomed. Opt.* **23** 037001

[42] Dua G and Mulaveesala R 2017 Infrared thermography for detection and evaluation of bone density variations by non-stationary thermal wave imaging *Biomed. Phys. Eng. Express* **3** 017006

[43] Dua G and Mulaveesala R 2013 Applications of Barker coded infrared imaging method for characterisation of glass fibre reinforced plastic materials *Electron. Lett.* **49** 1071–3

[44] Dua G, Mulaveesala R, Kher V and Rani A 2019 Gaussian windowed frequency modulated thermal wave imaging for non-destructive testing and evaluation of carbon fibre reinforced polymers *Infrared Phys. Technol.* **98** 125–31

[45] Dua G, Mulaveesala R, Mishra P and kaur J 2021 InfraRed image correlation for non-destructive testing and evaluation of delaminations in glass fibre reinforced polymer materials *Infrared Phys. Technol.* **116** 103803

[46] Dua G, Mulaveesala R and Siddique J A 2015 Effect of spectral shaping on defect detection in frequency modulated thermal wave imaging *J. Opt.* **17** 025604

[47] Dua G, Subbarao G V and Mulaveesala R 2015 Testing and evaluation of glass fiber reinforced polymers by thermal wave imaging *Int. Conf. on Signal Processing and Communication Engineering Systems – Proc. SPACES 2015, in Association with IEEE* pp 522–6

[48] Ghali V S, Jonnalagadda N and Mulaveesala R 2009 Three-dimensional pulse compression for infrared nondestructive testing *IEEE Sens. J.* **9** 832–3

[49] Ghali V S and Mulaveesala R 2011 Comparative data processing approaches for thermal wave imaging techniques for non-destructive testing *Sens. Imaging* **12** 15–33

[50] Ghali V S and Mulaveesala R 2010 Frequency modulated thermal wave imaging techniques for non-destructive testing *Insight: Non-Destr. Test. Cond. Monit.* **52** 475–80

[51] Ghali V S, Mulaveesala R and Takei M 2011 Frequency-modulated thermal wave imaging for non-destructive testing of carbon fiber-reinforced plastic materials *Meas. Sci. Technol.* **22** 104018

[52] Ghali V S, Panda S S B and Mulaveesala R 2011 Barker coded thermal wave imaging for defect detection in carbon fbre-reinforced plastics *Insight: Non-Destr. Test. Cond. Monit.* **53** 621–4

[53] Kaur K and Mulaveesala R 2021 Statistical post-processing approaches for active infrared thermography: a comparative study *2021 IEEE 11th Annual Computing and Communication Workshop and Conf., CCWC 2021* pp 1251–5

[54] Kaur K and Mulaveesala R 2020 Efficient selection of independent components for inspection of mild steel sample using infrared thermography *Electron. Lett.* **56** 990–3

[55] Kaur K and Mulaveesala R 2020 Active infrared imaging for estimation of sub-surface features in a steel material *Procedia Comput. Sci.* **171** 1204–11

[56] Kaur K and Mulaveesala R 2019 An efficient data processing approach for frequency modulated thermal wave imaging for inspection of steel material *Infrared Phys. Technol.* **103** 103083

[57] Kaur K and Mulaveesala R 2019 Experimental investigation on noise rejection capabilities of pulse compression favourable frequency-modulated thermal wave imaging *Electron. Lett.* **55** 352–3

[58] Kaur K, Sharma A, Rani A, Kher V and Mulaveesala R 2020 Physical insights into principal component thermography *Insight: Non-Destr. Test. Cond. Monit.* **62** 277–91

[59] Kher V and Mulaveesala R 2021 Probability of defect detection in glass fibre reinforced polymers using pulse compression favourable frequency modulated thermal wave imaging *Infrared Phys. Technol.* **113** 103616

[60] Kher V and Mulaveesala R 2020 Probability of defect detection in pulse compression favourable thermal excitation schemes for infra-red non-destructive testing *Electron. Lett.* **56** 998–1000

[61] Kher V and Mulaveesala R 2019 Probability of defect detection in pulse compression favourable frequency modulated thermal wave imaging *Electron. Lett.* **55** 789–91

[62] Morello R, Schena E, Oddo C M, Lay-Ekuakille A, Ruiz C, Mulaveesala R, Yuce M R and Pataranutaporn P 2021 Guest editorial special issue on advances and current trends in sensing physiological parameters for human wellness and patient monitoring *IEEE Sens. J.* **21** 13965–66

[63] Mukhopadhyay C K and Mulaveesala R 2021 Preface *Lecture Notes in Mechanical Engineering* (Berlin: Springer) pp v–vii

[64] Mulaveesala R and Arora V 2017 Complementary coded thermal wave imaging scheme for thermal non-destructive testing and evaluation *Quant. InfraRed Thermogr. J.* **14** 44–53

[65] Mulaveesala R, Arora V and Dua G 2022 Highly depth resolved coded thermal wave imaging technique for infrared non-destructive testing and evaluation *Lecture Notes in Mechanical Engineering* (Berlin: Springer) pp 233–41

[66] Mulaveesala R, Arora V and Dua G 2021 Pulse compression favorable thermal wave imaging techniques for non-destructive testing and evaluation of materials *IEEE Sens. J.* **21** 12789–97

[67] Mulaveesala R, Arora V, Dua G, Rani A, Kher V, Sharma A and Kaur K 2020 Pulse compression favorable thermal wave imaging methods for testing and evaluation of carbon fibre reinforced polymer *Proc. SPIE – Int. Soc. Opt. Eng.* **11409** 180–5

[68] Mulaveesala R, Arora V, Dua G, Vavilov V P, Guo X, Bison P and Avdelidis N P 2020 Guest editorial: non-destructive testing *Electron. Lett.* **56** 972–3

[69] Mulaveesala R, Arora V and Rani A 2019 Coded thermal wave imaging technique for infrared non-destructive testing and evaluation *Nondestruct. Test. Eval.* **34** 243–53

[70] Mulaveesala R, Arora V, Siddiqui J A and Muniyappa A 2014 Non-stationary thermal wave imaging for nondestructive testing and evaluation *Thermosense: Thermal Infrared Applications XXXVI* **9105** pp 163–8

[71] Mulaveesala R, Awasthi S and Tuli S 2008 Infrared non-destructive characterization of boiler tube *Sens. Lett.* **6** 312–8

[72] Mulaveesala R and Dua G 2016 Non-invasive and non-ionizing depth resolved infra-red imaging for detection and evaluation of breast cancer: a numerical study *Biomed. Phys. Eng. Express* **2** 055004

[73] Mulaveesala R, Dua G and Arora V 2022 Digitized frequency modulated thermal wave imaging for testing and evaluation of steel materials *Lecture Notes in Mechanical Engineering* (Berlin: Springer) pp 159–66

IOP Publishing

Advanced Signal Processing for Industry 4.0, Volume 1
Evolution, communication protocols, and applications in manufacturing systems
Irshad Ahmad Ansari and Varun Bajaj

Chapter 5

Customer-driven healthcare through mission-focused approach in 4IR

Marjo Rissanen, Antti Rissanen and Kari Rannaste

Socio-technical change is underway in the fields of health service, health technology, and informatics design, with a goal to optimize processes and to improve professionalism in healthcare. In healthcare, Industry 4.0 technologies change service process structures in many ways. Artificial intelligence, big data, robotics, the Internet of Things, and the development of sensor technology bring new perspectives to the field of health service design. Design and evaluation which underline mission-strategic patterns in customer-centric health technology development play a significant role in this change. What matters is the importance of products and services to the customer, as well as the cost-effectiveness and benefits of the innovation policy. Socially responsible Industry 4.0 challenges not only design activities and designers, but also involve stakeholders. Due to interrelated ethical issues, this is an educational challenge for healthcare providers as well as health consumers and stakeholders as part of the organization's social responsibility. Mission-strategic design makes customer-centric healthcare more achievable. This phenomenon and related trends are considered in this chapter.

5.1 Introduction

Socio-technical change and digital transformation in healthcare are present challenges for evaluative approaches. In socio-technical systems theory, people, tasks, structures, and technologies form the major components of an organizational system [1] but the dimensions also cover areas of goals, culture, infrastructure, processes, stakeholders, economic circumstances, and regulatory frameworks [2]. Therefore, new technologies also mean decentralized leadership challenges in organizations [3]. Artificial intelligence, digital twin domains, and development in the field of health monitoring wearables bring new perspectives to the field of health informatics and technology design in the fourth Industrial Revolution in healthcare (4IR) [4]. The content of

  5-1

customer-driven design practices will also expand in multiple contexts. The importance of design strategies in healthcare is increasing with the growth of an aging population. In the USA, for example, about 80% of the elderly population has a chronic illness which exposes this growing population to other morbidities as well [5], indicating a need for cost-effective, mission-emphasized design. Older adults also have the most frustration and anxiety with digitalization and therefore also towards digital health informatics [6]. However, successful emergent technologies can lead to positive outcomes in later life and smart technologies are increasingly used in their homes [7, 8]. The success of emergent technologies among the elderly depends largely on efficient techniques of design, implementation, and evaluation [7]. Elderly are often considered frail and passive, but their inspirational value as codesigners at a certain level can be an asset for technology development [8].

Healthcare organizations seek to solve problems through new practices, technologies, and harmonization in process management. Mission-centric design requires identification of the primary processes, which also guides resource allocation and the criteria for genuine customer orientation. Do new solutions and practices support professionalism and patient safety successfully in care? Are these renewed processes supported by appropriate training for designers, producers, researchers, healthcare staff, and clients? Both the technical subsystem (physical structures, buildings, technologies) and the functions of the social subsystem (people, values, norms) [9] in organizations and society play an essential role in the realization of a customer-centric socio-technical service system.

5.2 Fourth Industrial Revolution in healthcare

5.2.1 Possibilities and challenges

The need for resource efficiency, flexibility, and individualized products are triggers for the fourth revolution in healthcare [4]. From products, we expect adaptation to human needs, sustainability, social responsibility, cost savings, enhanced performance, innovativeness, and smart manufacturing [10]. Big data, cloud-based health information systems, telemedicine and disease monitoring, personalized healthcare, assisted living, and novel systems for medication, rehabilitation, and self-management represent new service lines in healthcare [11]. Augmented reality, modeling and simulation, nanotechnology, and platforms for decentralized transaction systems (block-chains) are part of design and manufacturing systems in the 4.0 era [10]. Internet of Things (IoT) networks connect medical and health-related services, wearable devices, and m-health technologies, allowing various connections (thing-to-human, human-to-human, thing-to-thing) and have a role in continuous health monitoring and remote healthcare [10, 11].

In healthcare 4.0 (HC 4.0) however, security, data privacy, extreme heterogeneity of big data, formats and standards [11], and the analysis of information [12] are known concerns. Problems in interoperability, network limitations, lack of quality data, unnecessary computational workload, and integrated legal and managerial issues are areas of difficulty [13]. Social responsibility and science–management collaboration generally need a wider set of ethical concepts to capture current demands in this transdisciplinary area [14–16]. Human-related challenges also cover knowledge and educational factors in organizations and thus leadership and organizational culture play a remarkable role [14].

5.2.2 Digital twins in healthcare

The digital twin (DT) concept integrated in the data and service context is seen as a promising tool in HC 4.0, even if it still faces challenges. DT refers to a virtual model of a physical entity, a digital corresponding item, and a link that draws these segments together [17–19]. Two-way flow of information between the physical and virtual creations allows digital representations of the physical entities [19]. Health consumers' baseline and clinical data can be utilized to produce a clinical state for a digital twin [20]. A key idea of DTs is their effective bidirectional mapping, which can cover healthy or diseased objects and their comparisons and representations (e.g., a diseased cell, a diseased organ, a disease or syndrome) [19]. DT instances of the same person are suitable for comparisons of different interventions or medications, for example [19]. DTs can offer transparency in a person's molecular and physiological structure [21]. DTs also support healthcare organizations in intensifying their monitoring and planning activities. These DTs support in intensifying monitoring and design activities. DTs integrate big data, artificial intelligence (AI), machine learning (ML), development of sensor technology and the IoT in health technology in Industry 4.0. However, data of any size can be utilized alone or combined with big data. DT is one method for ensuring smoother human–machine connections in healthcare. Interaction, communication, and exchange of information take place between the system, services, data, and digital model [19, 22].

In healthcare DTs are applied in precision medicine and patient-centered care in various medical specialty areas. In healthcare, DT applications can be integrated into different areas of design, such as platform, diagnosis, prognosis, simulation, and optimization. DTs provide support in educational interventions, hypothesis generation, testing procedures, and new knowledge acquisition [19]. Testing on the digital human body in a virtual environment gives better patient safety and clinical accuracy [20]. Besides diagnostics, monitoring and preventive medicine can take advantage of connected concept models [22, 23].

Challenges typical of modern AI and big data analytics also concern DT technologies, such as data quality and management, algorithm design, data privacy, and ethical concerns [24]. Challenges are linked to incomplete knowledge of DT modeling and connected implementation models. The common data collection and processing methodology may not always satisfy the requirements of DTs. Currently, shared methods, standards, and norms for the development of DTs are also lacking [22, 23].

5.2.3 Wearable sensors in continuous health monitoring

Sensors produce signals that can be utilized as inputs to electronic circuits which are used to measure, record, or control systems based on gathered values. Sensors are a subsection of transducers. These devices are used to produce input signals for various measurement, instrumentation or control systems. Transducers change energy in the form of sound, light, heat etc., into equal electrical signals. Transducers are utilized both as inputs to electronic circuits or as outputs or actuators from electronic circuits. Sensors are classified as passive, active, digital, or analog ones [25]. Accelerometers, strain gauges, and microphones are good examples of passive sensors. Piezoelectric ultrasonic sensors produce acoustic views and get data based on measured echoes [26]. This straightforward principle describes nicely how active sensors work.

Diseases like coronary heart disease, bronchitis, diabetes, and Parkinson's disease are common senile diseases without early recognizable signs. Moreover, elderly people often suffer simultaneously from several geriatric diseases. Constantly evolving wearable sensors mean promising options for convenient and noninvasive remote home monitoring, post care, and mobility assistance for geriatric patients [27–29]. This type of almost continuous measurement can be helpful in early identification of many serious health problems [28].

In the HC 4.0 era the aim is to develop sensor types which disturb consumers' daily life as little as possible and still serve health practices in a cost-effective way. A good example of tiny sensors are smart rings on the finger which measure body temperature, heart rate, sleep patterns, and fitness activity. Most wearable devices can be used without medical or technical assistance and can be worn almost anywhere on the body surface or implanted in clothing [28]. Wearables with auxiliary computation, sensor networks, signal processing, machine learning, and AI techniques and associated interfaces and devices can enhance care intensity levels in healthcare and thus quality of life [30, 31]. Electrodermal thin, skin-soft electrodes called e-tattoos (e-skin) are examples of light and compact ECG electrodes [32]. Electronic skin adapts well to any shape in the human body and can in principle obtain energy from the electrophysiological processes of the body [33].

New wearables cover various kinds of sensor types which serve continuous health monitoring. *Electrochemical biosensors* allow constant monitoring of the concentration of biomarkers in various body fluids (e.g., electrolyte levels, proteins, exogenous medicines in human sweat, urine, saliva) as well as work in monitoring cardiovascular health [27]. *Optical sensors* have a crucial role in the advancement of medical devices in diagnostics. They are basically detectors that can capture light or its variants. Quite often optical sensors are becoming an alternative to traditional electrical and mechanical sensors because in some cases direct contact with the person is no longer required. *Photoplethysmography* is utilized to control heart rate, blood oxygen and glucose, respiration, blood pressure, and signals from intact muscles in robotics and prosthetics [34]. *Radiation sensors* cover thermometers, thermal cameras, and daysimeters. Due to the nature of radiation sensors they are used to measure the body temperature in general or in a specific part of the body. Mass control for infectious diseases is a good example of utilization of these sensor types. *Biochemical optical sensors* (e.g. fluorescence, luminescence, and colorimetry) are useful also in measuring human fluids without traditional sample collection. *Optical fiber sensors* are commonly used in wearables (chest belts, smart clothing, and other textiles) [35]. They are suitable for measurement of cardiac activity, respiration, temperature, mobility, and electrolyte balance. Due to cost control and the need for automatic contact with the human body, the use of innovative materials is being studied in the field of wearable electrochemical biosensors (e.g., carbon-based or metallic compounds). Even magnetic nanoparticles have been proposed for sensor structure [27]. The advancement of paper-based equipmentation modified with nanomaterials offers an interesting future research and affordable development area [35].

Globally, *cardiovascular diseases* (CVDs) form the primary cause of death. Advances in wearable technologies change management of CVDs and may give good indicators for early interventions. At the moment bioelectric, mechano-electric, optoelectric, and ultrasonic wearables are available for CVD monitoring. With rapid development flexible wearables can serve the diagnostics and treatment of cardiovascular diseases [29]. Nowadays common commercial wearables like smart wristbands, patches, and smartwatches monitor heart rate, blood oxygen saturation, and even electrocardiogram data [33]. Compact, low-cost single-chip microcontrollers and advanced wireless communication technologies have inspired new equipment prototypes. They can be used to process the real-time signals generated by the biosensors and transmit the cleaned signals via the consumer's phone or home network to the server of the health center. Monitoring benefits those patients already diagnosed with heart failures, hypertension, or diabetes, as well as home-care elderly patients with general frailty [36].

Traditionally *blood glucose* is daily measured by invasive skin puncture methods and blood samples. Skin-like electrochemical sensors on the surface of the skin and biochemical optical sensors (e.g., colorimetric detection) represent noninvasive, almost continuous method of sweat-based glucose monitoring. Correlation between sweat and blood glucose has been proved in studies [37, 38].

Cancer morbidity and mortality are major health problems in the world. Electrochemical cytosensors, for example, have raised attention for tumor cell detection mainly because of the sensitivity of the techniques, low cost, and simple equipmentation when compared to traditional cancer detection techniques [39]. The fiber optic surface plasmon resonance (SPR) biosensor is one candidate for the early diagnosis of different types of cancers [40]. Due to the late diagnosis of prostate cancer, the mortality of this cancer type is high. Development of sensors for diagnosis of this condition may change the diagnostic possibilities in this field [41]. Breast cancer is the main cause of death among women and better, more practical early aids in diagnostics are important. Existing microwave sensors open a new option for breast cancer diagnostics with improved image resolution [42].

Drug concentration monitoring has a crucial role in health monitoring and pharmaceutical testing. Traditionally drug monitoring is based on blood or urine analysis. In early testing wearable microneedle sensor arrays provided functional Levodopa drug management among patients with Parkinson's disease [43]. Despite aspirin's wide applicability, its possible side effects require careful dosing. Electrochemical sensors are utilized for the detection of acetysalic concentration in tablets and human fluids [44, 45].

Wearables have been used even for *mental health* monitoring. Data from multimodal wearable sensors and connected AI powered data analytics enhance monitoring of mental conditions [46]. Sympathetic nervous system is activated during stressful situations for example, due to stressors in work, environment, events, thinking, or due to learned behavior models. This activation produces consequently physiological measurable reactions like variations in neural activity, changes in skin temperature, heartbeat, or respiration rate. Other observable issues are activation in perspiration, hormonal effects, and changes in sleep patterns [32, 47]. Wearables have also enhanced the precision in depression diagnostics [48].

Due to the novelty of the recent sensor invention policy, the experimentation of new applications is also extensive. Sensor technology has been tested among various patient groups, e.g., patients who suffer chronic obstructive pulmonary diseases [28], autism [49], or in seizure prediction [38]. Sensors are common in physical activity monitoring, sports medicine, and weight control [38]. They also work in the study of sleep disturbances or sleep disorders such as sleep apnea [50–52]. Wearable sensors have also been developed for detecting human movement, for example [53].

Despite all the progress, many wearables are still in their prototype phase [38]. Skin-friendliness, compactness, and comfort are challenges in artificial skin sensors [54]. Intrusion of ambient light with signal measurements, the small infiltration into the skin and other bioliquids are known disadvantages of optical biosensors [35]. Electrochemical sensors need also regular calibrations and sensor replacements which increases the overall cost. In constant body contact electromagnetic exposure to tissues needs more investigation [55]. The development of electrochemical wearable sensors also requires more stable and robust sensing mechanisms and pieces to allow correct measurements [56]. Accuracy, accessibility, repeatability, cost, and acceptance are issues to be clarified before wearables can be widely applied in healthcare [27, 57]. Nanomaterials are a promising area in sensor technology but further research on the properties and biocompatible characters of these materials is needed [27, 35, 56]. Excessive reliance on wearables and correspondingly neglecting lifestyle changes may reduce their efficiency. Data safety and ethics are essential concerns; such as security vulnerabilities, protection of users' personal information and physiological data from hacking, and ownership of the data. Large amounts of data from different sensors may also produce irrelevant data [35, 38, 55].

Health consciousness, performance expectancy, perceived usefulness, social influence, facilitating conditions, ease of use, compatibility, and social influence are issues that enhance adoption of health wearables [58–60]. Independent and consumer-led health monitoring of the elderly population requires attention and challenges user training and customer-oriented design. Older user groups can also actively participate in the development process of novel health innovations. Wearables represent an innovation type that usually reaches the user groups that are already interested in monitoring their own health status first. Users below the age of 44 form the main customer group and the spread to older users is limited [55]. The goal is therefore to reach the groups that would benefit the most from new innovations.

5.3 Customer-centric design in the context of Industry 4.0

5.3.1 Targeting and evaluating customer-centered service design

User-centered, cocreative, holistic, and evidence-based design targets are integrated to well-functioning service contexts [61]. Better customer orientation in the design of services and products [62, 63] are tried to enhance by communication between various interest groups [64], and through the equality of designers, different stakeholders [65], and consumers [66] in the planning process.

Customer-centricity is generally associated with how the application meets general usability requirements, supports user-centeredness in work, utilizes user participation, and adjusts to the needs of the user and environment [67]. Notification of customers' wishes or the suitability of the product for its intended use (fitness for use) [68–70] are widely emphasized. User-centered design strives from the notification of user experience and 'customers' voice' towards close user collaboration. Collaborative and participatory planning activities have been noteworthy in, for example, intensive care, oncology, and mental health [65]. Terms like 'empathic design' and 'socio-emotional intelligence' indicate eagerness to achieve a more customer-centric touch in design [71]. However, the need for a better understanding of customer-driven innovation practices challenges designers [72].

Digital services for growing older age groups mean not only opportunities but also challenges for design activities [73, 74]. Digital and health technology innovations are increasingly being produced from the perspective of inclusive design orientation [75]. Human–computer interaction (HCI) policy seeks constructive theoretical frames to support consumer-oriented design. In addition to empirical validity the focus is also on other constructive possibilities [76]. 'Out of the box' methodology means design flexibility, in which learning plays a central role, and potential theoretical frameworks are exploited in a coherent way, according to individual needs [77, 78]. Customer-centered design and evaluation activities in healthcare may require quite new types of scenarios and perspectives [78, 79], and designers who favor tangible representations typically have fewer design fixations [80]. System designers can influence service processes and embedded value frames in a remarkable way [81].

Professionalism of care, patient safety, and optimal cost–quality ratios are the key goals in healthcare and targets in the health technology area [82]. Creative innovation can be simultaneously useful for customers, industry, and society [83]. Customers' choices and acceptance are still ethical principles that regulate the utilization of novel applications. In health technology assessment, the primary interest typically focuses on embedded features of artifacts, and on their impact on clinical practice, health service systems, and patient outcomes. Therefore, evidence-based medicine plays a crucial role in assessment. In health technology and informatics assessment, well-known evaluation approaches are needs-based evaluation theory, behavior change theory, results-oriented evaluation (health outcomes and impact assessment), technology exploitation models (technology evaluation), socio-technical analysis, economic assessment, ethical considerations, and approaches that underline political and legal evaluation frames [84].

Product design is guided by standards and legislation that support ethical responsibility in a certain development area; e.g., international ISO standards (ISO/IEC 25022:2016, ISO/IEC 25023:2016, ISO/IEC 25010:2011) [85] focus on specific software quality issues. The *quality in use* standard focuses on aspects of effectiveness, efficiency, satisfaction, freedom from risk, and context coverage. The *product quality* standard underlines aspects such as functional suitability, performance efficiency, compatibility, usability, reliability, security, maintainability, and portability (ISO/IEC 25010:2011) [86].

As well the overall technological development of communication networks in healthcare (like Internet, sensor networks, machine-to-machine protocols) has produced architectures and standards to ascertain existing eHealth challenges [87]. These evaluation approaches in healthcare are significant and integrate customer-centric design and its evaluation. Specific feature-specific assessment categories are also relevant when systems are introduced as customer-centric systems. Ethical responsibility is integrated to a certain degree in all of these presented and well-known evaluation categories. However, the *mission-strategic evaluation perspective* in healthcare systems and application development focuses merely on the basic fundamentals and aspects of customer-centric design. This means justificative pondering and the basis for all other evaluations.

5.3.2 Customer-centric healthcare with a mission-strategic emphasis

Intelligent technologies may enable operations and analysis without human intervention. This brings new perspectives to the assessment of customer orientation. Mission quality is linked to issues of customer and product quality and ethical responsibility in healthcare. The product or service must be useful, safe, and cost-effective in an area where resources are limited. The importance of hearing, understanding, and codesigning with customers is typically emphasized in customer-centric design. However, the mission-strategic emphasis as an ideological starting point for design and customer-oriented implementation takes a wider view with more efficiency and professionalism in healthcare.

The main task is to evaluate how health consumers can get real value from an innovation or service. Does the planned product or service concept significantly improve safety and professionalism in patient care? What kinds of products and service packages represent genuinely customer-centric design in mission-strategic visions? The primary focus is on what is produced, rather than just on how it is produced and how the integration between the application and the user works. The question of methods is however also relevant. In evidence-based health informatics, the essential question is which system design and development methods can lead to safe and operative systems [88]. Mission-centeredness in production informs the nature and significance of innovations clearly to the customer (mission quality frame). In this view, the primary evaluation perspective focuses on the results and significance of the innovation. A change toward 'better' is underlined, which deepens the nature of customer-centric design.

Consequently, the customer perspective and mission-strategic viewpoint will focus on the product's and service models' ability to improve customer rights, customer safety, and professionalism in healthcare. The mission-strategic customer perspective needs to be examined also from the perspective of educational targets, as customer orientation challenges especially designers' and developers' training. However, the technologies of Industry 4.0 cover the life cycle of the product from early plans and design to deliverance and use, and thus evaluation is a common and shared responsibility for all involved stakeholders. Therefore, the question is about a versatile training challenge; health consumers, health professionals, researchers, stakeholders, and other interest groups are target groups for training. Service is a combination of technology and service processes, and thus training sessions should be sufficiently practice-oriented. Expertise, experience, and understanding of environmental characteristics should be integrated into the mentoring policy (see figure 5.1).

The purpose of translational medicine and design is the betterment of human health and customer-centric health services and innovations. Health-related socio-technical systems thinking has much in common with translational health design [89, 90]. Intensive theory and practice integration and orientation in research and clinical practice [91] as well as notification of essential ethical frames are underlined features of translational medicine and design.

Customer-centric design challenges the overall healthcare system, and evaluation takes place at different stages of the development process. The customer perspective integrates frameworks (knowledge bases) that support design activities, choices in design practice, and impact assessment in the target environment [70, 92]. In healthcare, translational research and research agendas focus also on the utilization and adoption of medical innovations as well as technology features [93]. In health technology-related evaluation, many sub-areas have their underlined aspects (e.g., product maturation, healthcare management, quality control, usability research and human factors in design). Aspects that are generally emphasized in translational health design are also valuable in the 4.0 era when aiming to improve customer-centeredness. In innovation policy the well-known principles of health-related translational design and related mission-strategic emphasis in the health technology area are identified in table 5.1.

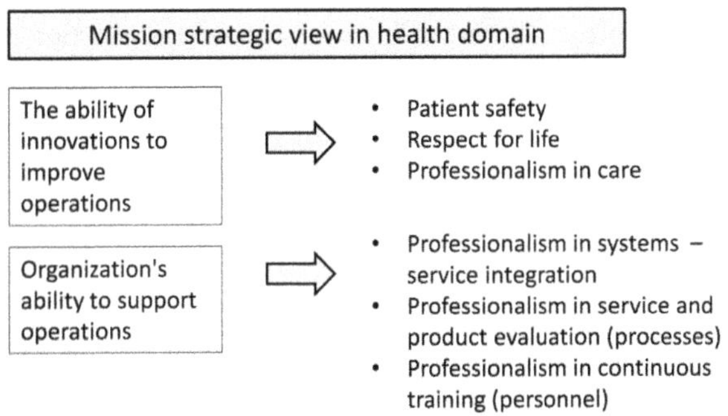

Figure 5.1. Mission-strategic view in translational health technology domain.

Table 5.1. Translational design principles in the 4.0 era.

Translational design principle	Mission-strategic perspective
Mission-strategic awareness	Mission value and identification as a priority task
Ethical awareness	Ethical acceptability and justification as a starting point in all innovation and service policies
Synergy thinking	Wide understanding and awareness of connected synergy effects
Collaborative forums for action	Versatile, multidisciplinary expertise and strategies that support cooperative aims
Intense praxis and theory integration	Awareness of practice-related questions in balance with theoretical, multifaceted knowledge
Sound clinical practices for betterment of human health	Patient safety, respect for life, and professionalism of care as main aims
Creative education and training	Integration of value frames in staff, stakeholder, and consumer training
Enhanced utilization of medical innovations	Enhancement of wide utilization of safe, ethically acceptable, and consumer-centered innovations

The evaluative approach in the mission-strategic view focuses on healthcare, particularly on the following themes:

- How intensively the product or sample of products integrated in the service process can enhance patient safety, customer rights, and professionalism of care. Theme: Mission-strategic justification analysis starting from the early phases of design.
- How the mission-strategic view should be noted in continuous service and application assessment in various quality views. Theme: Mission-centric customer orientation in application quality audit.
- How the challenges of mission-centric customer orientation should be taken into account in the training programs of designers and developers. Theme: Mission-centric customer orientation as an educational challenge.
- How mission-centric customer orientation will be integrated into all various stakeholders' training. Theme: Mission-strategic customer-centric design as a shared challenge.

5.3.3 Mission-strategic perspective, ethics, and overall quality in healthcare

Mission-strategic questioning and identification in the early phases of design are essential [94] because many crucial choices are made in the early stages of the product scheme [81]. Mission-focused ethical justification analysis at the very beginning of a development project forms a foundational layer for all other evaluations. Then, mission identification and awareness integrate into ethical views throughout the product or service design in healthcare. Social, environmental, ethical, and human rights-related evaluation views and measures should get more space in general in the 4.0 era [95]. Consequently, health-related business models should reflect human-centered, ethical, sustainable, and transparent values [95, 96]. Understanding the relationship between customer-centeredness, mission value, ethical principles, legal aspects, and different quality categories enhances operative intelligence in the field of health technology and service policy [89, 97].

In healthcare, the professionalism of the offered services is a main target. Novel innovations are connected to change. The essential question connects change policy in primary design plans. Is change needed, and why? [98]. The aim of emergent technologies is to aid in achieving a professional care intensity level in patient care which would reflect ethical responsibility in the system. Ethical aspects are thus also related to intensity thinking in healthcare [89]. An appropriate care intensity level is one guarantee of quality and can significantly support cost-effective care with enhanced health outcomes [99]. In well-organized service systems too low intensity level in care means neglected professionalism and produces diagnostic errors, delays in care, complications, fatal consequences, and consequently, more indirect costs [100]. Thus, besides ethical frames, economic and political factors are integrated into the evaluative focus on a mission-strategic basis. Novel technologies should reflect values that could promote the image value of the organization. Issues of real image quality connect trustworthiness in communication and are thus not far from ethical issues either [97]. This mission-strategic questioning connects the main quality aspects at the macro level (see table 5.2).

Table 5.2. Mission-strategic questioning in main quality areas.

Quality view	Mission-strategic questioning
Mission-strategic awareness	Is the mission justified, obvious, and shared?
Ethical justification	Ethical acceptability and justification as a basis in innovation and service policy
Customer	Patient safety, respect for life, professional intensity level of care as aims Customers' acceptance
Product	Is the technology highly beneficial? Is change needed, and why?
Process	Does the product support key processes intensively?
Efficiency	Is the innovation or service concept cost-effective?
Production	Production policy and maturation policy in balance with ethical guidelines
Image	Transparency and trustworthiness in communication Image creation based on facts Innovation in image value enhancement

5.3.4 Customer-centric design through metadesign

Customer-centric design means adaptive and sometimes even customized products and services. Cocreation, bottom-up design, and an adaptive design process are integrated in the metadesign approach [101]. Metadesign empowers users' creative competences in an open environment and forms an intrinsic tension between standardization and improvisation that allows users to design for themselves [102]. Metadesign addresses the following essentials: (a) products or designs must be elastic because they cannot be designed entirely prior to use; (b) these advance and evolve partly by users; and (c) these must be planned for progress. In this process, users can be designers throughout the entire life cycle of the system. Consumers and users can create their own visions and targets through their own customizations and extensions and thus make a product fit for their purposes. Metadesign is not restricted to end-user enlargement, but can be elaborated by software professionals or system developers as well [102].

Metadesign involves new cultures and mindsets. The coparticipation of different actors with their distinctive skills and desires nurtures transdisciplinary dialog in the design process [103]. In healthcare, metadesign approaches can offer novel possibilities by empowering clients' independent health management [89]. New technologies like AI and DTs can bring more options to metadesign kits for designers. In particular, higher levels of cognition, such as synthesis, evaluation, and analytical ponderings in self-health management, can become more personal for consumers with easy-use metadesign options in consumer targeted applications. Consumer could focus on such areas in one's health status which bother most and one could enlarge these focus areas with background material of one's own health-related history or daily monitored data. Self-diagnostics and health-related self-monitoring e.g., sensors and connected smartphone applications and rehabilitative activities could find more usable tactics with intense metadesign approaches.

5.4 Enhancing customer orientation in organizations

5.4.1 Customer-centric service policy through comprehensive training

Motivation for mission-related training integrates development challenges in healthcare. Ethical issues and mission-strategic guidelines are linked to customer quality dimensions, and knowledge and educational factors play a role in enabling socially responsible actions in organizational culture [14]. Long-term growth in organizations requires ethical, sustainable, and transparent business models and values that are coherent with these ideas [95]. AI can support service design, diagnostics, and decision making but is not able to make ethical or moral choices. Health professionals are needed for constructive interaction; to face the consumer and choose a treatment together with the patient [104].

In healthcare, training is aimed at enhancing service processes and improving human–technology integration. The mission value of the services and novel innovations must be recognizable, transparent, and acceptable in view of producers, health consumers, and stakeholders. Customer-centric product design aspires to intense service entities—deep integration of products, target processes, health consumers, and health professionals. Educational and tutorial interventions in the health context can contain elements of new knowledge, understanding, skills, values, and attitudes [105] (e.g., clinical skills, medical knowledge, cooperative and communication skills, and information managerial- and service-related issues and related value frames).

Ethically focused training and guidance should be sufficiently pragmatic and tailored to meet the problem areas of the target group or organization. The reflection process involves identifying the goal, process, and object of action in connected processes in clinical practice or operation environment [106]. This requires nowadays in healthcare abilities for lifelong learning and openness to new kinds of training interventions in which health professionals and health consumers can engage in dialog under teaching intercessions. Ethics-related educational and mentoring support for healthcare clients can be relegated to a side role in busy clinical practice unless both healthcare professionals and consumers are encouraged and supported through interactive training portals and renewed forms of tutorial collaboration. This would also increase discussion and the exchange of ideas about values. Ethically oriented training with deeper mission recognition is even more relevant in healthcare organizations as healthcare technology evolves. It can prevent mechanical work attitudes that are easily encountered in a busy work atmosphere and may thus promote intellectual reflection in work. Ethically oriented tutoring and education can help both healthcare professionals and patients make socially responsible choices by increasing principled awareness, transparency, and mutual communication.

Educational interventions alone, however, do not necessarily develop thinking skills and thereby improve problem-solving skills [107], especially in challenging questions. 'Cleaner production' covers also smarter action models in which there is enough space for the transparent exchange of ideas and conversations about questions and problematic areas. *Mission-strategic educational interventions should focus particularly on wicked problems in healthcare.* The main mission of customer-centric healthcare is to protect human life in all its aspects, and this is especially important in the most fragile stages of the human life cycle. This reflects genuine social responsibility and ethical quality in healthcare and is thus the best indicator of highly relevant healthcare. The healthcare of elderly people is insufficient and disorganized in many countries [108]. Furthermore, in old age, the need for intense care increases, but in practice, at the same time, care intensity levels typically decrease among elderly patient groups [109–111] and also despite technological developments that allow safer and less invasive monitoring and treatment options. Underutilization of healthcare resources or emergent technologies in the case of the need for treatment also means waste of resources [89]. Also, ineffective care protocols in healthcare mean rising total costs. Equally, abortion rates are still high, even in countries with high levels of living [112], and even in spite of available educational resources, consultative aids, and social support systems.

Best practices in healthcare also concern ethical protocols and choices. In an educational process, the conditions of learning, content, methods, and tools are

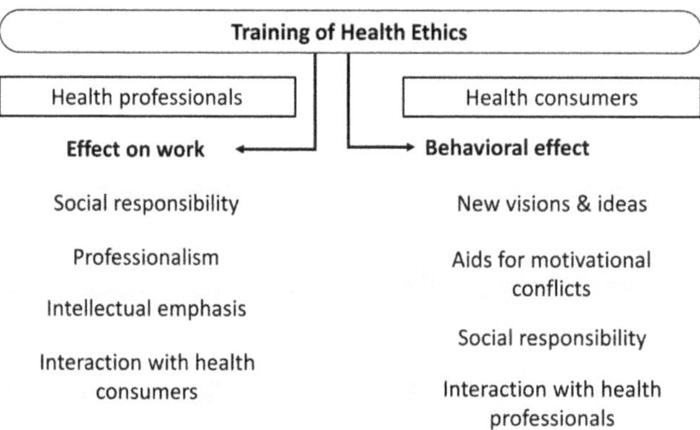

Figure 5.2. Organizational education challenges in wicked problems.

essential [105]. Artificial intelligence-based applications and combining data, as well as DTs, can provide new kinds of tooling support for training design [19] also on ethics- and mission-focused themes. The DT concept as an instrument for deeper analytical investigations could bring interesting background information about identified wicked and problematic target areas by allowing, for example, informative comparisons between different sizes of datasets. It is also possible to collect the necessary background material for training purposes in a more specific and targeted way. Similarly, more informative solutions are available for the visual presentation of data (see figure 5.2).

5.4.2 Shared mission in multidisciplinary design and action

Today, health informatics and technology design inspires working groups to practice interdisciplinarity more intensively, as multidisciplinary cooperative work is typical in various health technology subsectors. Cooperation and communication play key roles in complex organizational environments in healthcare. These social subsystems cover people from different research and discipline backgrounds, with their specific values, attitudes, and visionary scenes.

Different disciplines often operate with their own languages and theories, integrated with their own theoretical frames and bases, and this also applies to aspects of overall evaluation mechanisms. Translational healthcare encourages different disciplines to strive for a common debate, for example, on ethical issues [113, 114]. Interdisciplinary dialog generally opens up interpretive links between different disciplines, allowing room for more mature analysis in challenging problem areas.

Creativity is valued in all fields. Multidisciplinary thinking can also contribute to critical thinking and create a new kind of dialog that is less tied to existing frameworks. However, consensus seeking and individual emphases should both be given enough creativity space in multidisciplinary work environments [115]. As application or system development and testing often require multidisciplinary expertise in healthcare, this collaboration can also contribute to more diverse assessment activities. Cooperative functioning and continuous multidisciplinary team working are needed inside the production group, and often with the users, stakeholders, and other interest groups. Cooperation between different groups is affected by the prevailing operating methods, structures, and cultures in organizations. These frames and structures influence how successfully collaboration can finally enhance creativity and, ultimately, product innovation. Intense rigidity in an operating environment can easily hinder creative minds. Even if more specialists are needed in the health domain, also the importance of generalists in complex decision making is growing as well. AI and evolving technologies in healthcare contribute also to the need for more multi-experts in the healthcare domain.

5.5 Process reengineering with renewed mission

5.5.1 Case: customer-centric dental care

Dentistry 4.0 offers inventive solutions with new manufacturing chances and automation, data exchange, and developed technologies [116]. AI, the Internet of Dental Things (IoDT), computer-aided design (CAD), dental 3D scanners, cone beam computed tomography (CBCT), and biosensors can make dental treatments more sophisticated [117]. 3D skeletal dental examination and evaluation for treatment procedures serve in orthodontic, prosthesis, and orthognathic surgery [118]. DTs have been utilized, for example, when providing suitable orthodontic treatment to adult females with aid of facial scans and computed tomography [119]. CAD/CAM software applications allow dentists to perform procedures faster and more accurately. Intraoral scanning enables remote dental consultations and producing fixed ceramic porcelain crowns in a one-day visit instead of two weeks. Digital reproduction and milling machines are used in the restorative care. Patients can use a dental monitoring ScanBox device and their smartphone and take intraoral images from anywhere and send it to the dentist through a patient app [52]. Light-induced fluorescence (QLF) technologies are helpful in the detection of pathologic lesions and plaque by noninvasive actions [118].

Data dentistry allows more precise and personalized dental care [120]. Dental care is not limited to activities in the office; teledentistry and online care, online visits and prescriptions, remote help, and advice change practices in oral health [121]. Online interviews and information sharing before consultation in the office can reduce consumers' anxiety. Chat services bring more flexibility to doctor–patient interaction. Informative guides can be helpful by informing consumers about relevant topics, such as the etiologies of common symptoms [121]. Chair-side tissue biopsies bring more opportunities for diagnostics. Such tests are already in use for example in periodontology [104].

Instead of only treating a single tooth, the patient as a whole is in focus. Geriatric dentistry forms its own specialty area. Oral health contributes to one's wellbeing and physical and psychological status [122]. Regular dental check-ups are essential for all patients, but particularly for special groups like the elderly. Appropriate dental follow-up is also crucial in preventing more serious diseases in overall health status [121]. Therefore, geriatric dentistry should diversify its focus areas [123].

Customer-centric dental care means greater flexibility in communication between the patient and the dentist. A person's dental care profile should also have a place in the electronic health record. The condition of teeth is relevant, for example, in many preoperative anesthetic consultations [124]. Common information systems also will make dental treatment easier when the patient's general background information and previous treatment history are better understood [104].

In dentistry, digital care and monitoring systems have evolved, but the potential for digitalization in this area is also available with low-cost and easy-use operations which can make dental care more accurate and efficient. The exemplar DT system model map contains three actors: the patient, the dentist, and the administration. Patient activities include appointment reservations and pre-information delivery, and payment. The dentist takes care of pre-diagnosis, patient feedback, diagnosis, and treatment at the office, and the recording of procedures applied to patient. Administration takes care of resource allocation and billing. In this model, the client informs the dentist before the consultation at the office. Through the customer interface, the patient locates the treatment site in the area of the dentition as well as symptoms via the digital channel and consumer's individual, visual dentition map before the doctor's consultation.

A visual DT map of the person's individual dental condition also helps the patient more easily report their dental condition, mark possible areas of pain or dental damage, and gives an opportunity to add patient's own suggestions for their individual treatment (e.g., degree of suitable anesthetic). Pre-information could also contain issues of pain intensity, related disabilities, and connected stress reactions, as well as the DT map visualization. Aesthetic dentistry represents interactive activity in which hearing the client's views and wishes is important. The patient's individual dental map is currently used by the physician, but the patient most often makes suggestions without any aids.

An informative online connection before consultation time helps the doctor and the patient better interact and plan treatments. Through this prospective information, the doctor can make a preplan for diagnosis, investigations, and treatment, which will be rechecked during consultancy. Prospective activities that aid communication between administration and information automation decrease the workload of the dentist. The patient learns to understand and anticipate the development of their dentistry. The application could also automatically inform the patient of their upcoming annual follow-up visits (see table 5.3 and figure 5.3).

5.5.2 Process reengineering with mission-strategic vision

Intense possibilities to enhance and streamline activities with technological innovations and opening options require thorough mission-strategic ponderings. This advances innovation policy with a sustainable ethical basis. Even if the innovation's high-level requirements are typically part of the mission statement of the project [125], system-related mission strategic re-evaluations should be widen to system entity conception. This system entity covers innovations, users, health professionals, and connected service concepts. This entity should finally fulfill its targeted

Table 5.3. Renewed dentist–patient model.

Old model

	Step 1	Step 2	Step 3	Step 4	Step 5
Patient	Appointment reservation	Arrival at dentist			Payment
Dentist	Scheduling		Diagnosis	Treatment	
Database	Coded data from previous visit (for dentist)			Doctor's report	Update

New model

	Step 1	Step 2	Step 3	Step 4	Step 5
Patient	Inquiry through patient interface/ pre-information	Getting preliminary estimation	Acceptance of given slot	Arrival for treatment	Payment
Multilevel Database	New entry to patient's record, inquiry for appointment	Care resource allocation and usage estimate			Updates, billing, follow-up if necessary
Dentist	Scheduling workload, planning	Pre-diagnosis to database	Resource reservation	Diagnosis, treatment	Documentation

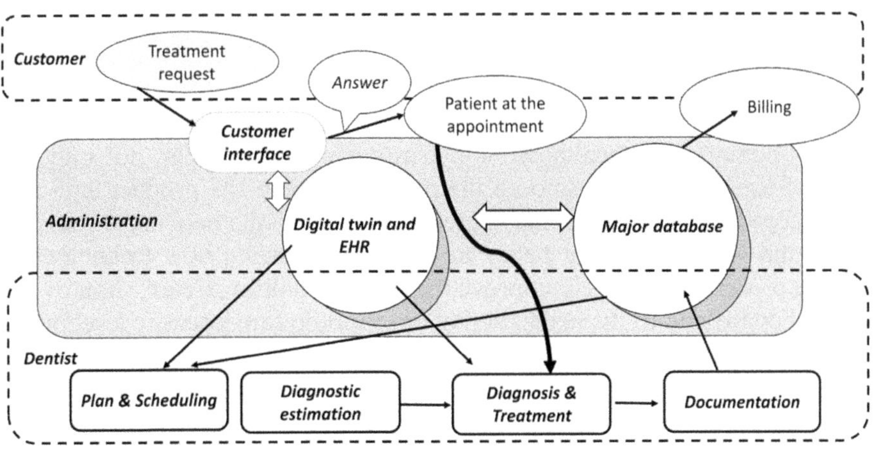

Figure 5.3. Renewed model for dentist consultation.

mission. Mission-strategic customer-centric evaluation should focus in healthcare, particularly in the following tasks and issues:

- Mission in healthcare means emphasis on patient safety and the overall professionalism of care. Ponderings around these theme areas should start all health technology innovation projects. This requires a mission-strategic justification analysis in the early stages of design.

- Mission-strategic view should concentrate on these three key areas with versatile quality thinking and approach to guarantee mission-centric customer orientation in quality auditing of innovations.
- Mission-centric customer orientation should be integrated especially in the training programs of health technology designers and developers to guarantee more sophisticated products.
- Mission-centric customer orientation should represent a shared challenge and thus be integrated also into all stakeholders' training including health consumers.
- Mission-centric customer orientation is, in healthcare, intensively integrated in ethical awareness. Ethical consciousness should increase with the technological advancement of the healthcare sector.
- In addition to technical development and process reengineering, multidisciplinarity must also be increased in ethically weighted cooperation. Interdisciplinary dialog opens up interpretive links between disciplines, allowing room for a more mature analysis on challenging problem areas.
- Multidisciplinary thinking can contribute to critical thinking and create a new kind of dialog, less tied to existing frameworks. The pursuit of consensus and individual emphases should both be given enough creativity space also in multidisciplinary work environments. Evaluating and maintaining the creativity space of the individual and the environment can open eyes to realize more possibilities [115].

5.6 Conclusions

A mission-strategic point of view guides the redesign and comprehension of customer orientation in healthcare 4.0. Customer-centric design not only means the ease of use of products, smooth interaction between the product and user, or products according to the customer's preferences. The issue here is the relevance of products and well-functioning policy in the service system; how technologies and healthcare processes ultimately improve the professionalism of care, support patient safety, and contribute to the realization of appropriate care intensity level in patient care and disease prevention. A doctor's work is most cost-effective when she treats the patient as professionally as possible. In fact, patients' professional care intensity level, an intensive focus on prevention, diagnostic accuracy, and elimination of medical errors are all powerful tools for intended savings in healthcare [89]. Emergent and existing health technologies in the early detection and prevention of diseases play a key role in the scheme of professional care. As a consequence this means reducing healthcare costs by reducing the need and demand for medical services [126].

Underdiagnosis and undertreatment multiply mortality specifically among elderly patient groups. Respiratory dysfunctions [127], diabetes [128], and heart problems [109] are well-known examples of underdiagnosis and undertreatment among older people. Polypharmacy among elderly people increases risk of adverse drug reactions which means complications that add significant additional healthcare costs [89, 129].

Recent pandemics have caused high mortality, especially among the elderly population. Relative new sensor technology represents new possibilities for elderly care. As a good example, early signals about health problems can give boosts for preventive actions and increase the cost-effectiveness of modern healthcare. Preventive healthcare is also known to be effective for the elderly population [130]. The development and implementation of sensor technology in, for example, the aforementioned areas can promote the intensity level in future healthcare.

With new technological challenges, the conceptual frames of customer-oriented design are expanding. Mission-strategic visions will typically be born in the ideas of the development teams in industry. However, consumers should also have a role in accepting and valuing these ideas. Also, consumers may produce new innovative design ideas. However, when pondering customers' wishes, attention must also be paid to the ethics and practicality of these signals [89].

The significance of the intended product innovation, the product's mission, is essential. Thus, applications and novelties that support the safety and professionalism of services can be considered customer-oriented in terms of their product mission. However, if IT investments and advanced technology policy proceed in an overly technology-oriented way, the outcome may be less successful [131], especially in healthcare. Advanced innovations, service concepts, health professionals, and health consumers form the service entity, which should function seamlessly and fulfill its mission [89]. Along with the assessment of health technology, the assessment of the strength of this synthesis entity is significant. The mission-centered perspective brings a useful primary dimension to the customer-centric evaluation context in the health technology area. This poses challenges for the development of training and auditing methods in this field and will sharpen design and evaluation frames in healthcare and service sectors. A mission-strategic evaluation focus integrated with ethical awareness represents a development policy that can make customer-centric design more achievable in healthcare.

Acknowledgments

We thank Villa Ensi Foundation, Finland, for the financial support of this research.

References

[1] Leavitt H J 1965 Applied organizational change ed I J G Industry *Handbook of Organizations* (Chicago, IL: Rand McNally)1144 March p 1170

[2] Münch C, Marx E, Benz L, Hartmann E and Matzner M 2022 Capabilities of digital servitization: evidence from the socio-technical systems theory *Technol. Forecast. Soc. Change* **176** 121361

[3] Spillane J P, Halverson R and Diamond J B 2001 Investigating school leadership practice: a distributed perspective *Educ. Res.* **30** 23–8

[4] Lasi H, Fettke P, Kemper H-G, Feld T and Hoffmann M 2014 Industry 4.0 *Bus. Inf. Syst. Eng.* **6** 239–42

[5] Pettigrew K A 2013 Senior community centers of San Diego as a preventive care model: a perspective *Am. J. Prev. Med.* **44** S34–8

[6] Nurgalieva L, Laconich J J J, Baez M, Casati F and Marchese M 2019 A systematic literature review of research-derived touchscreen design guidelines for older adults *IEEE Access* **7** 22035–58

[7] Neves B B and Vetere F 2019 Ageing and digital technology *Designing and Evaluating Emerging Technologies for Older Adults* (Singapore: Springer) pp 1–14

[8] Nicenboim I, Giaccardi E and Kuijer L 2018 Designing connected resources for older people *Proc. 2018 Designing Interactive Systems Conf* pp 413–25

[9] Troxler P and Lauche K 2015 Assessing, Creating and Sustaining Knowledge Culture in Organisations http://www.trox.net/km-cult/content/

[10] Abdullah F M, Saleh M, Al-Ahmari A M and Anwar S 2022 The impact of Industry 4.0 technologies on manufacturing strategies: proposition of technology-integrated selection *IEEE Access* **10** 21574–83

[11] Aceto G, Persico V and Pescapé A 2020 Industry 4.0 and health: Internet of Things, big data, and cloud computing for healthcare 4.0 *J. Ind. Inf. Integr.* **18** 100129

[12] da Silveira F, Neto I R, Machado F M, Silva da M P and Amaral F G 2019 Analysis of Industry 4.0 technologies applied to the health sector: systematic literature review *Occupational and Environmental Safety and Health* (Berlin: Springer) 701–9

[13] Rehman M U, Andargoli A E and Pousti H 2019 Healthcare 4.0: trends, challenges and benefits *Australasian Conf. on Information Systems (Perth)* pp 556–64

[14] Mukhuty S, Upadhyay A and Rothwell H 2022 Strategic sustainable development of Industry 4.0 through the lens of social responsibility: the role of human resource practices *Bus. Strategy Environ.* **31** 2068–81

[15] Wilmer H, Meadow A M, Brymer A B, Carroll S R, Ferguson D B, Garba I, Greene C, Owen G and Peck D E 2021 Expanded ethical principles for research partnership and transdisciplinary natural resource management science *Environ. Manage.* **68** 453–67

[16] Keleko A T, Kamsu-Foguem B, Ngouna R H and Tongne A 2022 Artificial intelligence and real-time predictive maintenance in industry 4.0: a bibliometric analysis *AI Ethics* **2** 553–77

[17] Grieves M 2014 Digital twin: manufacturing excellence through virtual factory replication *White Paper* **1** 1–7

[18] Li J, Tao F, Cheng Y and Zhao L 2015 Big data in product lifecycle management *Int. J. Adv. Manuf. Technol.* **81** 667–84

[19] Kamel Boulos M N and Zhang P 2021 Digital twins: from personalised medicine to precision public health *J. Pers. Med.* **11** 745

[20] Lal A, Li G, Cubro E, Chalmers S, Li H, Herasevich V, Dong Y, Pickering B W, Kilickaya O and Gajic O 2020 Development and verification of a digital twin patient model to predict specific treatment response during the first 24 h of sepsis *Crit. Care Explor.* **2** e0249

[21] Bruynseels K, Santoni de Sio F and Van den Hoven J 2018 Digital twins in health care: ethical implications of an emerging engineering paradigm *Front. Genet.* **31** 1–11

[22] Khan S, Arslan T and Ratnarajah T 2022 Digital twin perspective of fourth industrial and healthcare revolution *IEEE Access* **10** 25732–54

[23] Voigt I, Inojosa H, Dillenseger A, Haase R, Akgün K and Ziemssen T 2021 Digital twins for multiple sclerosis *Front. Immunol.* **12** 1556

[24] World Health Organization 2021 *Ethics and Governance of Artificial Intelligence for Health: WHO Guidance* (Geneva: World Health Organization)

[25] Tooley M 2006 *Electronic Circuits: Fundamentals and Applications* 3rd edn (London: Taylor and Francis)

[26] Malekloo A, Ozer E, AlHamaydeh M and Girolami M 2022 Machine learning and structural health monitoring overview with emerging technology and high-dimensional data source highlights *Struct. Health Monit.* **21** 1906–55

[27] Huang Z, Xu Y, Cheng Y, Xue M, Deng M, Jaffrezic-Renault N and Guo Z 2022 Recent advances in skin-like wearable sensors: sensor design, health monitoring, and intelligent auxiliary *Sens. Diagn.* **1** 686–708

[28] Kazanskiy N L, Butt M A and Khonina S N 2022 Recent advances in wearable optical sensor automation powered by battery versus skin-like battery-free devices for personal healthcare—a review *Nanomaterials* **12** 334

[29] Chen S, Qi J, Fan S, Qiao Z, Yeo J C and Lim C T 2021 Flexible wearable sensors for cardiovascular health monitoring *Adv. Healthcare Mater.* **10** 2100116

[30] Pozdin V A and Dieffenderfer J 2022 Towards wearable health monitoring devices *Biosensors* **12** 322

[31] Costin H-N and Sanei S 2022 Intelligent biosignal processing in wearable and implantable sensors *Biosensors* **12** 396

[32] Kang M and Chai K 2022 Wearable sensing systems for monitoring mental health *Sensors* **22** 994

[33] Prieto-Avalos G, Cruz-Ramos N A, Alor-Hernández G, Sánchez-Cervantes J L, Rodríguez-Mazahua L and Guarneros-Nolasco L R 2022 Wearable devices for physical monitoring of heart: a review *Biosensors* **12** 292

[34] Bansal A K, Hou S, Kulyk O, Bowman E M and Samuel I D 2015 Wearable organic optoelectronic sensors for medicine *Adv. Mater.* **27** 7638–44

[35] Vavrinsky E, Esfahani N E, Hausner M, Kuzma A, Rezo V, Donoval M and Kosnacova H 2022 The current state of optical sensors in medical wearables *Biosensors* **12** 217

[36] Khandare V V, Funde P M, Ingavale M A, Shinde S M and Saini S P 2022 Real time healthcare monitoring system *Int. J. Res. Publ. Rev.* **3** 227–33

[37] Bariya M, Nyein H Y Y and Javey A 2018 Wearable sweat sensors *Nat. Electron.* **1** 160–71

[38] Wu M and Luo J 2019 Wearable technology applications in healthcare: a literature review *Online J. Nurs. Inform* **23** 1 https://www.himss.org/resources/wearable-technology-applications-healthcare-literature-review

[39] Zhou H, Du X and Zhang Z 2021 Electrochemical sensors for detection of markers on tumor cells *Int. J. Mol. Sci.* **22** 8184

[40] Kaur B, Kumar S and Kaushik B K 2021 2D materials-based fiber optic SPR biosensor for cancer detection at 1550 nm *IEEE Sens. J.* **21** 23957–64

[41] Dejous C and Krishnan U M 2021 Sensors for diagnosis of prostate cancer: looking beyond the prostate specific antigen *Biosens. Bioelectron.* **173** 112790

[42] Wang L 2018 Microwave sensors for breast cancer detection *Sensors* **18** 655

[43] Goud K Y, Moonla C, Mishra R K, Yu C, Narayan R, Litvan I and Wang J 2019 Wearable electrochemical microneedle sensor for continuous monitoring of levodopa: toward Parkinson management *ACS Sens.* **4** 2196–204

[44] Lu H, He B and Gao B 2021 Emerging electrochemical sensors for life healthcare *Eng. Regen.* **2** 175–81

[45] Kalambate P K, Noiphung J, Rodthongkum N, Larpant N, Thirabowonkitphithan P, Rojanarata T, Hasan M, Huang Y and Laiwattanapaisal W 2021 Nanomaterials-based

electrochemical sensors and biosensors for the detection of non-steroidal anti-inflammatory drugs *TrAC Trends Anal. Chem.* **143** 116403

[46] Gedam S and Paul S 2021 A review on mental stress detection using wearable sensors and machine learning techniques *IEEE Access* **9** 84045–66

[47] Long N, Lei Y, Peng L, Xu P and Mao P 2022 A scoping review on monitoring mental health using smart wearable devices *Math. Biosci. Eng.* **19** 7899–919

[48] Roh T, Hong S and Yoo H-J 2014 Wearable depression monitoring system with heart-rate variability *2014 36th Annual Int. Conf. of the IEEE Engineering in Medicine and Biology Society* (Piscataway, NJ: IEEE) pp 562–5

[49] Daniels J, Haber N, Voss C, Schwartz J, Tamura S, Fazel A, Kline A, Washington P, Phillips J and Winograd T 2018 Feasibility testing of a wearable behavioral aid for social learning in children with autism *Appl. Clin. Inform.* **9** 129–40

[50] Heikkilä N 2022 Anturidatan analyysi unen vaikutuksista ja vaikutuksista uneen *BA Thesis* (Oulu: University of Oulu)

[51] Imtiaz S A 2021 A systematic review of sensing technologies for wearable sleep staging *Sensors* **21** 1562

[52] Lee L 2021 Why digitalisation in dentistry is more important today than ever https://www.dpdental.com/digitalisation-in-dentistry/

[53] Liao X, Zhang Z, Kang Z, Gao F, Liao Q and Zhang Y 2017 Ultrasensitive and stretchable resistive strain sensors designed for wearable electronics *Mater. Horiz.* **4** 502–10

[54] Liu X, Wei Y and Qiu Y 2021 Advanced flexible skin-like pressure and strain sensors for human health monitoring *Micromachines* **12** 695

[55] Yetisen A K, Martinez-Hurtado J L, Ünal B, Khademhosseini A and Butt H 2018 Wearables in medicine *Adv. Mater.* **30** 1706910

[56] Hatamie A, Angizi S, Kumar S, Pandey C M, Simchi A, Willander M and Malhotra B D 2020 Textile based chemical and physical sensors for healthcare monitoring *J. Electrochem. Soc.* **167** 037546

[57] Johnston L, Wang G, Hu K, Qian C and Liu G 2021 Advances in biosensors for continuous glucose monitoring towards wearables *Front. Bioeng. Biotechnol.* **9** 1–17

[58] Gopinath K, Selvam G and Narayanamurthy G 2022 Determinants of the adoption of wearable devices for health and fitness: a meta-analytical study *Commun. Assoc. Inf. Syst.* **50** 23

[59] Binyamin S S and Hoque M R 2020 Understanding the drivers of wearable health monitoring technology: an extension of the unified theory of acceptance and use of technology *Sustainability* **12** 9605

[60] Yang Q, Al Mamun A, Hayat N, Salleh M F M, Jingzu G and Zainol N R 2022 Modelling the mass adoption potential of wearable medical devices *PLoS One* **17** e0269256

[61] Schneider J and Stickdorn M 2011 *This Is Service Design Thinking: Basics, Tools, Cases* (New York: Wiley)

[62] Richards T, Coulter A and Wicks P 2015 Time to deliver patient centred care *BMJ* **350** h530

[63] van Biesen W, Van Der Straeten C, Sterckx S, Steen J, Diependaele L and Decruyenaere J 2021 The concept of justifiable healthcare and how big data can help us to achieve it *BMC Med. Inf. Decis. Making* **21** 1–17

[64] Reinert D, Schueler K and Storf H 2021 Usability design in medical informatics: a prospective research project *Stud. Health Technol. Inform.* **281** 1015–6

[65] Harrison R, Ni She E and Debono D 2022 Implementing and evaluating co-designed change in health *J. R. Soc. Med.* **115** 48–51

[66] Ferrucci F, Jorio M, Marci S, Bezenchek A, Diella G, Nulli C, Miranda F and Castelli-Gattinara G 2021 A web-based application for complex health care populations: user-centered design approach *JMIR Hum. Factors* **8** e18587

[67] Iivari J and Iivari N 2006 Varieties of user-centeredness *Proc. 39th Annual Hawaii Int. Conf. on System Sciences (HICSS'06)* 8 (Piscataway, NJ: IEEE) pp 176a

[68] Feigenbaum A V 1991 *Total Quality Control* (New York: McGraw-Hill) pp 1–342

[69] Ghobadian A, Speller S and Jones M 1994 Service quality: concepts and models *Int. J. Qual. Reliab. Manage.* **11** 43–66

[70] Hevner A R, March S T, Park J and Ram S 2004 Design science in information systems research *MIS Q.* **28** 75–105

[71] Pallot M, Trousse B, Senach B and Scapin D 2010 Living lab research landscape: from user centred design and user experience towards user cocreation *Technol. Innov. Manag. Rev.* **1** 1–11

[72] Fuglsang L 2019 Human-centric service co-innovation in public services from a practice-based perspective: a case of elderly care *Human-Centered Digitalization and Services* (Berlin: Springer) pp 17–36

[73] Meinel M, Eismann T T, Baccarella C V, Fixson S K and Voigt K-I 2020 Does applying design thinking result in better new product concepts than a traditional innovation approach? An experimental comparison study *Eur. Manag. J.* **38** 661–71

[74] Bakken S 2021 Patient safety and quality of care: a key focus for clinical informatics *J. Am. Med. Inform. Assoc.* **28** 1603–4

[75] Fishman B J, Marx R W, Best S and Tal R T 2003 Linking teacher and student learning to improve professional development in systemic reform *Teach. Teach. Educ.* **19** 643–58

[76] Oulasvirta A and Hornbæk K 2022 Counterfactual thinking: what theories do in design *Int. J. Hum.–Comput. Interact.* **38** 78–92

[77] Hevner A R 2007 A three cycle view of design science research *Scand. J. Inf. Syst.* **19** 4

[78] Winter R 2009 Interview with Alan R. Hevner on 'Design Science' *Bus. Inf. Syst. Eng* **1** 126–9

[79] Hevner A, vom Brocke J and Maedche A 2019 Roles of digital innovation in design science research *Bus. Inf. Syst. Eng.* **61** 3–8

[80] Youmans R J 2011 The effects of physical prototyping and group work on the reduction of design fixation *Des. Stud.* **32** 115–38

[81] Koo Y 2016 The role of designers in integrating societal value in the product and service development processes *Int. J. Des.* **10** 49–65

[82] Interviewed by Graber M L 2020 2019 John M. Eisenberg patient safety and quality awards: an interview with Gordon D. Schiff *Jt Comm J Qual Patient Saf* **46** 371–80

[83] Berkun S 2010 *The Myths of Innovation* (Sebastopol, CA: O'Reilly Media, Inc)

[84] Khoja S, Durrani H, Scott R E, Sajwani A and Piryani U 2013 Conceptual framework for development of comprehensive e-health evaluation tool *Telemed. e-Health* **19** 48–53

[85] Nakai H, Tsuda N, Honda K, Washizaki H and Fukazawa Y 2016 Initial framework for software quality evaluation based on ISO/IEC 25022 and ISO/IEC 25023 *2016 IEEE Int. Conf. on Software Quality, Reliability and Security Companion (QRS-C)* (Piscataway, NJ: IEEE) pp 410–1

[86] Estdale J and Georgiadou E 2018 Applying the ISO/IEC 25010 quality models to software product *European Conf. on Software Process Improvement* (Berlin: : Springer) pp 492–503

[87] Rubí J N S and Gondim P R L 2019 IoMT platform for pervasive healthcare data aggregation, processing, and sharing based on OneM2M and OpenEHR *Sensors* **19** 4283

[88] Wyatt J C 2016 Evidence-based health informatics and the scientific development of the field *Evidence-Based Health Informatics* (Amsterdam: IOS Press) pp 14–24

[89] Rissanen M 2021 Comprehending translational design scenarios and implications in consumer health informatics *Doctoral Thesis* (Espoo: Aalto University)

[90] Rissanen M 2020 Translational health technology and system schemes: enhancing the dynamics of health informatics *Health Inf. Sci. Syst.* **8** 39

[91] Verschuren P and Hartog R 2005 Evaluation in design-oriented research *Qual. Quant.* **39** 733–62

[92] Li J H, Land L P W, Chattopadhyay S and Ray P 2008 An approach for e-health system assessment and specification *HealthCom 2008-10th Int. Conf. on e-health Networking, Applications and Services* (Piscataway, NJ: IEEE) pp 134–9

[93] Wehling M 2008 Translational medicine: science or wishful thinking? *J. Transl. Med.* **6** 1–3

[94] Rissanen M 2014 Prioritizing quality attributes in health design *Leading Transformation to Sustainable Excellence Proc. Congress, 7th Quality Conf. in the Middle East* pp 133–42

[95] Machado C G, Winroth M P and Ribeiro da Silva E H D 2020 Sustainable manufacturing in Industry 4.0: an emerging research agenda *Int. J. Prod. Res.* **58** 1462–84

[96] Friedman B, Kahn P and Borning A 2002 Value sensitive design: theory and methods *University of Washington Technical Report* **2** 12

[97] Rissanen M 2015 Ethical quality in eHealth: a challenge with many facets *Health Information Science (Lecture Notes in Computer Science)* ed X Yin, K Ho, D Zeng, U Aickelin, R Zhou and H Wang (Cham: Springer International Publishing) pp 146–53

[98] Rissanen M K 2016 Integrating translational design ideology for consumer-targeted, informative eHealth *Int. J. Innov. Manage. Technol.* **7** 260

[99] Lerolle N, Trinquart L, Bornstain C, Tadié J-M, Imbert A, Diehl J-L, Fagon J-Y and Guérot E 2010 Increased intensity of treatment and decreased mortality in elderly patients in an intensive care unit over a decade *Crit. Care Med.* **38** 59–64

[100] Li C, Chen E, Savova G, Fraser H and Eickhoff C 2020 Mining misdiagnosis patterns from biomedical literature *AMIA Jt. Summits Transl. Sci. Proc.* **2020** 360–6

[101] Giaccardi E 2003 Principles of metadesign: processes and levels of co-creation in the new design space *PhD Thesis* (Plymouth, UK: University of Plymouth)

[102] Fischer G and Giaccardi E 2006 Meta-design: a framework for the future of end-user development *End User Development* (Berlin: Springer) 427–57

[103] Angelucci F 2021 Metadesigning the urban space/environmental system. Inter-and trans-disciplinary issues *Techne* **2** 53

[104] Pohjola V 2018 Digitalisaatio ja tekoäly—tulevaisuuden isäntiä vai renkejä? *Hammaslääkäri* **40** 1–3

[105] Targamadzė A and Petrauskienė R 2010 Impact of information technologies on modern learning *Inf. Technol. Control* **39** 169–75

[106] Procee H 2006 Reflection in education: a Kantian epistemology *Educ. Theory* **56** 237–53

[107] Montana-Hoyos C and Lemaitre F 2011 Systems thinking, disciplinarity and critical thinking in relation to creativity within contemporary arts and design education *Stud. Learn. Eval. Innov. Dev.* **8** 12–25

[108] Marcusson J, Nord M, Dong H-J and Lyth J 2020 Clinically useful prediction of hospital admissions in an older population *BMC Geriatr.* **20** 1–9

[109] Iung B, Cachier A, Baron G, Messika-Zeitoun D, Delahaye F, Tornos P, Gohlke-Bärwolf C, Boersma E, Ravaud P and Vahanian A 2005 Decision-making in elderly patients with severe aortic stenosis: why are so many denied surgery? *Eur. Heart J.* **26** 2714–20

[110] Boumendil A, Aegerter P, Guidet B and Network C-R 2005 Treatment intensity and outcome of patients aged 80 and older in intensive care units: a multicenter matched-cohort study *J. Am. Geriatr. Soc.* **53** 88–93

[111] Guidet B, Vallet H, Boddaert J, de Lange D W, Morandi A, Leblanc G, Artigas A and Flaatten H 2018 Caring for the critically ill patients over 80: a narrative review *Ann. Intensive Care* **8** 1–15

[112] Sedgh G, Finer L B, Bankole A, Eilers M A and Singh S 2015 Adolescent pregnancy, birth, and abortion rates across countries: levels and recent trends *J. Adolesc. Health* **56** 223–30

[113] Silva D S, Smith M J and Norman C D 2018 Systems thinking and ethics in public health: a necessary and mutually beneficial partnership *Monash Bioeth. Rev* **36** 54–67

[114] Kumar R, Singh R K and Dwivedi Y K 2020 Application of industry 4.0 technologies in SMEs for ethical and sustainable operations: analysis of challenges *J. Clean. Prod.* **275** 124063

[115] Rissanen A 2011 Enhancing physics education in the National Defence University: a design based research approach in the production of learning material *PhD Thesis* (Helsinki: University of Helsinki)

[116] Javaid M, Haleem A, Singh R P and Suman R 2021 Dentistry 4.0 technologies applications for dentistry during COVID-19 pandemic *Sustain. Oper. Comput.* **2** 87–96

[117] Liu Y-C, Shih M-C and Tu Y-K 2021 Using dental patient-reported outcomes (dPROs) in meta-analyses: a scoping review and methodological investigation *J. Evid. Based Dent. Pract.* **22** 101658

[118] Kim S-H, Kim K B and Choo H 2022 New Frontier in advanced dentistry: CBCT, intraoral scanner, sensors, and artificial intelligence in dentistry *Sensors* **22** 2942

[119] Cho S-W, Byun S-H, Yi S, Jang W-S, Kim J-C, Park I-Y and Yang B-E 2021 Sagittal relationship between the maxillary central incisors and the forehead in digital twins of korean adult females *J. Pers. Med.* **11** 203

[120] Schwendicke F and Krois J 2022 Data dentistry: how data are changing clinical care and research *J. Dent. Res.* **101** 21–9

[121] Shamsoddin E, DeTora L M, Tovani-Palone M R and Bierer B E 2021 Dental care in times of the COVID-19 pandemic: a review *Med. Sci.* **9** 13

[122] Ástvaldsdóttir Á, Boström A-M, Davidson T, Gabre P, Gahnberg L, Sandborgh Englund G, Skott P, Ståhlnacke K, Tranæus S and Wilhelmsson H 2018 Oral health and dental care of older persons—a systematic map of systematic reviews *Gerodontology* **35** 290–304

[123] Ky J, Scepanovic T, Senthilvadevel N, Mati S, Ming A L, Ng M, Nguyen D, Yeo P, Zhao T and Paolini R 2022 The effect of clinical interventions on the oral health-related quality of life in older adults *Aust. Dent. J.* **67** 302–13

[124] Gatt S P, Aurisch J and Wong K 2001 A standardized, uniform and universal dental chart for documenting state of dentition before anaesthesia *Anaesth. Intensive Care* **29** 48–51

[125] Bahill A T and Dean F 2009 Discovering system requirements *Handbook of Systems Engineering and Management* (New York: Wiley) pp 205–65

[126] Fries J F, Koop C E, Beadle C E, Cooper P P, England M J, Greaves R F, Sokolov J J and Wright DHealth Project Consortium the 1993 Reducing health care costs by reducing the need and demand for medical services *New Engl. J. Med.* **329** 321–5

[127] Dunn R M, Busse P J and Wechsler M E 2018 Asthma in the elderly and late-onset adult asthma *Allergy* **73** 284–94

[128] Ortiz-Martínez M, González-González M, Martagón A J, Hlavinka V, Willson R C and Rito-Palomares M 2022 Recent developments in biomarkers for diagnosis and screening of type 2 diabetes mellitus *Curr. Diabetes Rep.* **22** 95–115

[129] Tariq R A, Vashisht R, Sinha A and Scherbak Y 2020 *Medication Dispensing Errors And Prevention* (Treasure Island, FL: StatPearls)

[130] Goldman D P, Cutler D M, Shang B and Joyce G F 2006 The value of elderly disease prevention *Forum for Health Economics & Policy* 9 (Berlin: De Gruyter)

[131] Rannaste K 2011 Operatiivinen älykkyys ja event processing-teknologia *Tiede Ja Ase* **69** 161–77

IOP Publishing

Advanced Signal Processing for Industry 4.0, Volume 1

Evolution, communication protocols, and applications in manufacturing systems

Irshad Ahmad Ansari and Varun Bajaj

Chapter 6

The application of Industry 4.0 technologies for automated health monitoring and surveillance during pandemics and post-pandemic life

Hamid Reza Marateb, Mónica Rojas-Martínez, Shadi Zamani, Mehdi Shirzadi, Amirhossein Koochekian and Miguel Angel Mañanas

Industry 4.0 signifies a new step in the control and organization of the manufacturing value chain. It can provide better digital solutions for our day-to-day lives during pandemics and post-pandemic life, including better activity planning, global public health emergencies, and risk assessment. The Internet of Things (IoT), as a part of Industry 4.0, brings new opportunities in various applications, including smart healthcare and cities. The combination of an IoT system and artificial intelligence (AI) has the following capabilities in pandemic management: public security improvement, contact tracing to access public places, and automated diagnosis and post-pandemic prognosis. In this chapter, we discuss the contribution of Industry 4.0 (IoT, digitalization, Big Data, AI, and cloud-based computing) for automatic surveillance and health monitoring during pandemics and post-pandemic life. More than fifty papers published in 2020–2 were analyzed in this chapter.

6.1 Introduction

More than 551 million people were infected by COVID-19, among which more than six million died by July 6, 2022 [1]. The pandemic has been affecting entire countries, families, and communities. The world was changed by the COVID-19 disease. The implementation of digital and intelligent technologies has been accelerated by COVID-19 [2]. The preventive measures for COVID-19 are shown in figure 6.1.

In the 18th century, the first industrial revolution began. It initially focused on water and steam power. Mass production, standardization, and industrialization were related to the second revolution in the late 19th century into the early 20th

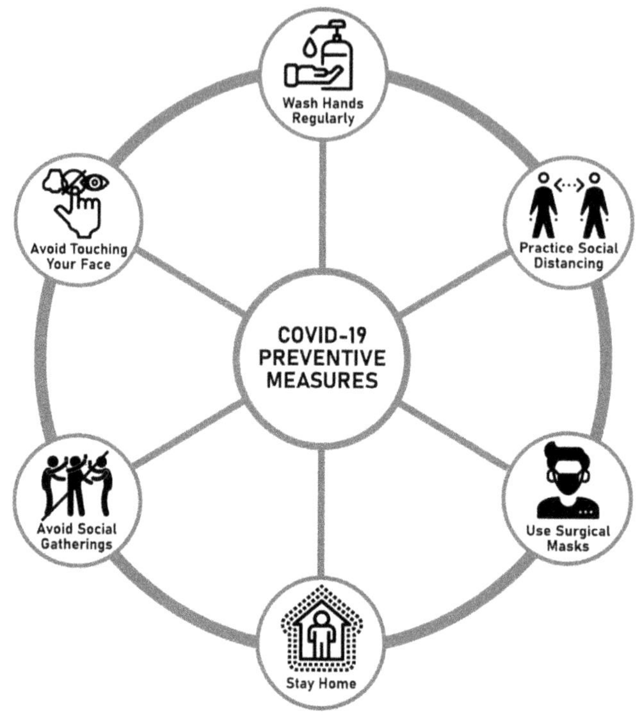

Figure 6.1. COVID-19 preventive measures (reproduced from [3], with permission).

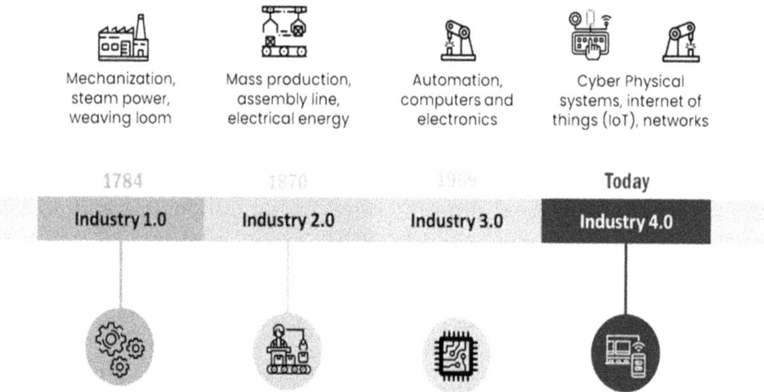

Figure 6.2. Fourth industrial revolution (Industry 4.0) (reproduced from [7], with permission).

century. Telecommunications, computers, and electronics were the key elements of the third revolution. Industry 4.0, the fourth industrial revolution, is a combination of the Internet of Things (IoT), artificial intelligence (AI), quantum computing, and robotics [4–6] (figure 6.2). The application of Industry 4.0's emerging knowledge has been accelerated after COVID-19.

IoT's efficiency and essential role have been proven in improving and developing healthcare facilities [8]. Also, AI has been used for various healthcare procedures throughout the COVID-19 pandemic [9]. In Industry 4.0, the entire production creates real-time information with the help of AI and IoT. Any medical system could benefit from this opportunity [10, 11]. Industry 4.0 can provide better solutions for pandemics, including COVID-19 [12–14]. Such benefits [15] can be shown in figure 6.3. In this chapter, we mainly concentrate on applying Industry 4.0 for automatic surveillance and health monitoring during pandemics and post-pandemic life.

6.2 Industry 4.0

Industry 4.0 has six design principles [16], including service orientation, real-time capability, modularity, decentralization, and virtualization. Many definitions have been proposed in the literature [17, 18]. In Industry 4.0, self-aware machines are used in highly automated manufacturing systems in which real-time data analytics and AI are used for decision-making [19]. The technology groups of Industry 4.0 are shown in figure 6.4.

In the following sections, the primary technology groups of Industry 4.0 are discussed, and their application for pandemic management is described.

6.2.1 IoT

Kevin Ashton used the term IoT for supply chain management for the first time in 1999 [21]. It has been mentioned with the term 'smartness' as the ability of a system to get information [22] independently. Thus, it refers to autonomous, intelligent, and addressable devices and sensors [22]. IoT can transform the world into a smart world, requiring less effort and time to access devices (figure 6.5).

Real-time health systems and remote monitoring are the primary applications of IoT for COVID-19. Using drones, an IoT system combined with AI could improve public security based on image processing and surveillance, contact tracing, and disinfection [23]. An example of using IoT for monitoring and surveillance in COVID-19 is COVID-SAFE. It monitors coughing patterns, blood oxygen saturation, heart rate, and body temperature. It also has a smartphone app and Bluetooth 4.0 technology for distance trackers and uses machine learning (ML) to send the required evidence to the users. The cloud-based ML could estimate the COVID-19 infection risk based on a Neuro-Fuzzy system with 15 rules [23]. The proposed system was trained and internally validated using the Khorshid COVID Cohort (KCC) dataset [24]. The risk of COVID-19 infection for two users, along with the input features, is shown in figure 6.6.

As a complete IoT system, it outperformed different technologies, including [25–30], in terms of monitoring COVID-19 symptoms, diagnosis, real-time updates, remote monitoring, and notifications. It also shows a map containing different zones (red: high risk of COVID-19 infection, yellow: moderate, and green: low). Using such real-time information, users' social activities could be managed with precautions. A sample of cone segmentation is provided in figure 6.7.

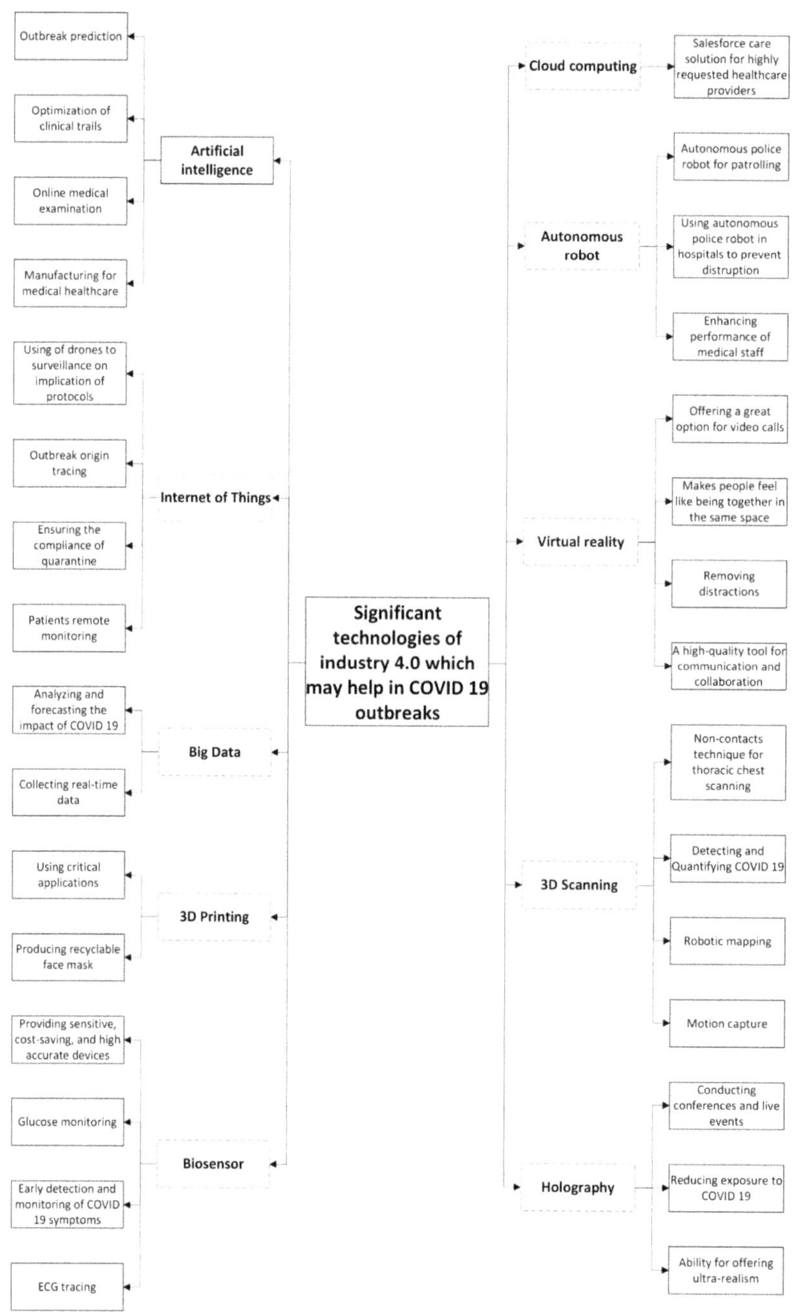

Figure 6.3. The application of Industry 4.0 for pandemic management (modified from [15], with permission).

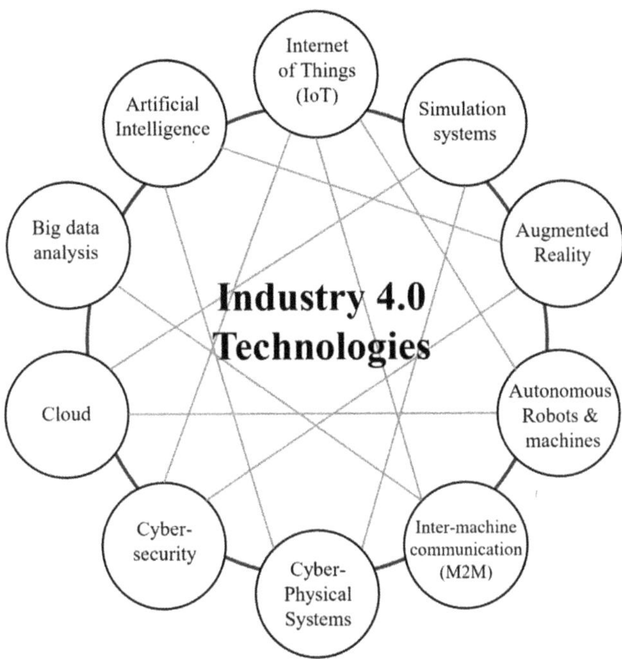

Figure 6.4. Ten technology groups of Industry 4.0 (reproduced from [20], with permission).

The applications of IoT in healthcare have seen an outbreak during recent years, especially with pandemics. As a result, patients, physicians, physicians, and even government entities have massively used several devices to measure body constants (e.g., smart wearables and thermometers) and monitor patients (e.g., telemedicine carts). IoT-based solutions like the Internet of Medical Things (IoMT) and the Cognitive Internet of Things (CIoT) collect real-time data from users at remote locations that can be later processed and interpreted for prediction and decision-making.

For example, Kinsa Health's network, initially intended to track the flu by smart thermometers deployed to households, was used by US authorities to classify possible COVID-19 hotspots during the pandemic [3]. Moreover, the practice of using IoMT, also known as telemedicine, has reported a high increase in demand, at least in countries like the US [31]. These practices, combined with the use of devices like smart wearables, drones, technologies like cloud storage and computing, 5G cellular networks, and AI, have helped reduce the virus spread and decrease the burden of hospital staff [32, 33].

Its consolidation can allow more dynamic healthcare ecosystems, enabling clinical care for mild COVID-19 cases and with difficulties accessing the healthcare system, like rural or underserved patients [31, 33]. Examples of applications of IoT for tackling COVID-19 pandemics are presented in figure 6.8.

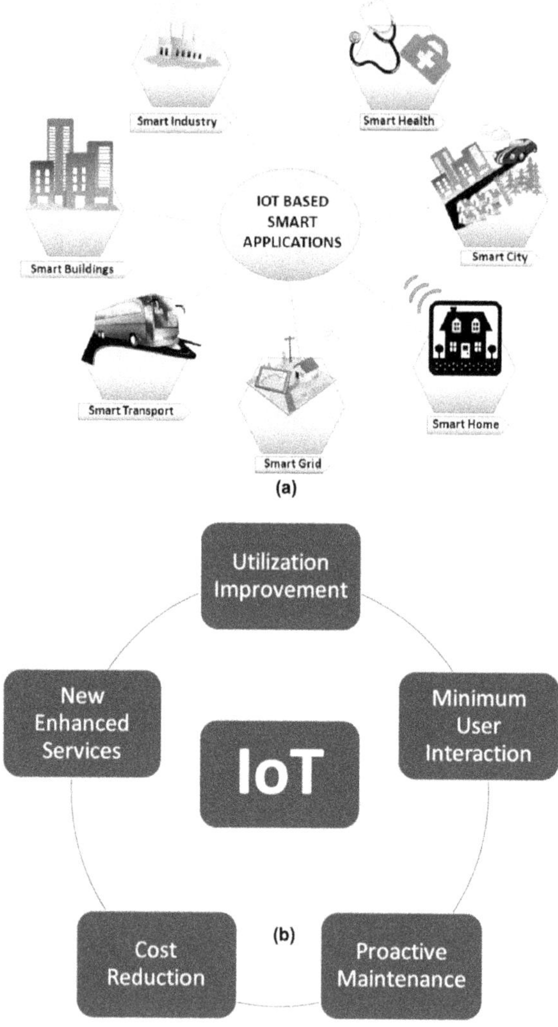

Figure 6.5. Block diagram including (a) smart applications based on IoT and (b) the cutting edge technology benefits of IoT (reproduced from [22] with permission).

IoT solutions have not only been used to monitor pandemics [32] but also to predict complications of patients with cardiovascular disease [33] and for the follow-up of chronic conditions like diabetes, cancer, and multiple sclerosis, among others (see [34] for a summary). Some studies advocated for the importance of tracking mental health, too, especially during lockdowns and homeschooling [35–37], where IoT and telehealth have a unique potential for exploitation given their capacity to rapidly detect people at risk of mental distress and dangerous behaviors [36, 38]. The incorporation of automatic processing, sensory information, and network communication is enabled by the Cognitive Internet of Medical Things (CIoMT) (figure 6.9).

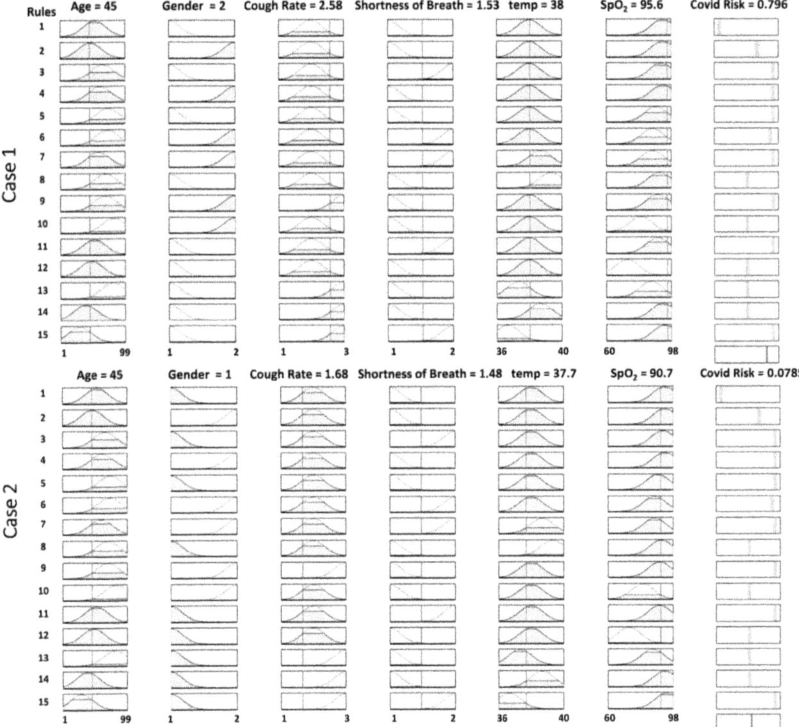

Figure 6.6. The designed Fuzzy rule-based system in COVID-SAFE. Two examples were provided, and their risk of COVID-19 infection was estimated. The system used six input variables and has 15 rules (reproduced from [23], with permission).

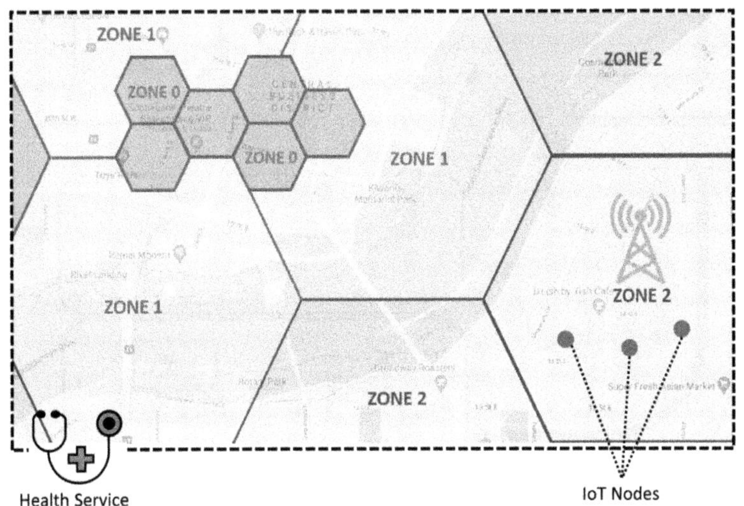

Figure 6.7. Real-time geolocation of hotspots from zone definition in COVID-SAFE. Zone two and zero have the lowest and highest risk of COVID-19 infection, respectively (reproduced from [23], with permission).

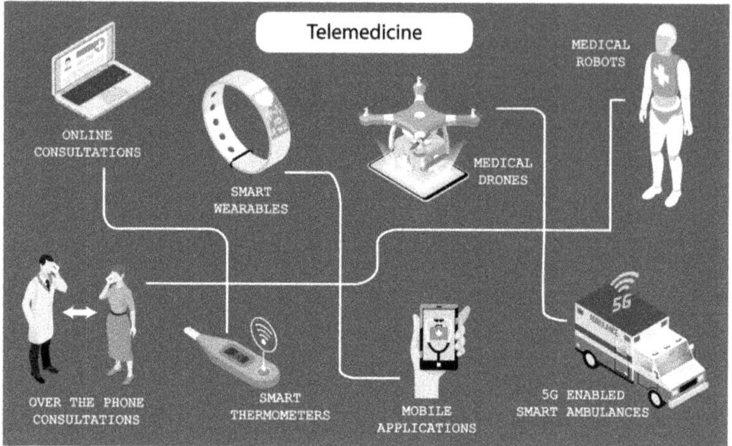

Figure 6.8. Examples of IoMT solutions for tackling COVID-19 pandemics (reproduced from [3], with permission).

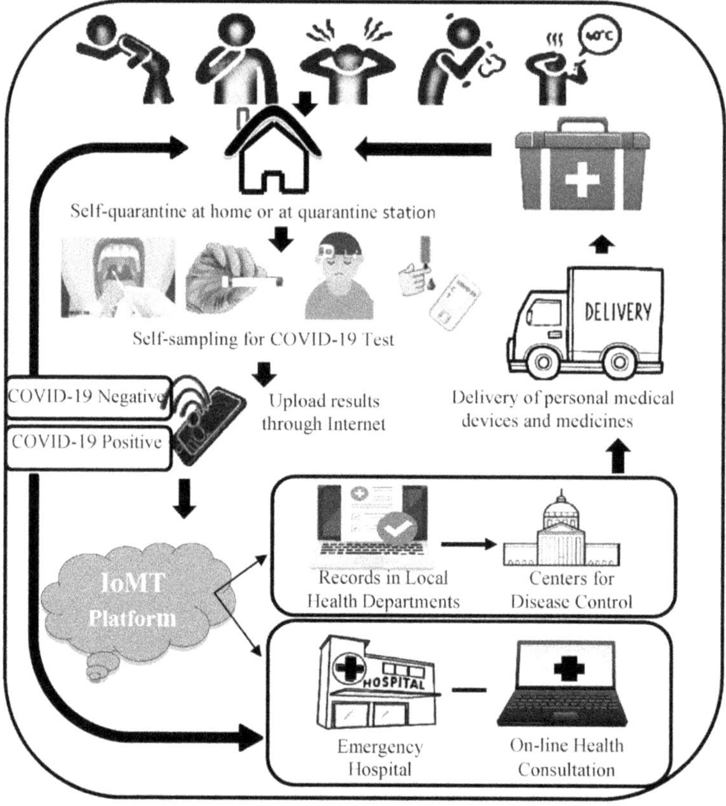

Figure 6.9. COVID-19 pandemic and Cognitive Internet of Medical Things (CIoMT) (reproduced from [32], with permission).

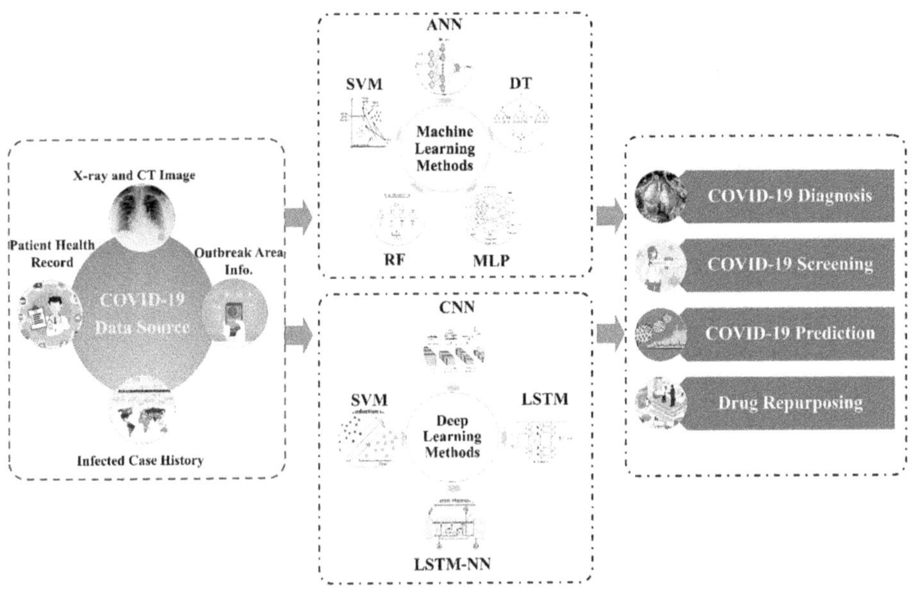

Figure 6.10. Fighting COVID-19 using machine learning (ML) (reproduced from [9], with permission).

6.2.2 AI, ML, big data, and cloud computing

AI is how machines mimic human intelligence, such as problem-solving. It enables systems to make decisions like a human [2]. ML is a subgroup of AI enabling the systems to learn automatically. Computer-aided diagnosis, prognosis, and clinical prediction models are parts of the ML methods. The ML's application in COVID-19 pandemics is shown in figure 6.10.

Big data, a large dataset usually automatically generated from different sources, is too complicated to analyze using traditional data analysis methods. Big data analysis transforms and visualizes such complicated data to analyze [39]. Cloud computing is Internet-based services, such as software, servers, and data storage [40]. It could support big data analysis. The key components of cloud-based computing are summarized in figure 6.11. Real-time cloud services are used by cloud clients. Medical staff use shared data using cloud apps. Infrastructure, software, and services are parts of a cloud platform [41, 42]. The patients' information, records, and reports are recorded in the cloud. They are shared with clinicians when necessary [43].

6.3 Comprehensive literature review

This chapter contains the literature review of the papers published in 2020, 2021, and 2022 focusing on COVID-19 health monitoring and surveillance. In table 6.1, we examine whether the published papers were related to Industry 4.0 technologies, including big data, IoT, Augmented Reality–Virtual Reality (AR–VR), digitalization, smart factories, ML, AI, and cloud computing.

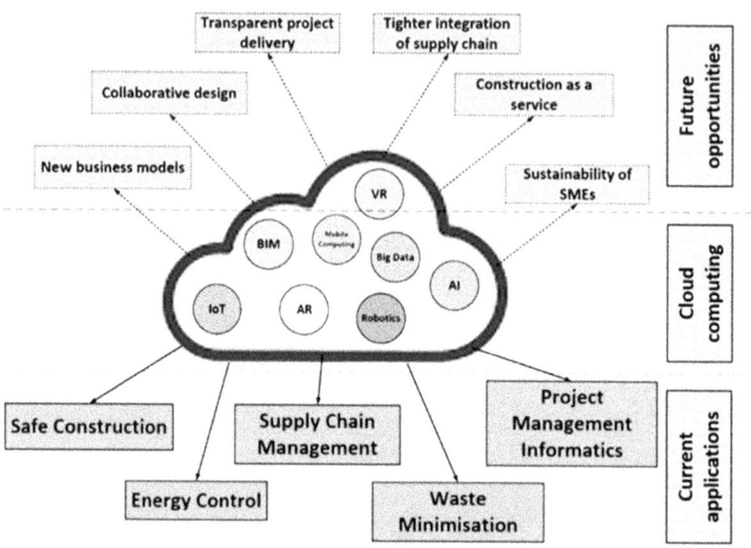

Figure 6.11. The critical components of cloud computing (reproduced from [44], with permission).

Table 6.1. COVID-19 health monitoring and surveillance methods concerning Industry 4.0 technologies.

Reference	Topic related to COVID-19	T1	T2	T3	T4	T5	T6	T7	T8
Chamola *et al* 2020 [3]	AI, IoT, drones, blockchain, and 5G applications	✓	✓	✓	✓	×	✓	✓	✓
Akhbarifar *et al* 2020 [33]	Remote health monitoring	✓	×	✓	×	×	×	✓	✓
Swayamsiddha and Mohanty 2020 [32]	Cognitive Internet of Medical Things	✓	×	×	×	×	×	×	×
Bai *et al* 2020 [28]	Internet of Things-aided diagnosis and treatment	✓	×	✓	✓	×	×	×	✓
Vedaei *et al* 2020 [23]	IoT-based health monitoring and surveillance	✓	×	×	×	×	×	✓	✓
Rahman *et al* 2020 [45]	IoT application	✓	×	✓	×	×	✓	✓	✓
Amin *et al* 2020 [46]	Edge intelligence and IoT	✓	×	✓	×	×	✓	✓	✓
Zhang *et al* 2020 [37]	Emotion-aware and intelligent IoMT	✓	×	×	×	×	×	×	✓
Singh *et al* 2020 [47]	IoMT for orthopedics in COVID-19	✓	✓	×	×	×	×	×	✓
Singh *et al* 2020 [48]	IoT applications	✓	×	✓	×	×	×	×	×
Nasajpour *et al* 2020 [49]	IoT applications	✓	×	×	×	×	×	×	✓
Kamal *et al* 2020 [50]	IoT applications	✓	✓	✓	×	×	✓	✓	✓
Kelly *et al* 2020 [51]	IoT and healthcare delivery	✓	✓	✓	×	×	✓	✓	✓
Yang *et al* 2021 [52]	IoT applications	✓	✓	×	×	×	×	×	✓
Pulicharla *et al* 2021 [53]	Wastewater monitoring	✓	✓	✓	×	✓	✓	×	×
Deshpande *et al* 2021 [54]	AI-based human audio processing	×	×	×	×	×	✓	✓	×
Lim *et al* 2021 [55]	Patient self-monitoring	×	✓	×	×	×	×	×	✓
Khalsa *et al* 2021[56]	AI and cardiac surgery	×	×	✓	×	×	✓	✓	×
Ni *et al* 2021 [57]	Automated monitoring of vital signs	×	✓	✓	×	×	×	✓	✓
Ganesh *et al* 2021 [58]	Automatic health monitoring	✓	×	×	×	×	✓	×	✓
Hao *et al* 2021 [59]	Control strategies	×	×	✓	×	×	×	×	×

		T1	T2	T3	T4	T5	T6	T7	T8
De Ridder *et al* 2021 [60]	Detection of spatio-temporal clusters	×	✓	×	×	×	×	×	×
Klonoff *et al* 2021 [61]	Digital connectivity	×	✓	×	×	×	×	×	×
Nogueira *et al* 2021 [62]	Surveillance on stroke care using AI	×	×	✓	×	×	✓	×	✓
Curtis *et al* 2021 [63]	Bluetooth low energy wearable tags	×	×	×	×	×	×	×	✓
Mukati *et al* 2021 [64]	IoT applications	✓	✓	×	×	×	×	✓	✓
Gallo *et al* 2021 [65]	Parenting training programs	×	×	×	×	×	×	✓	×
Pradhan *et al* 2021 [66]	IoT and robotics	✓	×	×	✓	×	✓	✓	✓
Muli *et al* 2021 [67]	Leveraging technology for health services	×	×	×	×	×	×	×	✓
Pasquel *et al* 2021 [68]	Management of diabetes and hyperglycemia in the hospital	×	×	×	×	×	×	×	✓
Dos Santos *et al* 2021 [69]	Newest diabetes-related technologies in pediatrics	×	✓	×	×	×	×	×	×
Sang M. Lee and DonHee Lee 2021 [70]	Contactless healthcare services	✓	✓	✓	✓	×	✓	✓	✓
Javaid *et al* 2021 [71]	Dentistry 4.0	✓	✓	✓	×	×	✓	×	✓
Jones *et al* 2021 [72]	Cloud-based computing and cancer surveillance	×	×	×	×	×	×	✓	×
Calatrava-Nicolás *et al* 2021 [73]	Robotic-based coaching system and well-being monitoring	×	×	×	×	×	✓	✓	✓
Sarker *et al* 2021 [74]	Robotics and AI in healthcare	✓	×	✓	×	×	✓	✓	✓
Efficace *et al* 2021 [75]	Digital health for hematologic malignancies	×	✓	×	×	×	×	×	×
Siriwardhana *et al* 2021 [76]	5G and healthcare	✓	✓	×	✓	✓	×	✓	✓
Wang *et al* 2021 [77]	Twitter data analysis study and health belief	×	×	×	×	×	×	✓	×
Deshpande *et al* 2022 [54]	AI-based human audio processing	×	×	×	×	×	✓	✓	✓
Deshpande *et al* 2022 [78]	Personal health detection from tweets	×	×	✓	×	×	×	✓	×
Agarwal *et al* 2022 [79]	Crowdsourced community support	×	✓	×	×	×	×	×	×
Sundaram *et al* 2022 [80]	Chest x-ray image diagnosis	×	×	×	×	×	×	✓	×
Kollu *et al* 2022 [81]	The application of AI and IoT	✓	✓	✓	×	×	✓	✓	✓
Eynstone-Hinkins and Moran 2022 [82]	Mortality data enhancement	×	✓	×	×	×	×	✓	×
Wen *et al* 2022 [83]	News reports analysis to predict public health interventions	×	×	×	×	×	×	✓	×
Pradeep Kumar Roy and Abhinav Kumar 2022 [84]	IoMT and early prediction	✓	✓	×	×	×	×	×	×
Lavric *et al* 2022 [85]	IoT and multi-sensors	✓	×	×	×	×	✓	✓	✓
Hasan *et al* 2022 [86]	Social Media analysis using natural language processing (NLP) pipelines	×	×	×	×	×	×	✓	×
Mann *et al* 2022 [87]	Reimagining connected care and digital medicine	✓	×	×	×	×	✓	✓	×
Van Vooren *et al* 2022 [88]	Reliable and scalable qPCR testing	×	✓	×	×	×	×	×	✓
Chen *et al* 2022 [89]	Big data technology during the pandemic	×	✓	✓	×	×	×	×	×

T1: IoT; T2: Digitalization; T3: Big data; T4: AR-VR; T5: Smart factory; T6: AI; T7: ML; T8: Cloud. VR: Virtual Reality; AR: Augmented Reality; ML: Machine Learning; AI: Artificial Intelligence.

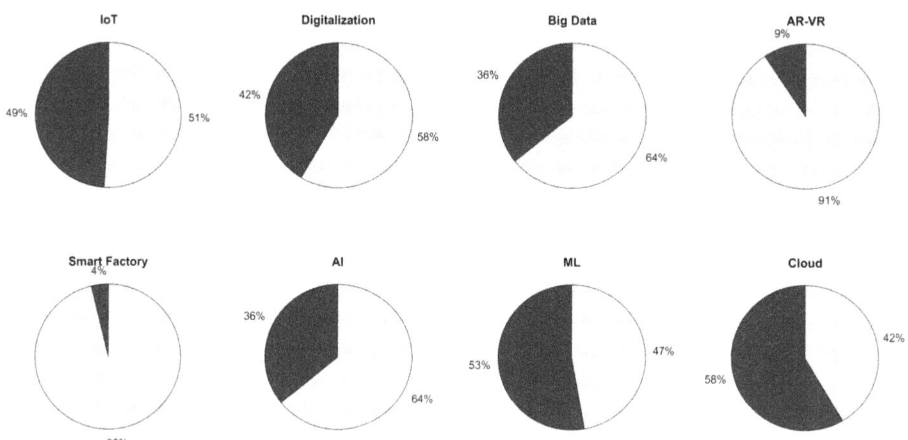

Figure 6.12. Industry 4.0 technologies in fighting COVID-19 ('yes' category in blue, and 'no' in yellow).

The Industry 4.0 technology application is summarized in figure 6.12. Fifty-two papers reviewed in table 6.1 were used to calculate the percentage of Industry 4.0 technologies in fighting COVID-19.

6.4 Automated monitoring and surveillance techniques

In patients' critical situations, an automated health monitoring system reacts or creates an alarm. The data is usually analyzed using microcontrollers. The condition is sent to medical experts and recorded for earlier diagnosis information [90]. The hardware for a typical monitoring system in healthcare is shown in figure 6.13.

A typical block diagram of an automated smart health system is provided in figure 6.14. It is different from the standard healthcare system. It is mainly based on the embedded sensor system, in which electronic data signals are recorded. The detectors, sensors, and monitors are connected synchronously in such systems. The analog signals, such as pulse rate, blood pressure, temperature, and oxygen level, are transformed into digital signals using the analog-to-digital converter (ADC). The microcontroller unit (MCU) then uses the recorded data for risk stratification or diagnosis [23]. Fast calculations could be performed on the mobile phone, while algorithms with high computational complexity are performed on the cloud server. Some of the results are shown on the display unit, and detailed information is provided to medical experts. In an automated smart health system, emergencies such as ST-segment elevation myocardial infarction (STEMI) [91] are also considered [92].

Non-invasive blood pressure (BP) measurement and monitoring are essential for predicting cardiovascular diseases. Cuffless BP measurement supports vital BP monitoring at night, which is not the case for cuff-based devices. Moreover, the prognostic role of such systems is more critical than daytime measurement [94]. Thus, cuffless BP monitoring is part of automated wearable monitoring systems [95]. Recently, bio-impedance (Bio-Z) sensors were used in the literature to estimate BP

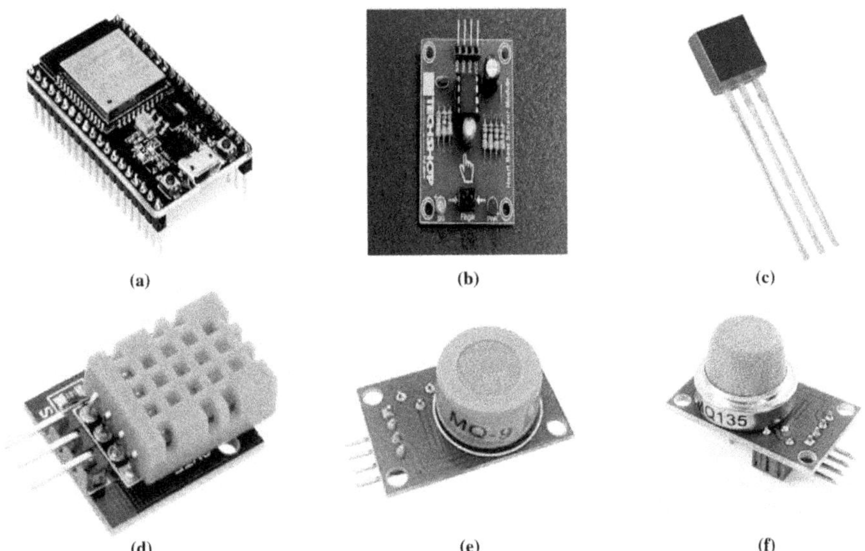

Figure 6.13. A typical monitoring system in healthcare includes (a) ESP32 (a very low-cost Linux system), (b) a heartbeat sensor using plethysmography, (c) body temperature sensor (e.g., LM35 series), (d) room temperature (e.g., DHT11 as a room humidity and temperature sensor), (e) CO (e.g., MQ-9), and (f) CO2 sensor (e.g., MQ-135) (reproduced from [90], with permission).

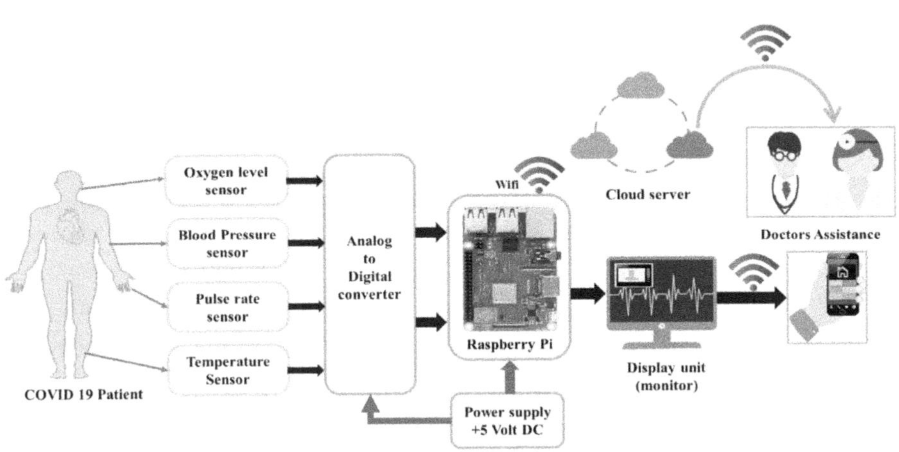

Figure 6.14. Block diagram of a typical IoT smart health system (reproduced from [93], with permission).

based on the activity of the arteries in the wrist. The combination of six Bio-Z sensor arrays, the convolutional neural network autoencoder, and the AdaBoost regression model was used by Ibrahim and Jafar [95] to estimate diastolic and systolic BP (figure 6.15).

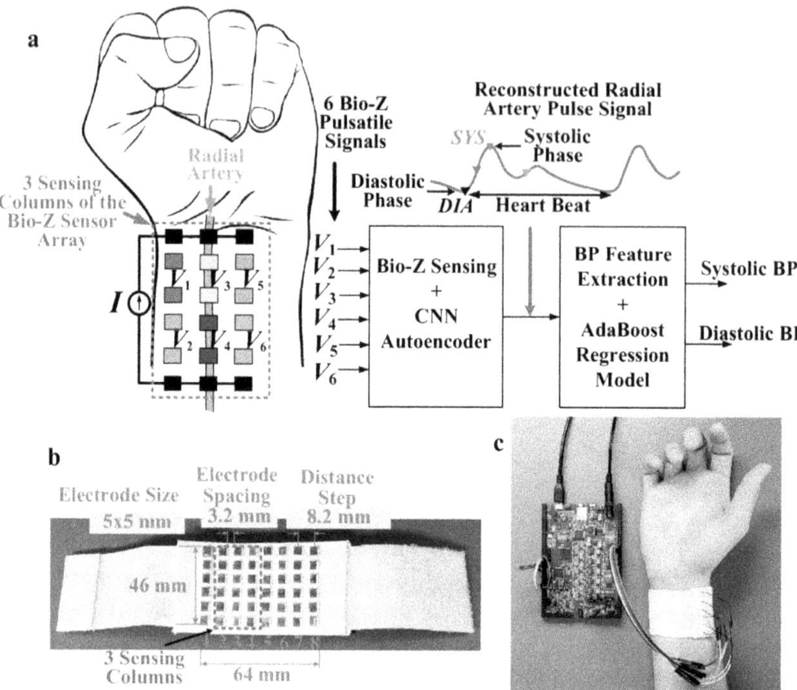

Figure 6.15. (a) A cuffless blood pressure (BP) monitoring device using the (Bio-Z) sensor array. An AdaBoost regression model and convolutional neural network autoencoder estimate the diastolic and systolic BP. (b) The implemented sensor array contains six by eight silver electrodes and (c) the combination of the Bio-Z sensing hardware and the wristband (reproduced from [95], with permission).

6.5 Conclusion and future scope

Undoubtedly, Industry 4.0 has various capacities for pandemic management. Health monitoring and surveillance are two critical issues of any pandemic. Most post-pandemic issues could be identified with a proper health monitoring system. More than 50 long-term properties of COVID-19 were described in the literature [96].

On a large scale, this task is only possible if Industry 4.0 technological advancements are used. However, a primary challenge of Industry 4.0 is security [97]. A data-driven system relying on information sharing could be highly susceptible to cyber-attacks [97]. Such risk themes must be taken into account.

Smart digital information continues from Industry 4.0 to Industry 5.0. COVID-19 pandemic challenges could be addressed by Industry 5.0, which focuses on personalized therapy [98]. Such smart systems signify the application of AI, ML, IoT, and cloud-based computing. It is also aligned with the digital health mega-trends and sustainable development goals (SDGs), especially SDG.1, SDG.9, and SDG.11. Much research is required to consider the validity measures of automatically personalized therapies to clinicians.

Acknowledgments

This work was supported by the Beatriu de Pinós post-doctoral program from the Office of the Secretary of Universities and Research from the Ministry of Business and Knowledge of the Government of Catalonia (Grant number: #2020 BP 00261) and the TecnioSpring Industry fellowship (ACCIO, H2020-EU-EXCELLENT SCIENCE—Marie Skłodowska-Curie Actions, #ACE026/21/000035) and the Ministry of Science and Innovation (MICINN), Spain, under contract PID2020-117751RB-I00.

References

[1] Dong E, Du H and Gardner L 2020 An interactive web-based dashboard to track COVID-19 in real time *Lancet Infect. Dis.*- **20** 533–4

[2] Moosavi J, Bakhshi J and Martek I 2021 The application of Industry 4.0 technologies in pandemic management: literature review and case study *Healthc. Anal.* **1** 100008

[3] Chamola V, Hassija V, Gupta V and Guizani M 2020 A comprehensive review of the COVID-19 pandemic and the role of IoT, drones, AI, blockchain, and 5G in managing its impact *IEEE Access* **8** 90225–65

[4] Kumar S, Suhaib M, Asjad M and Salah B 2022 Industry 4.0 and its suitability in post COVID-19 *J. Ind. Integr. Manag* 2250012

[5] Sigov A, Ratkin L, Ivanov L A and Xu L D 2022 Emerging enabling technologies for Industry 4.0 and beyond *Inf. Syst. Front.*

[6] Duan L and Da X L 2021 Data analytics in Industry 4.0: a survey *Inf. Syst. Front.* 1–17

[7] Silva F, Resende D, Amorim M and Borges M 2020 A field study on the impacts of implementing concepts and elements of Industry 4.0 in the biopharmaceutical sector *J. Open Innov* **6** 175

[8] Fagroud F Z, Toumi H, Ben Lahmar E H, Talhaoui M A, Achtaich K and Filali S E 2021 Impact of IoT devices in E-Health: a review on IoT in the context of COVID-19 and its variants *Procedia Comput. Sci.* **191** 343–8

[9] Khan M, Mehran M T, Haq Z U, Ullah Z, Naqvi S R, Ihsan M and Abbass H 2021 Applications of artificial intelligence in COVID-19 pandemic: a comprehensive review *Expert Syst. Appl.* **185** 115695

[10] Zeng J, Huang J and Pan L 2020 How to balance acute myocardial infarction and COVID-19: the protocols from Sichuan Provincial People's Hospital *Intensive Care Med* **46** 1111–3

[11] Manogaran G, Thota C, Lopez D and Sundarasekar R 2017 Big data security intelligence for healthcare Industry 4.0 *Springer Series in Advanced Manufacturing* (Cham: Springer International Publishing), pp 103–26

[12] Ahmed S F, Quadeer A A and McKay M R 2020 Preliminary identification of potential vaccine targets for the COVID-19 Coronavirus (SARS-CoV-2) based on SARS-CoV immunological studies *Viruses* **12** 254

[13] Cheng G-J, Liu L-T, Qiang X-J and Liu Y 2016 Industry 4.0 development and application of intelligent manufacturing *2016 Int. Conf. on Information System and Artificial Intelligence (ISAI)* (Piscataway, NJ: IEEE), pp 407–10

[14] Grasselli G, Pesenti A and Cecconi M 2020 Critical care utilization for the COVID-19 outbreak in Lombardy, Italy: early experience and forecast during an emergency response *JAMA* **323** 1545–6

[15] Javaid M, Haleem A, Vaishya R, Bahl S, Suman R and Vaish A 2020 Industry 4.0 technologies and their applications in fighting COVID-19 pandemic *Diabetes Metab. Syndr* **14** 419–22

[16] Kumar S, Suhaib M and Asjad M 2021 Narrowing the barriers to Industry 4.0 practices through PCA-Fuzzy AHP-K means *J. Adv. Manag. Res* **18** 200–26

[17] Moeuf A, Pellerin R, Lamouri S, Tamayo-Giraldo S and Barbaray R 2018 The industrial management of SMEs in the era of Industry 4.0 *Int. J. Prod. Res.* **56** 1118–36

[18] Buer S-V, Strandhagen J O and Chan F T S 2018 The link between Industry 4.0 and lean manufacturing: mapping current research and establishing a research agenda *Int. J. Prod. Res.* **56** 2924–40

[19] Kumar S, Asjad M and Suhaib M 2022 A labelling system and automation comparison index for Industry 4.0 system *Ind. Rob.* **49** 415–27

[20] Rosin F, Forget P, Lamouri S and Pellerin R 2022 Enhancing the decision-making process through Industry 4.0 technologies *Sustainability* **14** 461

[21] Oriwoh E, Sant P and Epiphaniou G 2013 Guidelines for internet of things deployment approaches—the thing commandments *Procedia Comput. Sci.* **21** 122–31

[22] Shafique K, Khawaja B A, Sabir F, Qazi S and Mustaqim M 2020 Internet of things (IoT) for next-generation smart systems: a review of current challenges, future trends and prospects for emerging 5G-IoT scenarios *IEEE Access* **8** 23022–40

[23] Vedaei S S, Fotovvat A, Mohebbian M R, Rahman G M E, Wahid K A, Babyn P, Marateb H R, Mansourian M and Sami R 2020 COVID-SAFE: an IoT-based system for automated health monitoring and surveillance in post-pandemic life *IEEE Access* **8** 188538–51

[24] Sami R *et al* 2020 A one-year hospital-based prospective COVID-19 open-cohort in the Eastern Mediterranean region: the Khorshid COVID Cohort (KCC) study *PLoS One* **15** e0241537

[25] Ting D S W, Carin L, Dzau V and Wong T Y 2020 Digital technology and COVID-19 *Nat. Med.* **26** 459–61

[26] Imran A, Posokhova I, Qureshi H N, Masood U, Riaz M S, Ali K, John C N, Hussain M I and Nabeel M 2020 AI4COVID-19: AI enabled preliminary diagnosis for COVID-19 from cough samples via an app *Inform. Med. Unlocked* **20** 100378

[27] Mei X *et al* 2020 Artificial intelligence-enabled rapid diagnosis of patients with COVID-19 *Nat. Med.* **26** 1224–8

[28] Bai L *et al* 2020 Chinese experts' consensus on the Internet of Things-aided diagnosis and treatment of coronavirus disease 2019 (COVID-19) *Clin. eHealth* **3** 7–15

[29] Allam Z and Jones D S 2020 On the Coronavirus (COVID-19) outbreak and the smart city network: universal data sharing standards coupled with artificial intelligence (AI) to benefit urban health monitoring and management *Healthcare (Basel)* **8** 46

[30] Qiu G, Gai Z, Tao Y, Schmitt J, Kullak-Ublick G A and Wang J 2020 Dual-functional plasmonic photothermal biosensors for highly accurate severe acute respiratory syndrome Coronavirus 2 detection *ACS Nano* **14** 5268–77

[31] Gajarawala S N and Pelkowski J N 2021 Telehealth benefits and barriers *J. Nurse Pract.* **17** 218–21

[32] Swayamsiddha S and Mohanty C 2020 Application of cognitive Internet of Medical Things for COVID-19 pandemic *Diabetes Metab. Syndr* **14** 911–5

[33] Akhbarifar S, Javadi H H S, Rahmani A M and Hosseinzadeh M 2020 A secure remote health monitoring model for early disease diagnosis in cloud-based IoT environment *Pers. Ubiquitous Comput.* 1–17

[34] Shamsabadi A *et al* 2022 Internet of Things in the management of chronic diseases during the COVID-19 pandemic: a systematic review *Health Sci. Rep.* **5** e557

[35] Awais M, Raza M, Singh N, Bashir K, Manzoor U, Islam S U and Rodrigues J J P C 2021 LSTM-based emotion detection using physiological signals: IoT framework for healthcare and distance learning in COVID-19 *IEEE Internet Things J.* **8** 16863–71

[36] Meng W, Cai Y, Yang L T and Chiu W-Y 2021 Hybrid emotion-aware monitoring system based on brainwaves for internet of medical things *IEEE Internet Things* J **8** 16014–22

[37] Zhang T, Liu M, Yuan T and Al-Nabhan N 2021 Emotion-aware and intelligent Internet of Medical Things toward emotion recognition during COVID-19 pandemic *IEEE Internet Things* J **8** 16002–13

[38] Torous J, Jän Myrick K, Rauseo-Ricupero N and Firth J 2020 Digital mental health and COVID-19: using technology today to accelerate the curve on access and quality tomorrow *JMIR Ment. Health* **7** e18848

[39] Pépin J L, Bruno R M, Yang R-Y, Vercamer V, Jouhaud P, Escourrou P and Boutouyrie P 2020 Wearable activity trackers for monitoring adherence to home confinement during the COVID-19 pandemic worldwide: data aggregation and analysis *J. Med. Internet Res.* **22** e19787

[40] Velte A T, Velte T J and Elsenpeter R 2010 *Cloud Computing: A Practical Approach* (New York: McGraw-Hill)

[41] Arora R, Arora P K, Kumar H and Pant M 2020 Additive manufacturing enabled supply chain in combating COVID-19 *J. Ind. Integr. Manag.* **5** 495–505

[42] Bahl S, Javaid M, Bagha A, Singh R, Haleem A, Vaishya R and Suman R 2020 Biosensors applications in fighting COVID-19 pandemic *Apollo Med.* **17** 221–3

[43] Alhomdy S, Thabit F, Abdulrazzak F H, Haldorai A and Jagtap S 2021 The role of cloud computing technology: a savior to fight the lockdown in COVID 19 crisis, the benefits, characteristics and applications *Int. J. Intell. Netw.* **2** 166–74

[44] Bello S A, Oyedele L O, Akinade O O, Bilal M, Davila Delgado J M, Akanbi L A, Ajayi A O and Owolabi H A 2021 Cloud computing in construction industry: use cases, benefits and challenges *Autom. Constr.* **122** 103441

[45] Rahman M S, Peeri N C, Shrestha N, Zaki R, Haque U and Hamid S H A 2020 Defending against the Novel Coronavirus (COVID-19) outbreak: how can the Internet of Things (IoT) help to save the world? *Health Policy Technol* **9** 136–8

[46] Amin S U and Hossain M S 2021 Edge intelligence and internet of things in healthcare: a survey *IEEE Access* **9** 45–59

[47] Pratap Singh R, Javaid M, Haleem A, Vaishya R and Ali S 2020 Internet of Medical Things (IoMT) for orthopaedic in COVID-19 pandemic: roles, challenges, and applications *J. Clin. Orthop. Trauma* **11** 713–7

[48] Singh R P, Javaid M, Haleem A and Suman R 2020 Internet of things (IoT) applications to fight against COVID-19 pandemic *Diabetes Metab. Syndr* **14** 521–4

[49] Nasajpour M, Pouriyeh S, Parizi R M, Dorodchi M, Valero M and Arabnia H R 2020 Internet of Things for current COVID-19 and future pandemics: an exploratory study *J. Healthc. Inform. Res* **4** 325–64

[50] Kamal M, Aljohani A and Alanazi E 2020 IoT meets COVID-19: status, challenges, and opportunities *arXiv [cs.DC]*

[51] Kelly J T, Campbell K L, Gong E and Scuffham P 2020 The Internet of Things: impact and implications for health care delivery *J. Med. Internet Res.* **22** e20135

[52] Yang S-J, Xiao N, Li J-Z, Feng Y, Ma J-Y, Quzhen G-S, Yu Q, Zhang T, Yi S-C and Zhou X-N 2021 A remote management system for control and surveillance of echinococcosis: design and implementation based on internet of things *Infect. Dis. Poverty* **10** 50

[53] Pulicharla R, Kaur G and Brar S K 2021 A year into the COVID-19 pandemic: rethinking of wastewater monitoring as a preemptive approach *J. Environ. Chem. Eng.* **9** 106063

[54] Deshpande G, Batliner A and Schuller B W 2022 AI-based human audio processing for COVID-19: a comprehensive overview *Pattern Recognit.* **122** 108289

[55] Lim H M, Teo C H, Ng C J, Chiew T K, Ng W L, Abdullah A, Abdul Hadi H, Liew C S and Chan C S 2021 An automated patient self-monitoring system to reduce health care system burden during the COVID-19 pandemic in Malaysia: development and implementation study *JMIR Med. Inform.* **9** e23427

[56] Khalsa R K, Khashkhusha A, Zaidi S, Harky A and Bashir M 2021 Artificial intelligence and cardiac surgery during COVID-19 era *J. Card. Surg* **36** 1729–33

[57] Ni X *et al* 2021 Automated, multiparametric monitoring of respiratory biomarkers and vital signs in clinical and home settings for COVID-19 patients *Proc. Natl Acad. Sci. USA* **118** e2026610118

[58] Ganesh D, Seshadri G, Sokkanarayanan S, Bose P, Rajan S and Sathiyanarayanan M 2021 Automatic health machine for COVID-19 and other emergencies *2021 Int. Conf. on COMmunication Systems & NETworkS (COMSNETS)* (Piscataway, NJ: IEEE) pp 685–9

[59] Hao R, Zhang Y, Cao Z, Li J, Xu Q, Ye L, Guo X, Zheng T and Song H 2021 Control strategies and their effects on the COVID-19 pandemic in 2020 in representative countries *J. Biosaf. Biosecur.* **3** 76–81

[60] De Ridder D *et al* 2021 Detection of spatiotemporal clusters of COVID-19-associated symptoms and prevention using a participatory surveillance app: protocol for the @choum study *JMIR Res. Protoc* **10** e30444

[61] Klonoff D C, Shang T, Zhang J Y, Cengiz E, Mehta C and Kerr D 2021 Digital connectivity: the sixth vital sign *J. Diabetes Sci. Technol.* **16** 1303–8

[62] Nogueira R G *et al* 2021 Epidemiological surveillance of the impact of the COVID-19 pandemic on stroke care using artificial intelligence *Stroke* **52** 1682–90

[63] Curtis S J, Rathnayaka A, Wu F, Al Mamun A, Spiers C, Bingham G, Lau C L, Peleg A Y, Yuce M R and Stewardson A J 2022 Feasibility of Bluetooth Low Energy wearable tags to quantify healthcare worker proximity networks and patient close contact: a pilot study *Infect. Dis. Health* **27** 66–70

[64] Mukati N, Namdev N, Dilip R, Hemalatha N, Dhiman V and Sahu B 2021 Healthcare assistance to COVID-19 patient using internet of things (IoT) enabled technologies *Mater. Today*

[65] Gallo C G, Berkel C, Mauricio A, Sandler I, Wolchik S, Villamar J A, Mehrotra S and Brown C H 2021 Implementation methodology from a social systems informatics and engineering perspective applied to a parenting training program *Fam. Syst. Health* **39** 7–18

[66] Pradhan B, Bharti D, Chakravarty S, Ray S S, Voinova V V, Bonartsev A P and Pal K 2021 Internet of Things and robotics in transforming current-day healthcare services *J. Healthc. Eng* **2021** 9999504

[67] Muli E *et al* 2021 Leveraging technology for health services continuity in times of COVID-19 pandemic: patient follow-up, and mitigation of worse patient outcomes *J. Glob. Health* **11** 05024

[68] Pasquel F J, Lansang M C, Dhatariya K and Umpierrez G E 2021 Management of diabetes and hyperglycaemia in the hospital *Lancet Diabetes Endocrinol* **9** 174–88

[69] Dos Santos T J, Rodrigues T C, Puñales M, Arrais R F and Kopacek C 2021 Newest diabetes-related technologies for pediatric type 1 diabetes and its impact on routine care: a narrative synthesis of the literature *Curr. Pediatr. Rep.* **9** 142–53

[70] Lee S M and Lee D 2021 Opportunities and challenges for contactless healthcare services in the post-COVID-19 era *Technol. Forecast. Soc. Change* **167** 120712

[71] Javaid M, Haleem A, Pratap Singh R and Suman R 2021 Pedagogy and innovative care tenets in COVID-19 pandemic: an enhancive way through Dentistry 4.0 *Sens. Int.* **2** 100118

[72] Jones D E, Alimi T O, Pordell P, Tangka F K, Blumenthal W, Jones S F, Rogers J D, Benard V B and Richardson L C 2021 Pursuing data modernization in cancer surveillance by developing a cloud-based computing platform: real-time cancer case collection *JCO Clin. Cancer Inform* **5** 24–9

[73] Calatrava-Nicolás F M, Gutiérrez-Maestro E, Bautista-Salinas D, Ortiz F J, González J R, Vera-Repullo J A, Jiménez-Buendía M, Méndez I, Ruiz-Esteban C and Mozos O M 2021 Robotic-based well-being monitoring and coaching system for the elderly in their daily activities *Sensors (Basel)* **21** 6865

[74] Sarker S, Jamal L, Ahmed S F and Irtisam N 2021 Robotics and artificial intelligence in healthcare during COVID-19 pandemic: a systematic review *Rob. Auton. Syst* **146** 103902

[75] Efficace F, Breccia M, Fazi P, Cottone F, Holzner B and Vignetti M 2021 The GIMEMA-ALLIANCE digital health platform for patients with hematologic malignancies in the COVID-19 pandemic and postpandemic era: protocol for a multicenter, prospective, observational study *JMIR Res. Protoc* **10** e25271

[76] Siriwardhana Y, Gür G, Ylianttila M and Liyanage M 2021 The role of 5G for digital healthcare against COVID-19 pandemic: opportunities and challenges *ICT Express* **7** 244–52

[77] Wang H, Li Y, Hutch M, Naidech A and Luo Y 2021 Using tweets to understand how COVID-19-related health beliefs are affected in the age of social media: Twitter data analysis study *J. Med. Internet Res.* **23** e26302

[78] Luo L, Wang Y and Liu H 2022 COVID-19 personal health mention detection from tweets using dual convolutional neural network *Expert Syst. Appl.* **200** 117139

[79] Agarwal A K, Southwick L, Schneider R, Pelullo A, Ortiz R, Klinger E V, Gonzales R E, Rosin R and Merchant R M 2022 Crowdsourced community support resources among patients discharged from the emergency department during the COVID-19 pandemic: pilot feasibility study *JMIR Ment. Health* **9** e31909

[80] Sundaram S G, Aloyuni S A, Alharbi R A, Alqahtani T, Sikkandar M Y and Subbiah C 2022 Deep transfer learning based unified framework for COVID19 classification and infection detection from chest X-ray images *Arab. J. Sci. Eng.* **47** 1675–92

[81] Kollu P K, Kumar K, Kshirsagar P R, Islam S, Naveed Q N, Hussain M R and Sundramurthy V P 2022 Development of advanced artificial intelligence and IoT automation in the crisis of COVID-19 detection *J. Healthc. Eng* **2022** 1987917

[82] Eynstone-Hinkins J and Moran L 2022 Enhancing Australian mortality data to meet future health information demands *Int. J. Environ. Res. Public Health* **19** 603

[83] Wen Z, Powell G, Chafi I, Buckeridge D L and Li Y 2022 Inferring global-scale temporal latent topics from news reports to predict public health interventions for COVID-19 *Patterns (N Y)* **3** 100435

[84] Roy P K and Kumar A 2022 Early prediction of COVID-19 using ensemble of transfer learning *Comput. Electr. Eng.* **101** 108018

[85] Lavric A, Petrariu A I, Mutescu P-M, Coca E and Popa V 2022 Internet of things concept in the context of the COVID-19 pandemic: a multi-sensor application design *Sensors* **22** 503

[86] Hasan A, Levene M, Weston D, Fromson R, Koslover N and Levene T 2022 Monitoring COVID-19 on social media: development of an end-to-end natural language processing pipeline using a novel triage and diagnosis approach *J. Med. Internet Res.* **24** e30397

[87] Mann D M and Lawrence K 2022 Reimagining connected care in the era of digital medicine *JMIR mHealth uHealth* **10** e34483

[88] Van Vooren S *et al* 2022 Reliable and scalable SARS-CoV-2 qPCR testing at a high sample throughput: lessons learned from the Belgian initiative *Life (Basel)* **12** 159

[89] Mao Z, Zou Q, Yao H and Wu J 2021 The application framework of big data technology in the COVID-19 epidemic emergency management in local government—a case study of Hainan Province, China *BMC Public Health* **21** 2001

[90] Islam M M, Rahaman A and Islam M R 2020 Development of smart healthcare monitoring system in IoT environment *SN Comput. Sci.* **1** 185

[91] Jacobs A K *et al* 2021 Systems of care for ST-segment-elevation myocardial infarction: a policy statement from the American Heart Association *Circulation* **144** e310–27

[92] Krishnamoorthy P *et al* 2021 Mobile application to optimize care for ST-segment elevation myocardial infarction patients in a large healthcare system, STEMIcathAID: rationale and design *Eur. Heart J. Digit. Health* **2** 189–201

[93] Bhardwaj V, Joshi R and Gaur A M 2022 IoT-based smart health monitoring system for COVID-19 *SN Comput. Sci.* **3** 137

[94] Ogedegbe G and Pickering T 2010 Principles and techniques of blood pressure measurement *Cardiol. Clin.* **28** 571–86

[95] Ibrahim B and Jafari R 2022 Cuffless blood pressure monitoring from a wristband with calibration-free algorithms for sensing location based on bio-impedance sensor array and autoencoder *Sci. Rep.* **12** 319

[96] Lopez-Leon S, Wegman-Ostrosky T, Perelman C, Sepulveda R, Rebolledo P A, Cuapio A and Villapol S 2021 More than 50 long-term effects of COVID-19: a systematic review and meta-analysis *Sci. Rep.* **11** 16144

[97] Lezzi M, Lazoi M and Corallo A 2018 Cybersecurity for Industry 4.0 in the current literature: a reference framework *Comput. Ind.* **103** 97–110

[98] Javaid M, Haleem A, Singh R P, Haq M I U, Raina A and Suman R 2020 Industry 5.0: potential applications in COVID-19 *J. Ind. Integr. Manag* **5** 507–30

IOP Publishing

Advanced Signal Processing for Industry 4.0, Volume 1
Evolution, communication protocols, and applications in manufacturing systems
Irshad Ahmad Ansari and Varun Bajaj

Chapter 7

A novel computational intelligence approach to making efficient decisions under parametric uncertainty of practical models and its applications to Industry 4.0

N A Nechval, G Berzins and K N Nechval

A fundamental issue in statistics is the decision making of expected outcomes under parametric uncertainty based on past and present knowledge. This issue occurs in various situations and has a variety of solutions. In this research, a novel computational intelligence approach with applicability to Industry 4.0 is proposed to make effective judgments under parametric stochastic model uncertainty. Ancillary statistics and crucial variables, whose distributions do not depend on the unknown parameters, are utilized since it is expected that only the functional form of the underlying distributions is defined but that some or all of its parameters are unknown. The unique computational intelligence approach efficiently isolates and removes unidentified factors from the underlying models. The proposed approach is innovative in the theory of statistical decisions since it is independent of prior choice, in contrast to the Bayesian approach, which depends on prior choice. It enables the removal of unknown parameters from the issue and the discovery of effective statistical decision rules, which frequently carry lower risk than any other known decision rules. Examples from the real world are provided to demonstrate the suggested strategy.

7.1 Introduction

Statistical predictive inferences and optimization for anticipated random quantities under parametric uncertainty, as well as signal detection and classification, are widely used in many industries including Industry 4.0 [1–4]. In this paper, an innovative statistical technique of computational intelligence (based on pivotal quantities and ancillary statistics to eliminate unknown parameters from the underlying models) is proposed for constructing efficient statistical decisions under parametric uncertainty.

The primary goal of the study under discussion is to demonstrate the applicability of the suggested statistical approach in the specific situation of building precise statistical prediction or tolerance limits and maximizing expected random quantities under parametric uncertainty. There are typically two distinct categories of statistical prediction limits: (1) statistical 'γ-content tolerance limits' with expected '$(1-\alpha)$-confidence' in expected results (also known as 'γ-content $(1-\alpha)$-tolerance limits'), and (2) statistical 'prediction limits' with expected '$(1-\alpha)$-confidence' in expected results (or simply $(1-\alpha)$-prediction limits).

The statistical prediction limit with predicted $(1-\alpha)$-confidence appears to be less effective than the statistical γ-content tolerance limit, but it is more challenging to create. The following categories of prediction issues exist:

1. *New-sample prediction:* The data from a previous sample of size n is utilized in this situation to forecast one or more expected units in a subsequent sample of size m from the same process or population [4]
2. *Within-sample prediction:* The issue at hand is predicting anticipated occurrences in a sample or process using the sample's or process's early-failure data [4].

7.2 Exponential distribution

Applications of the exponential distribution can be found in a fairly broad range of statistical techniques. Since the math relating to the exponential distribution is frequently of a straightforward character, explicit formulas in terms of fundamental functions can be obtained. One of the most significant and popular distributions in statistical practice is this one. It demonstrates excellent mathematical tractability while having a number of significant statistical features. One of the most-often utilized models in various life-testing and reliability study scenarios is the exponential distribution.

Given a sample of size m drawn from an exponential distribution with a probability density function, let $\mathbf{Z} = (Z_1 \leqslant \cdots \leqslant Z_k)$ represent the first k ordered observations (order statistics) in the sample.

$$f_\lambda(x) = \lambda^{-1} \exp(-x/\lambda), \ \lambda > 0, \ x > 0, \tag{7.1}$$

and the cumulative probability distribution function

$$F_\lambda(x) = 1 - \exp(-x/\lambda), \tag{7.2}$$

where λ represents the parameter of scale. The parameter is thought to be unknowable. The number of survivors is fixed in Type II censoring, which is the main topic of discussion here, whereas Z_k is a random variable. The likelihood function in this instance is provided by

$$L(\lambda) = \prod_{i=1}^{k} f_\lambda(z_i)(\overline{F}_\lambda(z_k))^{m-k} = \frac{1}{\lambda^k} \exp\left(-\left[\sum_{i=1}^{k} z_i + (m-k)z_k\right]\bigg/\lambda\right) = \frac{1}{\lambda^k}\exp\left(-\frac{S_k}{\lambda}\right), \quad (7.3)$$

where

$$S_k = \sum_{j=1}^{k} Z_j + (m-k)Z_k \tag{7.4}$$

is the complete sufficient statistic for λ. The probability density function of S_k is given by

$$f_\lambda(s_k) = \frac{\dfrac{1}{\lambda^k}\exp\left(-\dfrac{s_k}{\lambda}\right)}{\dfrac{1}{s_k^{k-1}}\displaystyle\int_0^\infty \dfrac{s_k^{k-1}}{\lambda^{k-1}}\exp\left(-\dfrac{s_k}{\lambda}\right)d\left(\dfrac{s_k}{\lambda}\right)} = \frac{\dfrac{1}{\lambda^k}\exp\left(-\dfrac{s_k}{\lambda}\right)}{\dfrac{\Gamma(k)}{s_k^{k-1}}} = \frac{s_k^{k-1}}{\Gamma(k)\lambda^k}$$

$$\exp\left(-\frac{s_k}{\lambda}\right), \ s_k \geqslant 0, \tag{7.5}$$

where

$$V = S_k/\lambda \tag{7.6}$$

is the pivotal quantity, the probability density function of which is given by

$$f(v) = \frac{1}{\Gamma(k)} v^{k-1} \exp(-v), \ v \geqslant 0. \tag{7.7}$$

The probability density function of the lth order statistic Z_l in a sample of size m is given by

$$g_\lambda(z_l; m) = \frac{m!}{(l-1)!(m-l)!}[F_\lambda(z_l)]^{l-1}[1 - F_\lambda(z_l)]^{m-l} f_\lambda(z_l)$$

$$= \frac{1}{B(l, m-l+1)}\sum_{j=0}^{l-1}\binom{l-1}{j}(-1)^j \tag{7.8}$$

$$\exp\left(-\frac{z_l}{\lambda}(m-l+1+j)\right)\frac{1}{\lambda}, \ z_l \in (0, \infty).$$

The cumulative probability distribution function of the lth order statistic Z_l in a sample of size m is given by

$$G_\lambda(z_l; m) = \int_0^{z_l} g_\lambda(z)dz = \int_0^{z_l} \frac{1}{B(l, m-l+1)}\sum_{j=0}^{l-1}\binom{l-1}{j}(-1)^j \exp\left(-\frac{z}{\lambda}(m-l+1+j)\right)\frac{1}{\lambda}dz$$

$$= 1 - \frac{1}{B(l, m-l+1)}\sum_{j=0}^{l-1}\binom{l-1}{j}(-1)^j \frac{\exp\left(-\frac{z_l}{\lambda}(m-l+1+j)\right)}{m-l+j+1} \tag{7.9}$$

The cumulative probability distribution function of the ancillary statistic

$$W_l = Z_l/S_k \qquad (7.10)$$

is given by

$$G(w_l) = \int_0^\infty G_\lambda(z_l; m) f(v_k) dv_k = \int_0^\infty G_\lambda\left(\frac{z_l}{s_k}\frac{s_k}{\lambda}; m\right) f(v) dv = \int_0^\infty G_\lambda(w_l v; m) f(v) dv$$

$$= 1 - \frac{1}{B(l, m - l + 1)} \sum_{j=0}^{l-1} \binom{l-1}{j} (-1)^j \frac{[1 + (m - l + j + 1)w_l]^{-k}}{m - l + j + 1}, \qquad (7.11)$$

where S_k is given by (7.4).

$$\frac{dG(w_l)}{dw_l} = g(w_l) = \frac{k}{B(l, m - l + 1)} \sum_{j=0}^{l-1} \frac{\binom{l-1}{j}(-1)^j}{[1 + (m - l + j + 1)w_l]^{k+1}}. \qquad (7.12)$$

is the probability density function of the ancillary statistic W_l.

Let $Z_1 \leqslant \cdots \leqslant Z_k$ be the first k ordered observations (order statistics) in a sample of size m from the exponential distribution with the probability density function f_λ (y) and distribution function F_λ (y), where λ is a parameter. Then it can be shown that the conditional probability distribution function of Z_l $(1 \leqslant k < l \leqslant m)$ given $Z_k = z_k$ is determined as

$$G_\lambda(z_l|z_k; m) = P_\lambda(Z_l \leqslant z_l|Z_k = z_k; m) = \sum_{j=l-k}^{m-k} \binom{m-k}{j}\left[1 - \frac{\overline{F}_\theta(y_l)}{\overline{F}_\theta(y_k)}\right]^j \left[\frac{\overline{F}_\theta(y_l)}{\overline{F}_\theta(y_k)}\right]^{m-k-j}$$

$$= 1 - \frac{1}{B(l - k, m - l + 1)} \sum_{j=0}^{l-k-1} \binom{l-k-1}{j}\frac{(-1)^j}{m - l + j + 1}\left[\frac{\overline{F}_\lambda(z_l)}{\overline{F}_\lambda(z_k)}\right]^{m-l+j+1}$$

$$= 1 - \frac{1}{B(l - k, m - l + 1)} \sum_{j=0}^{l-k-1} \binom{l-k-1}{j}\frac{(-1)^j}{m - l + 1 + j} \exp \qquad (7.13)$$

$$\left(-\frac{z_l - z_k}{\lambda}(m - l + 1 + j)\right)$$

The conditional probability density function of Z_l $(1 \leqslant k < l \leqslant m)$ given $Z_k = z_k$ is determined as

$$g_\lambda(z_l|z_k; m) = \frac{dG_\lambda(z_l|z_k; m)}{dz_l}$$

$$= \frac{1}{B(l - k, m - l + 1)} \sum_{j=0}^{l-k-1} \binom{l-k-1}{j}(-1)^j \exp\left(-\frac{z_l - z_k}{\lambda}(m - l + 1 + j)\right)\frac{1}{\lambda}. \qquad (7.14)$$

It follows from (7.13) and (7.7) that the cumulative probability distribution function of the ancillary statistic

$$W_{lk} = \frac{Z_l - Z_k}{S_k} \qquad (7.15)$$

is given by

$$G(w_{lk}) = \int_0^\infty G_\lambda(z_l|z_k; m)f(v)dv$$

$$= 1 - \frac{1}{B(l - k, m - l + 1)} \sum_{j=0}^{l-k-1} \binom{l - k - 1}{j}(-1)^j \frac{[1 + (m - l + 1 + j)w_{lk}]^{-k}}{m - l + 1 + j}. \tag{7.16}$$

The probability density function of the ancillary statistic W_{lk} is given by

$$g(w_{lk}) = \frac{dG(w_{lk})}{dw_{lk}}$$

$$= \frac{k}{B(l - k, m - l + 1)} \sum_{j=0}^{l-k-1} \binom{l - k - 1}{j}(-1)^j[1 + (m - l + 1 + j)w_{lk}]^{-k-1}, \; w_{lk} \geq 0. \tag{7.17}$$

Pivot-based estimation for the exponential distribution (novel analytical approach).
Let us suppose that X is a future observation from the same distribution (7.1),
independent of $\mathbf{X} = (X_1 \leq \cdots \leq X_k)$. Then the pivot-based estimate of (7.2) can be
determined as follows:

    Step 1. *Invariant embedding of S_k in (7.2) to isolate the unknown parameter λ
from the problem through V,*

$$F_\lambda(x) = 1 - \exp(-x/\lambda) = 1 - \exp\left(-\frac{x}{s_k}\frac{s_k}{\lambda}\right) = 1 - \exp\left(-\frac{x}{s_k}v\right) = F_\lambda\left(\frac{x}{s_k}v\right) = F_\lambda(wv), \tag{7.18}$$

        where

$$W = X/S_k \tag{7.19}$$

        is an ancillary statistic.

    Step 2. *Averaging (7.18) over the probability distribution of the pivotal quantity
V to eliminate unknown parameter λ from the problem.* It follows from
(7.18) and (7.7) that the pivot-based estimate of the cumulative
distribution function (7.2) (obtained through the pivot-based method)
is given by

$$F_{S_k}(x) = \int_0^\infty F_\lambda\left(\frac{x}{s_k}v\right)f(v)dv$$

$$= \int_0^\infty \left[1 - \exp\left(-\frac{x}{s_k}v\right)\right]\frac{1}{\Gamma(k)}v^{k-1}\exp(-v)dv = 1 - \left(1 + \frac{x}{s_k}\right)^{-k}, \tag{7.20}$$

where

$$\overline{F}_{S_k}(x) = 1 - F_{S_k}(x) = \left(1 + \frac{x}{s_k}\right)^{-k}. \tag{7.21}$$

The pivot-based estimate of the probability density function (7.1) is given by

$$f_{s_k}(x) = \frac{dF_{s_k}(x)}{dx} = \frac{k}{s_k}\left(1 + \frac{x}{s_k}\right)^{-k-1}, \; x \geqslant 0, \tag{7.22}$$

It follows from (7.20) that the cumulative probability distribution function of the ancillary statistic (7.19) is given by

$$F(w) = 1 - \frac{1}{(1 + w)^k}. \tag{7.23}$$

The probability density function of the ancillary statistic (7.19) is given by

$$f(w) = \frac{dF(w)}{dw} = \frac{k}{(1 + w)^{k+1}}, \; w \geqslant 0. \tag{7.24}$$

Unbiased estimation for the exponential distribution (novel analytical approach). Let us suppose that X is a future observation from the same distribution (7.1), independent of $\mathbf{X} = (X_1 \leqslant \cdots \leqslant X_k)$. Then the unbiased estimate of (7.1) can be determined as follows:

Step 1. The following transformation is performed:

$$\frac{1}{f_\lambda(x)} \times f_\lambda(s_k) = \frac{1}{\lambda^{-1}\exp(-x/\lambda)} \times \frac{1}{\Gamma(k)\lambda^k}s_k^{k-1}\exp\left(-\frac{s_k}{\lambda}\right)$$

$$= \frac{s_k}{(k-1)\left(1 - \dfrac{x}{s_k}\right)^{k-2}} \times \frac{1}{\Gamma(k-1)}\left(\frac{s_k}{\lambda}\left[1 - \frac{x}{s_k}\right]\right)^{k-2}\exp\left(-\frac{s_k}{\lambda}\left[1 - \frac{x}{s_k}\right]\right)\frac{1}{\lambda}. \tag{7.25}$$

Step 2. It follows from (7.25) that the statistical unbiased estimate of (7.1) can be determined as follows:

$$f_{s_k}^{\bullet}(x) = \left[\int_0^\infty \frac{s_k}{(k-1)(1 - x/s_k)^{k-2}} \times \frac{1}{\Gamma(k-1)}\left(\frac{s_k}{\lambda}\left[1 - \frac{x}{s_k}\right]\right)^{k-2}\right.$$

$$\left. \exp\left(-\frac{s_k}{\lambda}\left[1 - \frac{x}{s_k}\right]\right)d\left(\frac{s_k}{\lambda}\left[1 - \frac{x}{s_k}\right]\right)\right]^{-1}$$

$$\tag{7.26}$$

$$= \left[\int_0^\infty \frac{s_k}{(k-1)(1 - x/s_k)^{k-2}} \times \frac{1}{\Gamma(k-1)}(v^{\bullet})^{k-2}\exp(-v^{\bullet})d(v^{\bullet})\right]^{-1}$$

$$= \frac{k-1}{s_k}\left(1 - \frac{x}{s_k}\right)^{k-2}, \; x \in (0, s_k),$$

where

$$v^{\bullet} = \frac{S_k}{\lambda}\left[1 - \frac{X}{S_k}\right] \tag{7.27}$$

is a pivotal quantity. It follows from (7.26) that the unbiased estimate of the probability distribution function (7.2) is given by

$$F_{S_k}^{\bullet}(x) = \int_0^x f_{S_k}^{\bullet}(u)du = \int_0^x \frac{k-1}{S_k}\left(1 - \frac{u}{S_k}\right)^{k-2} du = 1 - \left(1 - \frac{x}{S_k}\right)^{k-1}. \tag{7.28}$$

Verification that (7.28) is an unbiased estimate for (7.2) can be done as follows:

Step 1. The following transformation is performed:

$$F_{S_k}^{\bullet}(x) \times f_{\lambda}(s_k) = \left[1 - \left(1 - \frac{x}{S_k}\right)^{k-1}\right] \times \frac{1}{\Gamma(k)\lambda^k} s_k^{k-1} \exp\left(-\frac{s_k}{\lambda}\right)$$

$$= \frac{1}{\Gamma(k)\lambda^k}\left(\frac{s_k}{\lambda}\right)^{k-1} \exp\left(-\frac{s_k}{\lambda}\right)\frac{1}{\lambda} - \left(1 - \frac{x}{S_k}\right)^{k-1} \times \frac{1}{\Gamma(k)}\left(\frac{s_k}{\lambda}\right)^{k-1} \exp\left(-\frac{s_k}{\lambda}\right)\frac{1}{\lambda}$$

$$= \frac{1}{\Gamma(k)}\left(\frac{s_k}{\lambda}\right)^{k-1} \exp\left(-\frac{s_k}{\lambda}\right)$$

$$\frac{1}{\lambda} - \exp\left(-\frac{x}{\lambda}\right) \times \frac{1}{\Gamma(k)}\left(\frac{s_k}{\lambda}\left[1 - \frac{x}{S_k}\right]\right)^{k-1} \exp\left(-\frac{s_k}{\lambda}\left[1 - \frac{x}{S_k}\right]\right)\frac{1}{\lambda}$$

$$\tag{7.29}$$

Step 2. It follows from (7.29) that the expected parametric function is determined as follows:

$$\int_0^{\infty} \frac{1}{\Gamma(k)}\left(\frac{s_k}{\lambda}\right)^{k-1} \exp\left(-\frac{s_k}{\lambda}\right)d\left(\frac{s_k}{\lambda}\right)$$

$$- \int_0^{\infty} \exp\left(-\frac{x}{\lambda}\right) \times \frac{1}{\Gamma(k)}\left(\frac{s_k}{\lambda}\left[1 - \frac{x}{S_k}\right]\right)^{k-1} \exp\left(-\frac{s_k}{\lambda}\left[1 - \frac{x}{S_k}\right]\right)d\left(\frac{s_k}{\lambda}\left[1 - \frac{x}{S_k}\right]\right)$$

$$= \int_0^{\infty} \frac{1}{\Gamma(k)}v^{k-1}\exp(-v)dv - \int_0^{\infty} \exp\left(-\frac{x}{\lambda}\right) \times \frac{1}{\Gamma(k)}(v^{\bullet})^{k-1}\exp(-v^{\bullet})dv^{\bullet}$$

$$= 1 - \exp\left(-\frac{x}{\lambda}\right) = F_{\lambda}(x). \tag{7.30}$$

7.2.1 Two-parameter exponential distribution

Let $\mathbf{Z} = (Z_1 \leqslant \cdots \leqslant Z_r)$ be the first r ordered observations (order statistics) in a sample of size m from the two-parameter exponential distribution with the probability density function

$$f_\omega(x) = \vartheta^{-1} \exp\left(-\frac{x - \delta}{\vartheta}\right), \; \vartheta > 0, \; x > 0,$$ (7.31)

and the cumulative probability distribution function

$$F_\omega(x) = 1 - \exp\left(-\frac{x - \delta}{\vartheta}\right), \; \overline{F}_\omega(x) = 1 - F_\omega(x) = \exp\left(-\frac{x - \delta}{\vartheta}\right),$$ (7.32)

where $\omega = (\delta, \vartheta)$, δ is the shift parameter and ϑ is the scale parameter. It is assumed that these parameters are unknown. In Type II censoring, which is of primary interest here, the number of survivors is fixed and Z_k is a random variable. In this case, the likelihood function is given by

$$
\begin{aligned}
L(\delta, \vartheta) &= \prod_{i=1}^{r} f_\omega(z_i)(\overline{F}_\omega(z_r))^{m-r} = \frac{1}{\vartheta^r} \exp(-[\sum_{i=1}^{r}(z_i - \delta) + (m - r)(z_r - \delta)]/\vartheta) \\
&= \frac{1}{\vartheta^r} \exp(-[\sum_{i=1}^{r}(z_i - z_1 + z_1 - \delta) + (m - r)(z_r - z_1 + z_1 - \delta)]/\vartheta) \\
&= \frac{1}{\vartheta^{r-1}} \exp(-[\sum_{i=1}^{r}(z_i - z_1) + (m - r)(z_k - z_1)]/\vartheta) \times \frac{1}{\vartheta} \exp(-\frac{m(z_1 - \delta)}{\vartheta}) \\
&= \frac{1}{\vartheta^{r-1}} \exp(-\frac{S_r}{\vartheta}) \times \frac{1}{\vartheta} \exp(-\frac{m(s_1 - \delta)}{\vartheta}),
\end{aligned}
$$ (7.33)

where

$$\mathbf{S} = \left(S_1 = Z_1, \; S_r = \sum_{j=1}^{r} Z_j + (m - r)Z_r\right)$$ (7.34)

is the complete sufficient statistic for ω. The probability density function of $\mathbf{S} = (S_1, S_r)$ is given by

$$
\begin{aligned}
f_\omega(s_1, s_r) &= \frac{\dfrac{1}{\vartheta^{r-1}} \exp\left(-\dfrac{S_r}{\vartheta}\right) \times \dfrac{1}{\vartheta} \exp\left(-\dfrac{m(s_1 - \delta)}{\vartheta}\right)}{\dfrac{1}{s_r^{r-2}} \displaystyle\int_0^\infty \dfrac{s_r^{r-2}}{\vartheta^{r-1}} \exp\left(-\dfrac{S_r}{\vartheta}\right) ds_r \times \dfrac{1}{m} \displaystyle\int_0^\infty \dfrac{m}{\vartheta} \exp\left(-\dfrac{m(s_1 - \delta)}{\vartheta}\right) ds_1} \\
&= \frac{\dfrac{1}{\vartheta^{r-1}} \exp\left(-\dfrac{S_r}{\vartheta}\right) \times \dfrac{1}{\vartheta} \exp\left(-\dfrac{m(s_1 - \delta)}{\vartheta}\right)}{\dfrac{\Gamma(r - 1)}{s_r^{r-2}} \times \dfrac{1}{m}} \\
&= \frac{1}{\Gamma(r - 1)\vartheta^{r-1}} s_r^{r-2} \exp\left(-\dfrac{S_r}{\vartheta}\right) \times \dfrac{m}{\vartheta} \exp\left(-\dfrac{m(s_1 - \delta)}{\vartheta}\right) = f_\vartheta(s_r) f_\omega(s_1),
\end{aligned}
$$ (7.35)

where

$$f_\omega(s_1) = \frac{m}{\vartheta}\exp\left(-\frac{m(s_1 - \delta)}{\vartheta}\right), \; s_1 \geqslant \delta, \tag{7.36}$$

$$f_\vartheta(s_r) = \frac{1}{\Gamma(r-1)\vartheta^{r-1}}s_r^{r-2}\exp\left(-\frac{s_r}{\vartheta}\right), \; s_r \geqslant 0. \tag{7.37}$$

$$v_1 = \frac{S_1 - \delta}{\vartheta} \tag{7.38}$$

is the pivotal quantity, the probability density function of which is given by

$$f_1(v_1) = m\exp(-mv_1), \; v_1 \geqslant 0, \tag{7.39}$$

$$v_r = \frac{S_r}{\vartheta} \tag{7.40}$$

is the pivotal quantity, the probability density function of which is given by

$$f_r(v_r) = \frac{1}{\Gamma(r-1)}v_r^{r-2}\exp(-v_r), \; v_r \geqslant 0. \tag{7.41}$$

The pivot-based estimate of the cumulative distribution function (7.32) (obtained through the pivot-based method) is given by

$$F_S(x) = 1 - \left(1 + \frac{x - s_1}{s_r}\right)^{-(r-1)}, \; \overline{F}_S(x) = \left(1 + \frac{x - s_1}{s_r}\right)^{-(r-1)}, \; x \geqslant s_1. \tag{7.42}$$

7.3 Analytical inferences for constructing new-sample prediction limits

Theorem 7.1 (*Adequate transformation of the cumulative distribution function of order statistics*). Let us assume that there is a random sample of m ordered observations $Y_1 \leqslant \cdots \leqslant Y_m$ from a known distribution with probability density function $f_\theta(y)$ and probability distribution function $F_\theta(y)$, where θ is the parameter (in general, vector), then the adequate transformation of the cumulative distribution function of the lth order statistic Y_l, $l \in \{1, 2, \ldots, m\}$, is given by

$$G_\theta(y_l|m) = P_\theta(Y_l \leqslant y_l|m) = \sum_{j=l}^{m}\binom{m}{j}\left[F_\theta(y_l)\right]^j\left[1 - F_\theta(y_l)\right]^{m-j} = \int_0^{F_\theta(y_l)}f_{l,\,m-l+1}(u)du, \tag{7.43}$$

where

$$f_{l,m-l+1}(u) = \frac{1}{B(l,\,m-l+1)}u^{l-1}(1 - u)^{(m-l+1)-1}, \; 0 < u < 1, \tag{7.44}$$

is the probability density function of the beta distribution ($Beta(l, m-l+1)$) with shape parameters l and $m-l+1$.

Proof. On the one hand, it follows from (7.43) that

$$
\frac{d}{dy_l}G_\theta(y_l|m) = \frac{d}{dy_l}P_\theta(Y_l \leqslant y_l|m)
$$

$$
= \frac{d}{dy_l}\sum_{j=l}^{m}\binom{m}{j}\left[F_\theta(y_l)\right]^j\left[1 - F_\theta(y_l)\right]^{m-j} = \sum_{j=l}^{m}\binom{m}{j}\frac{d}{dy_l}\left[F_\theta(y_l)\right]^j
$$

$$
\left[1 - F_\theta(y_l)\right]^{m-j}
$$

$$
= \sum_{j=l}^{m}\binom{m}{j}\left[jF_\theta(y_l)^{j-1}\left(1 - F_\theta(y_l)\right)^{m-j}F'_\theta(y_l)\right.
$$

$$
\left. - (m - j)F_\theta(y_l)^j\left(1 - F_\theta(y_l)\right)^{m-j-1}F'_\theta(y_l)\right]
$$

$$
= \sum_{j=l}^{m}\binom{m}{j}\left[jF_\theta(y_l)^{j-1}\left(1 - F_\theta(y_l)\right)^{m-j}\right.
$$

$$
\left. - (m - j)F_\theta(y_l)^j\left(1 - F_\theta(y_l)\right)^{m-j-1}\right]f_\theta(y_l) \tag{7.45}
$$

$$
= \frac{m!}{(l - 1)!(m - l)!}F_\theta(y_l)^{l-1}\left(1 - F_\theta(y_l)\right)^{m-l}f_\theta(y_l)
$$

$$
+ \sum_{j=l+1}^{m}\frac{m!}{(j - 1)!(m - j)!}F_\theta(y_l)^{j-1}\left(1 - F_\theta(y_l)\right)^{m-j}f_\theta(y_l)
$$

$$
- \sum_{j=l}^{m-1}\frac{m!}{j!(m - j - 1)!}F_\theta(y_l)^j\left(1 - F_\theta(y_l)\right)^{m-j-1}f_\theta(y_l)
$$

$$
= \frac{1}{B(l, m - l + 1)}F_\theta(y_l)^{l-1}\left(1 - F_\theta(y_l)\right)^{m-l}f_\theta(y_l) = g_\theta(y_l|m).
$$

If $j = i+1$, we have that

$$
\sum_{j=l+1}^{m}\frac{m!}{(j - 1)!(m - j)!}F_\theta(y_l)^{j-1}\left(1 - F_\theta(y_l)\right)^{m-j}f_\theta(y_l)
$$

$$
= \sum_{i=l}^{m-1}\frac{m!}{i!(m - i - 1)!}F_\theta(y_l)^i\left(1 - F_\theta(y_l)\right)^{m-i-1}f_\theta(y_l). \tag{7.46}
$$

Thus, it follows from (7.43) that

$$
\frac{d}{dy_l}G_\theta(y_l|m) = \frac{d}{dy_l}P_\theta(Y_l \leqslant y_l|m)
$$

$$
= \frac{1}{B(l, m - l + 1)}F_\theta(y_l)^{l-1}\left(1 - F_\theta(y_l)\right)^{(m-l+1)-1}f_\theta(y_l) = g_\theta(y_l|m). \tag{7.47}
$$

On the other hand, it follows from (7.43) that

$$\frac{d}{dy_l}G_\theta(y_l|m) = \frac{d}{dy_l}P_\theta(Y_l \leq y_l|m)$$

$$= \frac{d}{dy_l}\int_0^{F_\theta(y_l)} f_{l,\,m-l+1}(u)du = \frac{d}{dy_l}\int_0^{F_\theta(y_l)} \frac{1}{B(l,\,m-l+1)}u^{l-1}(1-u)^{(m-l+1)-1}du \qquad (7.48)$$

$$= \frac{1}{B(l,\,m-l+1)}F_\theta(y_l)^{l-1}(1-F_\theta(y_l))^{(m-l+1)-1}f_\theta(y_l) = g_\theta(y_l|m).$$

Thus, $F_\theta(y_l)$ is the generalized pivotal quantity:

$$F_\theta(y_l) = u \sim f_{l,m-l+1}(u) = \frac{1}{B(l,\,m-l+1)}u^{l-1}(1-u)^{(m-l+1)-1},\ 0 < u < 1. \qquad (7.49)$$

This ends the proof.
    Corollary 7.1.

$$G_\theta(y_l|m) = P_\theta(Y_l \leq y_l|m) = \sum_{j=l}^m \binom{m}{j}\big[F_\theta(y_l)\big]^j\big[1-F_\theta(y_l)\big]^{m-j} =$$

$$\int_{1-F_\theta(y_l)}^1 f_{m-l+1,\,l}(u)du, \qquad (7.50)$$

where

$$f_{m-l+1,l}(u) = \frac{1}{B(m-l+1,\,l)}u^{(m-l+1)-1}(1-u)^{l-1},\ 0 < u < 1, \qquad (7.51)$$

is the probability density function (PDF) of the beta distribution ($Beta(m-l+1,\,l)$) with shape parameters $m-l+1$ and l.
    Proof.

$$\frac{d}{dy_l}G_\theta(y_l|m) = \frac{d}{dy_l}P_\theta(Y_l \leq y_l|m)$$

$$= \frac{d}{dy_l}\int_{1-F_\theta(y_l)}^1 f_{m-l+1,\,l}(u)du = \frac{d}{dy_l}$$

$$\int_{1-F_\theta(y_l)}^1 \frac{1}{B(m-l+1,\,l)}u^{(m-l+1)-1}(1-u)^{l-1}du \qquad (7.52)$$

$$= -\frac{1}{B(m-l+1,\,l)}(1-F_\theta(y_l))^{(m-l+1)-1}F_\theta(y_l)^{l-1}(-f_\theta(y_l))$$

$$= \frac{1}{B(l,\,m-l+1)}F_\theta(y_l)^{l-1}(1-F_\theta(y_l))^{m-l}f_\theta(y_l) = g_\theta(y_l|m).$$

Thus, $1-F_\theta(y_l)$ is the generalized pivotal quantity:

$$1-F_\theta(y_l) = u \sim f_{m-l+1,l}(u) = \frac{1}{B(m-l+1,\,l)}u^{(m-l+1)-1}(1-u)^{l-1},\ 0 < u \qquad (7.53)$$

$$< 1.$$

This ends the proof.

Corollary 7.2.

$$G_\theta(y_l|m) = P_\theta(Y_l \leqslant y_l|m) = \sum_{j=l}^{m} \binom{m}{j} [F_\theta(y_l)]^j [1 - F_\theta(y_l)]^{m-j}$$

$$= \int_0^{\frac{m-l+1}{l}\frac{F_\theta(y_l)}{1-F_\theta(y_l)}} \varphi_{l,\,m-l+1}(u)\,du,$$

(7.54)

where

$$\varphi_{l,\,m-l+1}(u) = \frac{1}{B(l,\,m-l+1)} \frac{\left[\frac{l}{m-l+1}u\right]^{l-1}}{\left[1 + \frac{l}{m-l+1}u\right]^{m+1}} \frac{l}{m-l+1}, \quad u \in (0,\,\infty).$$

(7.55)

is the PDF of the F distribution $(F(l,\,m-l+1))$ with parameters l and $m-l+1$, which are positive integers known as the degrees of freedom for the numerator and the degrees of freedom for the denominator, respectively.

Proof.

$$\frac{d}{dy_l}G_\theta(y_l|m) = \frac{d}{dy_l}P_\theta(Y_l \leqslant y_l|m) = \frac{d}{dy_l}\int_0^{\frac{m-l+1}{l}\frac{F_\theta(y_l)}{1-F_\theta(y_l)}} \varphi_{l,\,m-l+1}(u)\,du$$

$$= \frac{d}{dy_l}\int_0^{\frac{m-l+1}{l}\frac{F_\theta(y_l)}{1-F_\theta(y_l)}} \frac{1}{B(l,\,m-l+1)} \frac{\left[\frac{l}{m-l+1}u\right]^{l-1}}{\left[1 + \frac{l}{m-l+1}u\right]^{m+1}} \frac{l}{m-l+1}\,du$$

$$= \frac{1}{B(l,\,m-l+1)} \frac{\left[\frac{F_\theta(y_l)}{1-F_\theta(y_l)}\right]^{l-1}}{\left[1 + \frac{F_\theta(y_l)}{1-F_\theta(y_l)}\right]^{m+1}} \frac{1}{(1-F_\theta(y_l))^2}f_\theta(y_l)$$

(7.56)

$$= \frac{1}{B(l,\,m-l+1)} \frac{\left[\frac{F_\theta(y_l)}{1-F_\theta(y_l)}\right]^{l-1}}{\left[\frac{1}{1-F_\theta(y_l)}\right]^{m+1}} \frac{1}{(1-F_\theta(y_l))^2}f_\theta(y_l)$$

$$= \frac{1}{B(l,\,m-l+1)}F_\theta(y_l)^{l-1}(1-F_\theta(y_l))^{-(l-1)+m+1-2}f_\theta(y_l)$$

$$= \frac{1}{B(l,\,m-l+1)}F_\theta(y_l)^{l-1}(1-F_\theta(y_l))^{m-l}f_\theta(y_l) = g_\theta(y_l|m).$$

Thus, $\dfrac{m-l+1}{l}\dfrac{F_\theta(y_l)}{1-F_\theta(y_l)}$ is the generalized pivotal quantity:

$$\frac{m-l+1}{l}\frac{F_\theta(y_l)}{1-F_\theta(y_l)} = u \sim \varphi_{l,\,m-l+1}(u)$$

$$= \frac{1}{B(l,\,m-l+1)}\frac{\left[\dfrac{l}{m-l+1}u\right]^{l-1}}{\left[1+\dfrac{l}{m-l+1}u\right]^{m+1}}\frac{l}{m-l+1},\ u \in (0,\,\infty). \tag{7.57}$$

This ends the proof.

Corollary 7.3.

$$G_\theta(y_l|m) = P_\theta(Y_l \leqslant y_l|m) = \sum_{j=l}^{m}\binom{m}{j}\left[F_\theta(y_l)\right]^{j}\left[1-F_\theta(y_l)\right]^{m-j} \tag{7.58}$$

$$= \int_{\frac{l}{m-l+1}\frac{1-F_\theta(y_l)}{F_\theta(y_l)}}^{\infty}\varphi_{m-l+1,\,l}(u)du,$$

where

$$\varphi_{m-l+1,\,l}(u) = \frac{1}{B(m-l+1,\,l)}\frac{\left[\dfrac{m-l+1}{l}u\right]^{m-l}}{\left[1+\dfrac{m-l+1}{l}u\right]^{m+1}}\frac{m-l+1}{l},\ u \in (0,\,\infty), \tag{7.59}$$

is the probability density function of the F distribution ($F(l,\,m-l+1)$) with parameters l and $m-l+1$, which are positive integers known as the degrees of freedom for the numerator and the degrees of freedom for the denominator, respectively.

Proof.

$$\frac{d}{dy_l}G_\theta(y_l|m) = \frac{d}{dy_l}P_\theta(Y_l \leqslant y_l|m) = \frac{d}{dy_l}\int_{\frac{l}{m-l+1}\frac{1-F_\theta(y_l)}{F_\theta(y_l)}}^{\infty}\varphi_{m-l+1,\,l}(u)du$$

$$= \frac{d}{dy_l}\int_{\frac{l}{m-l+1}\frac{1-F_\theta(y_l)}{F_\theta(y_l)}}^{\infty}\frac{1}{B(m-l+1,\,l)}\frac{\left[\dfrac{m-l+1}{l}u\right]^{m-l}}{\left[1+\dfrac{m-l+1}{l}u\right]^{m+1}}\frac{m-l+1}{l}du$$

$$= -\frac{1}{B(m-l+1,\,l)}\frac{\left[\dfrac{1-F_\theta(y_l)}{F_\theta(y_l)}\right]^{m-l}}{\left[1+\dfrac{1-F_\theta(y_l)}{F_\theta(y_l)}\right]^{m+1}}\frac{-1}{F_\theta(y_l)^2}f_\theta(y_l) \tag{7.60}$$

$$= \frac{1}{B(m-l+1,\,l)}\frac{\left[\dfrac{1-F_\theta(y_l)}{F_\theta(y_l)}\right]^{m-l}}{\left[\dfrac{1}{F_\theta(y_l)}\right]^{m+1}}\frac{1}{F_\theta(y_l)^2}f_\theta(y_l)$$

$$= \frac{1}{B(m-l+1,\,l)}(1-F_\theta(y_l))^{m-l}F_\theta(y_l)^{m+1-m+l-2}f_\theta(y_l)$$

$$= \frac{1}{B(l,\,m-l+1)}F_\theta(y_l)^{l-1}(1-F_\theta(y_l))^{m-l}f_\theta(y_l) = g_\theta(y_l|m).$$

Thus, $\dfrac{l}{m-l+1}\dfrac{1-F_\theta(y_l)}{F_\theta(y_l)}$ is the generalized pivotal quantity:

$$\frac{l}{m-l+1}\frac{1-F_\theta(y_l)}{F_\theta(y_l)} = u \sim \varphi_{m-l+1,\,l}(u)$$

$$= \frac{1}{B(m-l+1,\,l)}\frac{\left[\dfrac{m-l+1}{l}u\right]^{m-l}}{\left[1+\dfrac{m-l+1}{l}u\right]^{m+1}}\frac{m-l+1}{l},\ u \in (0,\,\infty), \tag{7.61}$$

This ends the proof.

7.3.1 Example of constructing new-sample $(1-\alpha)$-prediction limits

Consider a life test of some sample $Z_1 \leqslant \cdots \leqslant Z_m$ where m units, whose lifetimes follow the same exponential distribution (7.1), are put on test simultaneously, and where all units are observed until failure. It is necessary, using the results derived above, give a prediction limit for the largest lifetime Z_m on the basis of the first k ordered observations (lifetimes) $Y_1 \leqslant \cdots \leqslant Y_k$ from the past sample of size m from the above exponential distribution (7.1); Z_m is in this case the total elapsed time required to complete the test. Then the exact lower one-sided $(1-\alpha)$-prediction limit $z_l(=L_{1-\alpha,\,l})$ (for the lth order statistic Z_l, $l \in \{1, \dots, m\}$, from a set of m future ordered observations $Z_1 \leqslant \cdots \leqslant Z_m$ also from the distribution (7.1)), has to satisfy

$$P_\lambda(Z_l > z_l;\, m) = \int_{z_l}^{\infty} g_\lambda(z;\, m)dz = \overline{G}_\lambda(z_l|;m) = 1-\alpha, \tag{7.62}$$

and the exact upper $(1-\alpha)$-prediction limit $z_l(=U_{1-\alpha;l})$ for Z_l, $l \in \{1, \dots, m\}$, has to satisfy

$$P_\lambda(Z_l \leqslant z_l;\, m) = \int_0^{z_l} g_\lambda(z;\, m)dz = G_\lambda(z_l;\, m) = 1-\alpha. \tag{7.63}$$

It follows from (7.10) and (7.11) that

$$L_{1-\alpha;l} = z_l = w_l s_k, \tag{7.64}$$

where

$$w_l = \underset{w_l}{\operatorname{argmin}}$$

$$\left(\frac{1}{B(l,\,m-l+1)}\sum_{j=0}^{l-1}\binom{l-1}{j}(-1)^j\frac{[1+(m-l+j+1)w_l]^{-k}}{m-l+j+1} - (1-\alpha)\right)^2, \tag{7.65}$$

In a similar way it can be shown that

$$U_{1-\alpha;l} = z_l = w_l s_k, \tag{7.66}$$

where

$$w_l = \underset{w_l}{\mathrm{argmin}}$$

$$\left(\frac{1}{B(l,\, m - l + 1)} \sum_{j=0}^{l-1} \binom{l-1}{j} (-1)^j \frac{[1 + (m - l + j + 1)w_l]^{-k}}{m - l + j + 1} - \alpha \right)^2, \qquad (7.67)$$

If, say, $\alpha = 0.05$, $k = 4$, $m = 10$, $l = m$,

$$S_k = \sum_{i=1}^{k} Y_i + (m - k)Y_k = 33 + 87 + 125 + 165 + 6 \times 165 = 1400, \qquad (7.68)$$

it follows from (7.64) that the exact lower $(1 - \alpha)$-prediction limit $z_l (= L_{1-\alpha;l})$ for Z_l is given by

$$L_{1-\alpha;l} = z_l = w_l s_k = 0.265 \times 1400 = 370.7, \qquad (7.69)$$

and it follows from (7.66) that the exact upper $(1 - \alpha)$-prediction limit $z_l (= U_{1-\alpha;l})$ for Y_l is given by

$$U_{1-\alpha;l} = y_l = w_l s_k = 2.418 \times 1400 = 3386. \qquad (7.70)$$

7.4 Analytical inferences for constructing within-sample prediction limits

Theorem 7.2 (*Adequate transformation of the lth order statistic's conditional cumulative distribution function*). Let us assume that there is a random sample of m ordered observations $Y_1 \leqslant \cdots \leqslant Y_m$ from a known distribution with probability density function $f_\lambda(y)$ and cumulative distribution function $F_\theta(y)$, where θ is the parameter (in general, vector), then the adequate transformation of the conditional cumulative distribution function of the lth order statistic Y_l $(1 \leqslant k < l \leqslant m)$ given $Y_k = y_k$ is determined as

$$G_\theta(y_l | y_k; m) = P_\theta(Y_l \leqslant y_l | Y_k = y_k; m)$$

$$= \sum_{j=l-k}^{m-k} \binom{m-k}{j} \left[1 - \frac{\overline{F}_\theta(y_l)}{\overline{F}_\theta(y_k)} \right]^j \left[\frac{\overline{F}_\theta(y_l)}{\overline{F}_\theta(y_k)} \right]^{m-k-j} \qquad (7.71)$$

$$= \int_0^{1 - \frac{\overline{F}_\theta(y_l)}{\overline{F}_\theta(y_k)}} f_{l-k,\, m-l+1}(u)\, du,$$

where $\overline{F}_\theta(y) = 1 - F_\theta(y)$,

$$f_{l-k,\, m-l+1}(u) = \frac{1}{B(l - k,\, m - l + 1)} u^{l-k-1}(1 - u)^{(m-l+1)-1} du, \quad 0 < u < 1, \qquad (7.72)$$

is the PDF of the beta distribution ($Beta(l - k, m - l + 1)$) with shape parameters $l - k$ and $m-k+1$.

Proof. On the one hand, it follows from (7.71) that

$$\frac{d}{dy_l} G_\theta(y_l|y_k; m) = \frac{d}{dy_l} P_\theta(Y_l \leqslant y_l | Y_k = y_k; m)$$

$$= \frac{d}{dy_l} \sum_{j=l-k}^{m-k} \binom{m-k}{j} \left[1 - \frac{\overline{F}_\theta(y_l)}{\overline{F}_\theta(y_k)} \right]^j \left[\frac{\overline{F}_\theta(y_l)}{\overline{F}_\theta(y_k)} \right]^{m-k-j}$$

$$= \sum_{j=l-k}^{m-k} (m - kj) \frac{d}{dy_l} \left(\left[1 - \frac{\overline{F}_\theta(y_l)}{\overline{F}_\theta(y_k)} \right]^j \left[\frac{\overline{F}_\theta(y_l)}{\overline{F}_\theta(y_k)} \right]^{m-k-j} \right)$$

$$= \sum_{j=l-k}^{m-k} \binom{m-k}{j} \left(j \left[1 - \frac{\overline{F}_\theta(y_l)}{\overline{F}_\theta(y_k)} \right]^{j-1} \left[\frac{\overline{F}_\theta(y_l)}{\overline{F}_\theta(y_k)} \right]^{m-k-j} \right.$$

$$\left. - (m - k - j) \left[1 - \frac{\overline{F}_\theta(y_l)}{\overline{F}_\theta(y_k)} \right]^j \left[\frac{\overline{F}_\theta(y_l)}{\overline{F}_\theta(y_k)} \right]^{m-k-j-1} \right) \frac{f_\theta(y_l)}{\overline{F}_\theta(y_k)}$$

$$= \sum_{j=l-k}^{m-k} \frac{(m-k)!}{(j-1)!(m-k-j)!} \left[1 - \frac{\overline{F}_\theta(y_l)}{\overline{F}_\theta(y_k)} \right]^{j-1} \left[\frac{\overline{F}_\theta(y_l)}{\overline{F}_\theta(y_k)} \right]^{m-k-j} \frac{f_\theta(y_l)}{\overline{F}_\theta(y_k)}$$

$$- \sum_{j=l-k}^{m-k-1} \frac{(m-k)!}{j!(m-k-j-1)!} \left[1 - \frac{\overline{F}_\theta(y_l)}{\overline{F}_\theta(y_k)} \right]^j \left[\frac{\overline{F}_\theta(y_l)}{\overline{F}_\theta(y_k)} \right]^{m-k-j-1} \frac{f_\theta(y_l)}{\overline{F}_\theta(y_k)}$$

$$= \frac{(m-k)!}{(l-k-1)!(m-l)!} \left[1 - \frac{\overline{F}_\theta(y_l)}{\overline{F}_\theta(y_k)} \right]^{l-k-1} \left[\frac{\overline{F}_\theta(y_l)}{\overline{F}_\theta(y_k)} \right]^{m-l} \frac{f_\theta(y_l)}{\overline{F}_\theta(y_k)}$$

$$+ \sum_{j=l-k+1}^{m-k} \frac{(m-k)!}{(j-1)!(m-k-j)!} \left[1 - \frac{\overline{F}_\theta(y_l)}{\overline{F}_\theta(y_k)} \right]^{j-1} \left[\frac{\overline{F}_\theta(y_l)}{\overline{F}_\theta(y_k)} \right]^{m-k-j} \frac{f_\theta(y_l)}{\overline{F}_\theta(y_k)}$$

$$- \sum_{j=l-k}^{m-k-1} \frac{(m-k)!}{j!(m-k-j-1)!} \left[1 - \frac{\overline{F}_\theta(y_l)}{\overline{F}_\theta(y_k)} \right]^j \left[\frac{\overline{F}_\theta(y_l)}{\overline{F}_\theta(y_k)} \right]^{m-k-j-1} \frac{f_\theta(y_l)}{\overline{F}_\theta(y_k)}$$

$$= \frac{(m-k)!}{(l-k-1)!(m-l)!} \left[1 - \frac{\overline{F}_\theta(y_l)}{\overline{F}_\theta(y_k)} \right]^{l-k-1} \left[\frac{\overline{F}_\theta(y_l)}{\overline{F}_\theta(y_k)} \right]^{m-l} \frac{f_\theta(y_l)}{\overline{F}_\theta(y_k)}$$

$$= \frac{1}{B(l-k, m-l+1)} \left[1 - \frac{\overline{F}_\theta(y_l)}{\overline{F}_\theta(y_k)} \right]^{l-k-1} \left[\frac{\overline{F}_\theta(y_l)}{\overline{F}_\theta(y_k)} \right]^{m-l} \frac{f_\theta(y_l)}{\overline{F}_\theta(y_k)},$$

$$\tag{7.73}$$

If $j = i+1$, we have that

$$\sum_{j=l-k+1}^{m-k} \frac{(m-k)!}{(j-1)!(m-k-j)!} \left[1 - \frac{\overline{F}_\theta(y_l)}{\overline{F}_\theta(y_k)} \right]^{j-1} \left[\frac{\overline{F}_\theta(y_l)}{\overline{F}_\theta(y_k)} \right]^{m-k-j} \frac{f_\theta(y_l)}{\overline{F}_\theta(y_k)}$$

$$= \sum_{i=l-k}^{m-k-1} \frac{(m-k)!}{i!(m-k-i-1)!} \left[1 - \frac{\overline{F}_\theta(y_l)}{\overline{F}_\theta(y_k)} \right]^i \left[\frac{\overline{F}_\theta(y_l)}{\overline{F}_\theta(y_k)} \right]^{m-k-i-1} \frac{f_\theta(y_l)}{\overline{F}_\theta(y_k)}.$$

$$\tag{7.74}$$

Thus, it follows from (7.71) that

$$
\frac{d}{dy_l} G_\theta(y_l|y_k; m) = \frac{d}{dy_l} P_\theta(Y_l \leqslant y_l | Y_k = y_k; m) = \frac{d}{dy_l}
$$

$$
\sum_{j=l-k}^{m-k} (m - kj)\left[1 - \frac{\overline{F_\theta(y_l)}}{\overline{F_\theta(y_k)}}\right]^j \left[\frac{\overline{F_\theta(y_l)}}{\overline{F_\theta(y_k)}}\right]^{m-k-j} \tag{7.75}
$$

$$
= \frac{1}{B(l - k, m - l + 1)}\left[1 - \frac{\overline{F_\theta(y_l)}}{\overline{F_\theta(y_k)}}\right]^{l-k-1}\left[\frac{\overline{F_\theta(y_l)}}{\overline{F_\theta(y_k)}}\right]^{m-l} \frac{f_\theta(y_l)}{\overline{F_\theta(y_k)}}
$$

$$
= g(y_l|y_k; m).
$$

On the other hand, it follows from (7.71) that

$$
\frac{d}{dy_l} G_\theta(y_l|y_k; m) = \frac{d}{dy_l} P_\theta(Y_l \leqslant y_l | Y_k = y_k; m)
$$

$$
= \frac{d}{dy_l}\int_0^{1-\frac{\overline{F_\theta(y_l)}}{\overline{F_\theta(y_k)}}} f_{k,\,m-k+1}(u)du = \frac{d}{dy_l}
$$

$$
\int_0^{1-\frac{\overline{F_\theta(y_l)}}{\overline{F_\theta(y_k)}}} \frac{1}{B(l - k, m - l + 1)} u^{l-k-1}(1 - u)^{(m-l+1)-1}du \tag{7.76}
$$

$$
= \frac{1}{B(l - k, m - l + 1)}\left[1 - \frac{\overline{F_\theta(y_l)}}{\overline{F_\theta(y_k)}}\right]^{l-k-1}\left[\frac{\overline{F_\theta(y_l)}}{\overline{F_\theta(y_k)}}\right]^{m-l} \frac{f_\theta(y_l)}{\overline{F_\theta(y_k)}}
$$

$$
= g(y_l|y_k; m).
$$

Thus, $1 - \dfrac{\overline{F_\theta(y_l)}}{\overline{F_\theta(y_k)}}$ is the generalized pivotal quantity:

$$
1 - \frac{\overline{F_\theta(y_l)}}{\overline{F_\theta(y_k)}} = u \sim f_{l-k,\,m-l+1}(u) = \frac{1}{B(l - k, m - l + 1)} u^{l-k-1}(1 - u)^{(m-l+1)-1} \tag{7.77}
$$

$$
du, \, 0 < u < 1.
$$

This ends the proof.

Corollary 7.4.

$$
G_\theta(y_l|y_k; m) = P_\theta(Y_l \leqslant y_l | Y_k = y_k; m)
$$

$$
= \sum_{j=l-k}^{m-k} \binom{m-k}{j}\left[1 - \frac{\overline{F_\theta(y_l)}}{\overline{F_\theta(y_k)}}\right]^j \left[\frac{\overline{F_\theta(y_l)}}{\overline{F_\theta(y_k)}}\right]^{m-k-j} \tag{7.78}
$$

$$
= \int_{\frac{\overline{F_\theta(y_l)}}{\overline{F_\theta(y_k)}}}^1 f_{m-l+1,\,l-k}(u)du,
$$

where

$$f_{l-k,\, m-l+1}(u) = \frac{1}{B(m-l+1,\, l-k)} u^{(m-l+1)-1}(1-u)^{l-k-1} du, \quad 0 < u < 1, \quad (7.79)$$

is the PDF of the beta distribution ($Beta(m-l+1,\, l-k)$) with shape parameters $m-l+1$ and $l-k$.

Proof.

$$\frac{d}{dy_l} G_\theta(y_l | y_k; m) = \frac{d}{dy_l} P_\theta(Y_l \leq y_l | Y_k = y_k; m)$$

$$= \frac{d}{dy_l} \int_{\frac{\overline{F_\theta(y_l)}}{\overline{F_\theta(y_k)}}}^{1} f_{k,\, m-k+1}(u) du = \frac{d}{dy_l}$$

$$\int_{\frac{\overline{F_\theta(y_l)}}{\overline{F_\theta(y_k)}}}^{1} \frac{1}{B(m-l+1,\, l-k)} u^{(m-l+1)-1}(1-u)^{l-k-1} du$$

$$= -\frac{1}{B(m-l+1,\, l-k)} \left[\frac{\overline{F_\theta(y_l)}}{\overline{F_\theta(y_k)}}\right]^{(m-l+1)-1} \left[1 - \frac{\overline{F_\theta(y_l)}}{\overline{F_\theta(y_k)}}\right]^{l-k-1} \qquad (7.80)$$

$$\frac{-f_\theta(y_l)}{\overline{F_\theta(y_k)}}$$

$$= \frac{1}{B(l-k,\, m-l+1)} \left[1 - \frac{\overline{F_\theta(y_l)}}{\overline{F_\theta(y_k)}}\right]^{l-k-1} \left[\frac{\overline{F_\theta(y_l)}}{\overline{F_\theta(y_k)}}\right]^{m-l} \frac{f_\theta(y_l)}{\overline{F_\theta(y_k)}}$$

$$= g(y_l | y_k; m).$$

Thus, $\dfrac{\overline{F_\theta(y_l)}}{\overline{F_\theta(y_k)}}$ is the generalized pivotal quantity:

$$\frac{\overline{F_\theta(y_l)}}{\overline{F_\theta(y_k)}} = u \sim f_{l-k,\, m-l+1}(u) = \frac{1}{B(m-l+1,\, l-k)} u^{(m-l+1)-1}(1-u)^{l-k-1} du$$

$$, \quad 0 < u < 1. \qquad (7.81)$$

This ends the proof.

Corollary 7.5.

$$G_\theta(y_l | y_k; m) = P_\theta(Y_l \leq y_l | Y_k = y_k; m)$$

$$= \sum_{j=l-k}^{m-k} \binom{m-k}{j} \left[1 - \frac{\overline{F_\theta(y_l)}}{\overline{F_\theta(y_k)}}\right]^{j} \left[\frac{\overline{F_\theta(y_l)}}{\overline{F_\theta(y_k)}}\right]^{m-k-j} \qquad (7.82)$$

$$= \int_0^{\frac{m-l+1}{l-k}\left(1-\frac{\overline{F_\theta(y_l)}}{\overline{F_\theta(y_k)}}\right) \Big/ \frac{\overline{F_\theta(y_l)}}{\overline{F_\theta(y_k)}}} \varphi_{l-k,\, m-l+1}(u) du,$$

where

$$\varphi_{l-k,\,m-l+1}(u) = \frac{1}{B(l-k,\,m-l+1)}\frac{\left[\dfrac{l-k}{m-l+1}u\right]^{l-k-1}}{\left[1+\dfrac{l-k}{m-l+1}u\right]^{m-k+1}}\frac{l-k}{m-l+1},\ u \in (0,\,\infty). \tag{7.83}$$

is the PDF of the F distribution ($F(k,\,m-k+1)$) with parameters k and $m-k+1$, which are positive integers known as the degrees of freedom for the numerator and the degrees of freedom for the denominator, respectively.

Proof.

$$\frac{d}{dy_l}G_\theta(y_l|m) = \frac{d}{dy_l}P_\theta(Y_l \leqslant y_l|m) = \frac{d}{dy_l}$$

$$\int_0^{\frac{m-l+1}{l-k}\left(1-\frac{\overline{F_\theta}(y_l)}{\overline{F_\theta}(y_k)}\right)\Big/\frac{\overline{F_\theta}(y_l)}{\overline{F_\theta}(y_k)}} \varphi_{l-k,\,m-l+1}(u)\,du$$

$$= \frac{d}{dy_l}\int_0^{\frac{m-l+1}{l-k}\left(1-\frac{\overline{F_\theta}(y_l)}{\overline{F_\theta}(y_k)}\right)\Big/\frac{\overline{F_\theta}(y_l)}{\overline{F_\theta}(y_k)}} \frac{1}{B(l-k,\,m-l+1)}\frac{\left[\dfrac{l-k}{m-l+1}u\right]^{l-k-1}}{\left[1+\dfrac{l-k}{m-l+1}u\right]^{m-k+1}}$$

$$\frac{l-k}{m-l+1}du$$

$$= \frac{1}{B(l-k,\,m-l+1)}\frac{\left[\left(1-\frac{\overline{F_\theta}(y_l)}{\overline{F_\theta}(y_k)}\right)\Big/\frac{\overline{F_\theta}(y_l)}{\overline{F_\theta}(y_k)}\right]^{l-k-1}}{\left[1+\left(1-\frac{\overline{F_\theta}(y_l)}{\overline{F_\theta}(y_k)}\right)\Big/\frac{\overline{F_\theta}(y_l)}{\overline{F_\theta}(y_k)}\right]^{m-k+1}}\frac{1}{\left(\frac{\overline{F_\theta}(y_l)}{\overline{F_\theta}(y_k)}\right)^2}\frac{f_\theta(y_l)}{\overline{F_\theta}(y_k)} \tag{7.84}$$

$$= \frac{1}{B(l-k,\,m-l+1)}\frac{\left[\left(1-\frac{\overline{F_\theta}(y_l)}{\overline{F_\theta}(y_k)}\right)\Big/\frac{\overline{F_\theta}(y_l)}{\overline{F_\theta}(y_k)}\right]^{l-k-1}}{\left[1\Big/\frac{\overline{F_\theta}(y_k)}{\overline{F_\theta}(y_l)}\right]^{m-k+1}}\frac{1}{\left(\frac{\overline{F_\theta}(y_l)}{\overline{F_\theta}(y_k)}\right)^2}\frac{f_\theta(y_l)}{\overline{F_\theta}(y_k)}$$

$$= \frac{1}{B(l-k,\,m-l+1)}\left(1-\frac{\overline{F_\theta}(y_l)}{\overline{F_\theta}(y_k)}\right)^{l-k-1}\left(\frac{\overline{F_\theta}(y_l)}{\overline{F_\theta}(y_k)}\right)^{m-k+1-l+k+1-2}\frac{f_\theta(y_l)}{\overline{F_\theta}(y_k)}$$

$$= \frac{1}{B(l-k,\,m-l+1)}\left(1-\frac{\overline{F_\theta}(y_l)}{\overline{F_\theta}(y_k)}\right)^{l-k-1}\left(\frac{\overline{F_\theta}(y_l)}{\overline{F_\theta}(y_k)}\right)^{m-l}\frac{f_\theta(y_l)}{\overline{F_\theta}(y_k)} = g(y_l|y_k;\,m).$$

Thus, $\dfrac{m-l+1}{l-k}\left(1-\dfrac{\overline{F_\theta}(y_l)}{\overline{F_\theta}(y_k)}\right)\Big/\dfrac{\overline{F_\theta}(y_l)}{\overline{F_\theta}(y_k)}$ is the generalized pivotal quantity:

$$\frac{m-l+1}{l-k}\left(1-\frac{\overline{F_\theta(y_l)}}{\overline{F_\theta(y_k)}}\right)\bigg/\frac{\overline{F_\theta(y_l)}}{\overline{F_\theta(y_k)}} = u \sim \varphi_{l-k,\,m-l+1}(u)$$

$$= \frac{1}{B(l-k,\,m-l+1)}\frac{\left[\dfrac{l-k}{m-l+1}u\right]^{l-k-1}}{\left[1+\dfrac{l-k}{m-l+1}u\right]^{m-k+1}}\frac{l-k}{m-l+1}\,,\; u \in (0,\infty). \tag{7.85}$$

This ends the proof.

Corollary 7.6.

$$G_\theta(y_l|y_k;m) = P_\theta(Y_l \leqslant y_l | Y_k = y_k;m)$$

$$= \sum_{j=l-k}^{m-k}\binom{m-k}{j}\left[1-\frac{\overline{F_\theta(y_l)}}{\overline{F_\theta(y_k)}}\right]^j\left[\frac{\overline{F_\theta(y_l)}}{\overline{F_\theta(y_k)}}\right]^{m-k-j} = \tag{7.86}$$

$$\int_{\frac{l-k}{m-l+1}\frac{\overline{F_\theta(y_l)}}{\overline{F_\theta(y_k)}}\big/\left(1-\frac{\overline{F_\theta(y_l)}}{\overline{F_\theta(y_k)}}\right)}^{\infty}\varphi_{m-l+1,\,l-k,}(u)du,$$

where

$$\varphi_{m-l+1,\,l-k}(u) = \frac{1}{B(m-l+1,\,l-k)}\frac{\left[\dfrac{m-l+1}{l-k}u\right]^{m-l+1}}{\left[1+\dfrac{m-l+1}{l-k}u\right]^{m-k+1}}\frac{m-l+1}{l-k}\,,\; u \in (0,\infty). \tag{7.87}$$

is the PDF of the F distribution $(F(m-l+1,\,l-k))$ with parameters $m-l+1$ and $l-k$, which are positive integers known as the degrees of freedom for the numerator and the degrees of freedom for the denominator, respectively.

Proof.

$$\frac{d}{dy_l}G_\theta(y_l|m) = \frac{d}{dy_l}P_\theta(Y_l \leqslant y_l|m) = \frac{d}{dy_l}\int_{\frac{l-k}{m-l+1}\frac{\overline{F_\theta(y_l)}}{\overline{F_\theta(y_k)}}\big/\left(1-\frac{\overline{F_\theta(y_l)}}{\overline{F_\theta(y_k)}}\right)}^{\infty}\varphi_{m-l+1,\,l-k}(u)du$$

$$= \frac{d}{dy_l}\int_{\frac{l-k}{m-l+1}\frac{\overline{F_\theta(y_l)}}{\overline{F_\theta(y_k)}}\big/\left(1-\frac{\overline{F_\theta(y_l)}}{\overline{F_\theta(y_k)}}\right)}^{\infty}\frac{1}{B(m-l+1,\,l-k)}\frac{\left[\dfrac{m-l+1}{l-k}u\right]^{m-l}}{\left[1+\dfrac{m-l+1}{l-k}u\right]^{m-k+1}}\frac{m-l+1}{l-k}du$$

$$= -\frac{1}{B(m-l+1,\,l-k)}\frac{\left[\dfrac{\overline{F_\theta(y_l)}}{\overline{F_\theta(y_k)}}\big/\left(1-\dfrac{\overline{F_\theta(y_l)}}{\overline{F_\theta(y_k)}}\right)\right]^{m-l}}{\left[1+\dfrac{\overline{F_\theta(y_l)}}{\overline{F_\theta(y_k)}}\big/\left(1-\dfrac{\overline{F_\theta(y_l)}}{\overline{F_\theta(y_k)}}\right)\right]^{m-k+1}}\frac{-1}{\left(1-\dfrac{\overline{F_\theta(y_l)}}{\overline{F_\theta(y_k)}}\right)^2}\frac{f_\theta(y_l)}{\overline{F_\theta(y_k)}} \tag{7.88}$$

$$= \frac{1}{B(m-l+1,\,l-k)}\frac{\left[\dfrac{\overline{F_\theta(y_l)}}{\overline{F_\theta(y_k)}}\big/\left(1-\dfrac{\overline{F_\theta(y_l)}}{\overline{F_\theta(y_k)}}\right)\right]^{m-l}}{\left[1\big/\left(1-\dfrac{\overline{F_\theta(y_l)}}{\overline{F_\theta(y_k)}}\right)\right]^{m-k+1}}\frac{1}{\left(1-\dfrac{\overline{F_\theta(y_l)}}{\overline{F_\theta(y_k)}}\right)^2}\frac{f_\theta(y_l)}{\overline{F_\theta(y_k)}}$$

$$= \frac{1}{B(m-l+1,\,l-k)}\left(1-\frac{\overline{F_\theta(y_l)}}{\overline{F_\theta(y_k)}}\right)^{m-k+1-m+l-2}\left(\frac{\overline{F_\theta(y_l)}}{\overline{F_\theta(y_k)}}\right)^{m-l}\frac{f_\theta(y_l)}{\overline{F_\theta(y_k)}}$$

$$= \frac{1}{B(l-k,\,m-l+1)}\left(1-\frac{\overline{F_\theta(y_l)}}{\overline{F_\theta(y_k)}}\right)^{l-k-1}\left(\frac{\overline{F_\theta(y_l)}}{\overline{F_\theta(y_k)}}\right)^{m-l}\frac{f_\theta(y_l)}{\overline{F_\theta(y_k)}} = g(y_l|y_k;m).$$

Thus, $\dfrac{l-k}{m-l+1}\dfrac{\overline{F}_\theta(y_l)}{\overline{F}_\theta(y_k)}\Big/\left(1-\dfrac{\overline{F}_\theta(y_l)}{\overline{F}_\theta(y_k)}\right)$ is the generalized pivotal quantity:

$$\dfrac{l-k}{m-l+1}\dfrac{\overline{F}_\theta(y_l)}{\overline{F}_\theta(y_k)}\Big/\left(1-\dfrac{\overline{F}_\theta(y_l)}{\overline{F}_\theta(y_k)}\right) = u \sim \varphi_{m-l+1,\,l-k}(u)$$

$$= \dfrac{1}{B(m-l+1,\,l-k)}\dfrac{\left[\dfrac{m-l+1}{l-k}u\right]^{m-l+1}}{\left[1+\dfrac{m-l+1}{l-k}u\right]^{m-k+1}}\dfrac{m-l+1}{l-k},\ u \in (0,\,\infty). \tag{7.89}$$

This ends the proof.

7.4.1 Example of constructing within-sample $(1-\alpha)$-prediction limits

Let $Z_1 \leqslant \cdots \leqslant Z_k$ be the first k ordered observations from the sample of size m from the one-parameter exponential distribution defined by the probability density function (7.1). Then the exact lower one-sided $(1-\alpha)$-prediction limit $z_l(L_{1-\alpha,\,l})$ (for the lth order statistic Z_l $(1 \leqslant k < l \leqslant m)$ of the above sample) given $Z_k=z_k$ has to satisfy

$$P_\lambda(Z_l > z_l | Z_k = z_k; m) = \int_{z_l}^{\infty} g_\lambda(z|z_k; m)dz = \overline{G}_\lambda(z_l|z_k; m) = 1 - \alpha, \tag{7.90}$$

and the exact upper $(1-\alpha)$-prediction limit $z_l(=U_{1-\alpha;l})$ for Z_l has to satisfy

$$P_\lambda(Z_l \leqslant z_l | Z_k = z_k; m) = \int_{z_k}^{z_l} g_\lambda(z|z_k; m)dz = G_\lambda(z_l|z_k; m) = 1 - \alpha. \tag{7.91}$$

It follows from (7.15) and (7.16) that

$$L_{1-\alpha;l} = z_l = z_k + w_{lk}s_k, \tag{7.92}$$

where

$$w_{lk} = \operatorname*{argmin}_{w_{lk}}$$

$$\left(\dfrac{1}{B(l-k,\,m-l+1)} \times \sum_{j=0}^{l-k-1}\binom{l-k-1}{j}(-1)^j\dfrac{[1+(m-l+1+j)w_{lk}]^{-k}}{m-l+1+j} - (1-\alpha)\right)^2. \tag{7.93}$$

In a similar way it can be shown that

$$U_{1-\alpha;l} = Z_l = Z_k + w_{lk}S_k, \tag{7.94}$$

where

$$w_{lk} = \underset{w_{lk}}{\text{argmin}} \left(\frac{1}{B(l-k, m-l+1)} \right.$$
$$\left. \sum_{j=0}^{l-k-1} \binom{l-k-1}{j} (-1)^j \frac{[1+(m-l+1+j)w_{lk}]^{-k}}{m-l+1+j} - \alpha \right)^2 \qquad (7.95)$$

If, say, $\alpha = 0.05$, $k = 4$, $m = 10$, $l = m$,

$$S_k = \sum_{i=1}^{k} Z_i + (m-k)Z_k = 33 + 87 + 125 + 165 + 6 \times 165 = 1400, \qquad (7.96)$$

it follows from (7.92) that the exact lower $(1 - \alpha)$-prediction limit $z_l(=L_{1-\alpha;l})$ for Z_l is given by

$$L_{1-\alpha;l} = y_l = y_k + w_{lk}s_k = 165 + 0.192433 \times 1400 = 434.4062, \qquad (7.97)$$

and it follows from (7.94) that the exact upper $(1 - \alpha)$-prediction limit $z_l(=U_{1-\alpha;l})$ for Y_l is given by

$$U_{1-\alpha;l} = z_l = z_k + w_{lk}s_k = 165 + 2.098182 \times 1400 = 3102.455. \qquad (7.98)$$

7.5 Optimization of anticipated inspection process

This section discusses the optimization of the intended control process for elements or structures that fall into one of two categories, with one state being preferred over the other. It is assumed that we do not know when the transition from state $S(u_0)$ (healthy at time $u_0 = 0$) to state $S(u_1)$ (some kind of 'damage' at a later time, u_1) will happen, and the only way to see the 'damage' is to check it.

One-parameter exponential distribution. In this case, the inspection strategy can be defined as follows. Let

$$u_1 = \text{arg}(\Pr\{Y > u_1\} = \Delta), \qquad (7.99)$$

where $\Delta \in (0,1)$ and

$$u_j = \text{arg}(\Pr\{Y > u_j | Y > u_{j-1}\} = \Delta), j \geqslant 2, \qquad (7.100)$$

where Y is a random variable that represents the component's (structure) lifetime and $\{u_j\}_{j=1,2,...}$ are inspection times of the same. Reliability-based inspection is the term used to describe this. When the structure was still in use at the time of the last inspection, the value of Δ can be thought of as the 'minimum reliability necessary' (also known as the 'reliability index') for the time duration, when the parameter remained (during last inspection) operational. If the structure lifetime distribution F_λ is exponential (7.1) and λ is unknown, the inspection technique is defined as follows:

$$u_j = \text{arg}(\overline{F}_{s_k}(u_j) = \Delta^j) = s_k(\Delta^{-j/k} - 1), \quad j \geq 1, \qquad (7.101)$$

where

$$\overline{F}_{s_k}(u_j) = 1 - F_{s_k}(u_j) = \left(1 + \frac{u_j}{s_k}\right)^{-k}. \tag{7.102}$$

The total anticipated cost per inspection cycle is given by equation (7.103), if it is known that each inspection costs q_1 and that the cost of leaving an undetected failure (some form of 'damage') costs q_2 per unit time.

$$Q(u(s_k, \Delta)) = \sum_{j=1}^{\infty} \int_{u_{j-1}}^{u_j} [jq_1 + q_2(u_j - y)] f_{s_k}(y) dy$$

$$= q_1 \sum_{j=1}^{\infty} j \, [F_{s_k}(u_j) - F_{s_k}(u_{j-1})] + q_2$$

$$\sum_{j=1}^{\infty} u_j [F_{s_k}(u_j) - F_{s_k}(u_{j-1})] - q_2 \int_0^{\infty} y f_{s_k}(y) dy$$

$$= q_1 \sum_{j=0}^{\infty} \overline{F}_{s_k}(u_j) + q_2 \sum_{j=1}^{\infty} u_j [\overline{F}_{s_k}(u_{j-1}) - \overline{F}_{s_k}(u_j)] - q_2 E_{s_k}\{Y\}$$

$$= \frac{q_1}{1 - \Delta} + q_2 \sum_{j=1}^{\infty} s_k(\Delta^{-j/k} - 1)[\Delta^{j-1} - \Delta^j] - q_2 E_{s_k}\{Y\}$$

$$= \frac{q_1}{1 - \Delta} + q_2 s_k \sum_{j=1}^{\infty} (\Delta^{-j/k}[\Delta^{j-1} - \Delta^j] - [\Delta^{j-1} - \Delta^j]) - q_2 E_{s_k}\{Y\}$$

$$= \frac{q_1}{1 - \Delta} + q_2 s_k \left(\sum_{j=1}^{\infty} \Delta^{-j/k}[\Delta^{j-1} - \Delta^j] - \sum_{j=1}^{\infty} [\Delta^{j-1} - \Delta^j]\right) - q_2 \int_0^{\infty} \overline{F}_{s_k}(y) dy \tag{7.103}$$

$$= \frac{q_1}{1 - \Delta} + q_2 s_k \left(\sum_{j=1}^{\infty} \Delta^{-j/k} \Delta^j [1/\Delta - 1] - \sum_{j=1}^{\infty} \Delta^{j-1}[1 - \Delta]\right)$$

$$- q_2 \int_0^{\infty} (1 + \frac{y}{s_k})^{-k} dy$$

$$= \frac{q_1}{1 - \Delta} + q_2 s_k \left(\frac{1 - \Delta}{\Delta} \sum_{j=1}^{\infty} [\Delta^{1-1/k}]^j - \sum_{j=1}^{\infty} \Delta^{j-1}[1 - \Delta]\right)$$

$$- q_2 s_k \int_0^{\infty} (1 + \frac{y}{s_k})^{-k} d(\frac{y}{s_k})$$

$$= \frac{q_1}{1 - \Delta} + q_2 s_k \left(\frac{1 - \Delta}{\Delta} \frac{\Delta^{1-1/k}}{1 - \Delta^{1-1/k}} - 1\right) - \frac{q_2 s_k}{k - 1}$$

$$= \frac{q_1}{1 - \Delta} + q_2 s_k [(\frac{1 - \Delta^{1/k}}{\Delta^{1/k} - \Delta}) - \frac{1}{k - 1}]$$

$$= q_1 [\frac{1}{1 - \Delta} + s_k \frac{q_2}{q_1} (\frac{1 - \Delta^{1/k}}{\Delta^{1/k} - \Delta} - \frac{1}{k - 1})].$$

If, say, $k=3$, $q_1=4$, $q_2=7$, $s_k=100$, it follows from (7.103) that the reliability index Δ minimizing the total estimated cost per inspection cycle is given by

$$\Delta^* = \arg\min_{0<\Delta<1} Q(u(s_k, \Delta)) = \arg\min_{0<\Delta<1} q_1$$
$$\left[\frac{1}{1-\Delta} + s_k\frac{q_2}{q_1}\left(\frac{1-\Delta^{1/k}}{\Delta^{1/k}-\Delta} - \frac{1}{k-1}\right)\right] = 0.863\ 946, \tag{7.104}$$

where

$$Q(u(s_k, \Delta^*)) = 55.84008. \tag{7.105}$$

The index of improvement percentage in effectiveness of the optimal inspection strategy (with$\Delta = \Delta^* = 0.863946$) as compared with the standard inspection strategy (with $\Delta = \Delta_{st} = 0.95$) is given by

$$I_{imp.per}(\Delta^*, \Delta_{st}) = \frac{Q(u(s_k, \Delta_{st})) - Q(u(s_k, \Delta^*))}{Q(u(s_k, \Delta_{st}))}100\%$$
$$= \frac{89.0793 - 55.840\ 08}{89.0793}100\% = 37.3\% \tag{7.106}$$

Two-parameter exponential distribution. If the two-parameter exponential distribution (7.31) is used, the total expected cost per inspection cycle is given by

$$Q(u(s_1, s_r, \Delta)) = \sum_{j=1}^{\infty} \int_{u_{j-1}}^{u_j} [jq_1 + q_2(u_j - y)] f_S(y) dy$$

$$= q_1 \sum_{j=1}^{\infty} j \ [F_S(u_j) - F_S(u_{j-1})] + q_2 \sum_{j=1}^{\infty} u_j [F_S(u_j) - F_S(u_{j-1})] - q_2 \int_0^{\infty} y f_S(y) dy$$

$$= q_1 \sum_{j=0}^{\infty} \overline{F}_S(u_j) + q_2 \sum_{j=1}^{\infty} u_j [\overline{F}_S(u_{j-1}) - \overline{F}_S(u_j)] - q_2 E_S\{Y\}$$

$$= \frac{q_1}{1 - \Delta} + q_2 \sum_{j=1}^{\infty} [s_1 + s_r(\Delta^{-j/(r-1)} - 1)][\Delta^{j-1} - \Delta^j] - q_2 E_S\{Y\}$$

$$= \frac{q_1}{1 - \Delta} + q_2 \sum_{j=1}^{\infty} [s_1[\Delta^{j-1} - \Delta^j] + s_r(\Delta^{-j/(r-1)}[\Delta^{j-1} - \Delta^j] - [\Delta^{j-1} - \Delta^j])] - q_2 E_S\{Y\}$$

$$= \frac{q_1}{1 - \Delta} + q_2[s_1 \sum_{j=1}^{\infty} [\Delta^{j-1} - \Delta^j] + s_r(\sum_{j=1}^{\infty} \Delta^{-j/(r-1)}[\Delta^{j-1} - \Delta^j] - \sum_{j=1}^{\infty}[\Delta^{j-1} - \Delta^j])]$$

$$- q_2 \int_{s_1}^{\infty} \overline{F}_S(y) dy$$

$$= \frac{q_1}{1 - \Delta} + q_2[s_1 \sum_{j=1}^{\infty} \Delta^{j-1}[1 - \Delta] + s_r(\sum_{j=1}^{\infty} \Delta^{-j/(r-1)}\Delta^j[1/\Delta - 1] - \sum_{j=1}^{\infty}\Delta^{j-1}[1 - \Delta])] \qquad (7.107)$$

$$- q_2 \int_{s_1}^{\infty} (1 + \frac{y - s_1}{s_r})^{-(r-1)} dy$$

$$= \frac{q_1}{1 - \Delta} + q_2[s_1 \frac{1 - \Delta}{1 - \Delta} + s_r(\frac{1 - \Delta}{\Delta} \sum_{j=1}^{\infty} [\Delta^{1-1/(r-1)}]^j - \frac{1 - \Delta}{1 - \Delta})]$$

$$= -q_2 s_r \int_{s_1}^{\infty} (1 + \frac{y - s_1}{s_r})^{-(r-1)} d(1 + \frac{y - s_1}{s_r})$$

$$= -q_2 s_r \int_{s_1}^{\infty} (1 + \frac{y - s_1}{s_r})^{-(r-1)} d(1 + \frac{y - s_1}{s_r})$$

$$= \frac{q_1}{1 - \Delta} + q_2[s_1 + s_r(\frac{1 - \Delta}{\Delta} \frac{\Delta^{1-1/(r-1)}}{1 - \Delta^{1-1/(r-1)}} - 1)] - \frac{q_2 s_r}{r - 2}$$

$$= \frac{q_1}{1 - \Delta} + q_2(s_1 + s_r[(\frac{1 - \Delta^{1/(r-1)}}{\Delta^{1/(r-1)} - \Delta}) - \frac{1}{r - 2}])$$

$$= q_1[\frac{1}{1 - \Delta} + \frac{q_2}{q_1}(s_1 + s_r(\frac{1 - \Delta^{1/(r-1)}}{\Delta^{1/(r-1)} - \Delta} - \frac{1}{r - 2}))].$$

where

$$u_j = \arg(\overline{F}_S(u_j) \ = \Delta^j) = s_1 + s_k(\Delta^{-j/(r-1)} - 1), \quad j \geq 1, \qquad (7.108)$$

$$\overline{F}_S(u_j) = 1 - F_S(u_j) = \left(1 + \frac{u_j - s_1}{s_r}\right)^{-(r-1)}. \qquad (7.109)$$

If, say, $r = 4$, $q_1 = 4$, $q_2 = 7$, $s_1 = 10$, $s_r = 100$, it follows from (7.107) that the reliability index Δ minimizing the total estimated cost per inspection cycle is given by

$$\Delta^* = \arg\min_{0<\Delta<1} \ Q(u(s_1, s_r, \Delta))$$

$$= \arg\min_{0<\Delta<1} \ q_1\left[\frac{1}{1-\Delta} + \frac{q_2}{q_1}\left(s_1 + s_r\left(\frac{1 - \Delta^{1/(r-1)}}{\Delta^{1/(r-1)} - \Delta} - \frac{1}{r-2}\right)\right)\right] \quad (7.110)$$

$$= 0.863946,$$

where

$$Q(u(s_1, s_r, \Delta^*)) = 125.8401. \quad (7.111)$$

The index of improvement percentage in effectiveness of the optimal inspection strategy (with $\Delta = \Delta^* = 0.863946$) as compared with the standard inspection strategy (with $\Delta = \Delta_{st} = 0.95$) is given by

$$I_{\text{imp.per}}(\Delta^*, \Delta_{st}) = \frac{Q(u(s_1, s_r, \Delta_{st})) - Q(u(s_1, s_r, \Delta^*))}{Q(u(s_1, s_r, \Delta_{st}))}100\%$$
$$= [(159.0793 - 125.8401)/159.0793]100\% = 20.9\%. \quad (7.112)$$

7.6 Optimization of single-period decision-making models

The Newsboy Problem, in which a newsboy chooses the best order quantity in order to maximize predicted profit, is how the single-period decision-making models are expressed. Demand for newspapers is unpredictable, so the newsboy who sells them must choose how many to get from his source. If he buys too many newspapers, he will end up with unsold copies that are worthless; if he buys too few, he will miss the chance to sell more copies at a bigger profit. The newsboy problem is a popular one-item, one-period inventory problem that asks for the following information:

Y: quantity demanded (random variable),

$f_\lambda(y)$: the exponential probability density function of Y (when the parameter λ is known),

$f_{s_k}(y)$: the statistical pivot-based estimate of $f_\lambda(y)$ (when the parameter λ is not known),

u: the order quantity (decision variable) to satisfy the demand Y,

$Q(Y|u)$: the profit (random variable) which depends on the demand Y and the order quantity u,

c_u: unit cost price,

c_y: unit selling price ($c_y > c_u$),

c_u^-: unit salvage cost ($c_u > c_u^-$),

c_y^-: unit shortage penalty.

When a quantity u is ordered and Y is the demand (random variable), the profit $Q(Y|u)$ (random variable) is determined by

$$Q(Y|u) = \begin{cases} c_y Y + c_u^-(u - Y) - c_u u, & Y \leqslant u \\ c_y u - c_y^-(Y - u) - c_u u, & Y \geqslant u \end{cases}$$

$$= \begin{cases} (c_y - c_u^-)Y - (c_u - c_u^-)u, & Y \leqslant u \\ (c_y + c_y^- - c_u)u - c_y^- Y, & Y \geqslant u, \end{cases} \tag{7.113}$$

where $Q(Y|u)$ increases for $Y < u$ and decreases for $Y > u$. To demonstrate how to solve a typical newsboy problem with continuous demand (random variable), we formulate a problem as follows.

Maximize the expected profit:

$$E_\lambda\{Q(Y|u)\} = \int_0^u \left[(c_y - c_u^-)y - (c_u - c_u^-)u\right]f_\lambda(y)d$$

$$y + \int_u^\infty \left[(c_y + c_y^- - c_u)u - c_y^- y\right]f_\lambda(y)dy$$

$$= (c_y - c_u^-)\int_0^u yf_\lambda(y)dy - (c_u - c_u^-)u\int_0^u f_\lambda(y)dy + (c_y + c_y^- - c_u)u$$

$$\int_u^\infty f_\lambda(y)dy - c_y^- \int_u^\infty yf_\lambda(y)dy$$

$$= (c_y - c_u^-)\int_0^u yf_\lambda(y)dy + c_y^- \int_0^u yf_\lambda(y)dy - c_y^- \int_0^u yf_\lambda(y)dy - c_y^-$$

$$\int_u^\infty yf_\lambda(y)dy$$

$$- (c_u - c_u^-)u\int_0^u f_\lambda(y)dy - (c_u - c_u^-)u\int_u^\infty f_\lambda(y)dy + (c_u - c_u^-)u$$

$$\int_u^\infty f_\lambda(y)dy + (c_y + c_y^- - c_u)u\int_u^\infty f_\lambda(y)dy \tag{7.114}$$

$$= (c_y + c_y^- - c_u^-)\int_0^u yf_\lambda(y)dy - c_y^- E_\lambda\{Y\} - (c_u - c_u^-)u$$

$$+ (c_y + c_y^- - c_u^-)u\int_u^\infty f_\lambda(y)dy$$

$$= (c_y + c_y^- - c_u^-)\int_0^u yf_\lambda(y)dy - c_y^- E_\lambda\{Y\} - (c_u - c_u^-)u$$

$$+ (c_y + c_y^- - c_u^-)u\int_u^\infty f_\lambda(y)dy$$

$$= (c_y + c_y^- - c_u^-)\left[-u\overline{F}_\lambda(u) + \int_0^u \overline{F}_\lambda(y)dy\right] - c_y^- E_\lambda\{Y\}$$

$$- (c_u - c_u^-)u + (c_y + c_y^- - c_u^-)u\overline{F}_\lambda(u)$$

$$= (c_y + c_y^- - c_u^-)\int_0^u \overline{F}_\lambda(y)dy - c_y^- E_\lambda\{Y\} - (c_u - c_u^-)u.$$

subject to

$$u > 0. \tag{7.115}$$

Solution:

$$\frac{dE_\lambda\{Q(Y|u)\}}{du} = (c_y + c_y^- - c_u^-)\overline{F}_\lambda(u) - (c_u - c_u^-) = 0. \tag{7.116}$$

It follows from (7.116) that

$$\overline{F}_\lambda(u) = \exp(-u/\lambda) = \frac{c_u - c_u^-}{c_y + c_y^- - c_u^-}. \tag{7.117}$$

It follows from (7.117) that the optimal order quantity is given by

$$u^* = \lambda \ln \frac{c_y + c_y^- - c_u^-}{c_u - c_u^-}. \tag{7.118}$$

It follows from (7.114) and (7.118) that the maximum expected profit, which can be realized, is given by

$$E_\lambda\{Q(Y|u^*)\} = (c_y + c_y^- - c_u^-)\int_0^{u^*} \overline{F}_\lambda(y)dy - c_y^- E_\lambda\{Y\} - (c_u - c_u^-)u^*. \tag{7.119}$$

If the parameter λ is unknown, then, using (7.114) and (7.21), we obtain

$$\frac{dE_{s_k}\{Q(Y|u)\}}{du} = (c_y + c_y^- - c_u^-)\overline{F}_{s_k}(u) - (c_u - c_u^-) = 0. \tag{7.120}$$

It follows from (7.120) and (7.21) that

$$\overline{F}_{s_k}(u) = \left(1 + \frac{u}{s_k}\right)^{-k} = \frac{c_u - c_u^-}{c_y + c_y^- - c_u^-}. \tag{7.121}$$

It follows from (7.121) that the optimal order quantity (based on S_k (7.4)) under parametric uncertainty is given by

$$u^\bullet = s_k\left[\left(\frac{c_y + c_y^- - c_u^-}{c_u - c_u^-}\right)^{1/k} - 1\right]. \tag{7.122}$$

It follows from (7.114) and (7.122) that the maximum expected profit (based on S_k (7.4)) under parametric uncertainty, which can be realized, is given by

$$E_{s_k}\{Q(Y|u^\bullet)\} = (c_y + c_y^- - c_u^-)\int_0^{u^\bullet} \overline{F}_{s_k}(y)dy - c_y^- E_{s_k}\{Y\} - (c_u - c_u^-)u^\bullet. \tag{7.123}$$

Generalized single-period decision-making model. The model described above can be generalized as follows. Consider n different items (say, newspapers) that are bought and sold. Here we have to determine the optimal order quantity for each of n different items in order to maximize the expected total profit. It is assumed that the volume level of the total order quantity is limited by the available monetary resource C. To demonstrate how to solve a generalized newsboy problem with n continuous demands (random variables), we formulate the problem as follows.

Maximize the expected total profit:

$$\sum_{i=1}^{n} E_{\lambda_i}\{Q(Y_i|u_i)\} = \sum_{i=1}^{n}\left((c_{y_i} + c_{y_i}^- - c_{u_i}^-)\int_0^{u_i} \overline{F}_{\lambda_i}(y_i)dy_i - c_{y_i}^- E_{\lambda_i}\{Y_i\} \right.$$
$$\left. - (c_{u_i} - c_{u_i}^-)u_i \right) \tag{7.124}$$

subject to

$$\sum_{i=1}^{n} c_{u_i} u_i = C, \; u_i \geqslant 0, \; i = 1,\dots, n. \tag{7.125}$$

Solution: Using the Lagrange multiplier method, the Lagrange function is determined as

$$L(u_1, \dots, u_n, \lambda_\bullet) = \sum_{i=1}^{n} E_{\lambda_i}\{Q(Y_i|u_i)\} - \lambda_\bullet\left(\sum_{i=1}^{n} c_{u_i} u_i - C \right), \tag{7.126}$$

where a new variable $\lambda_\bullet$ called a Lagrange multiplier is introduced. Then

$$\frac{\partial L}{\partial u_i} = (c_{y_i} + c_{y_i}^- - c_{u_i}^-)\overline{F}_{\lambda_i}(u_i) - (c_{u_i} - c_{u_i}^-) - \lambda_\bullet c_{u_i}$$
$$= (c_{y_i} + c_{y_i}^- - c_{u_i}^-)\overline{F}_{\lambda_i}(u_i) - \left[(\lambda_\bullet + 1)c_{u_i} - c_{u_i}^- \right] = 0, \; i = 1,\dots, n, \tag{7.127}$$

$$\frac{\partial L}{\partial \lambda_\bullet} = \sum_{i=1}^{n} c_{u_i} u_i - C = 0. \tag{7.128}$$

It follows from (7.127) that

$$u_i = \beta_i \ln \frac{c_{y_i} + c_{y_i}^- - c_{u_i}^-}{(\lambda + 1)c_{u_i} - c_{u_i}^-}, \; i = 1,\dots, n. \tag{7.129}$$

Substituting (7.129) into (7.128), we have that

$$\lambda_\bullet^* = \underset{\lambda_\bullet \in (-\infty,\infty)}{\arg}\left(\sum_{i=1}^{n} c_{u_i}\lambda_i \ln \frac{c_{y_i} + c_{y_i}^- - c_{u_i}^-}{(\lambda_\bullet + 1)c_{u_i} - c_{u_i}^-} = C \right)$$

or

$$\lambda^* = \underset{\lambda_\bullet \in (-\infty,\infty)}{\arg} \min \left(\sum_{i=1}^{n} c_{u_i}\lambda_i \ln \frac{c_{y_i} + c_{y_i}^- - c_{u_i}^-}{(\lambda_\bullet + 1)c_{u_i} - c_{u_i}^-} - C \right)^2. \tag{7.130}$$

Substituting (7.130) into (7.129), we determine the optimal order quantity for each from n different items to maximize the expected total profit,

$$u_i^* = \lambda_i \ln \frac{c_{y_i} + c_{y_i}^- - c_{u_i}^-}{(\lambda_\bullet^* + 1)c_{u_i} - c_{u_i}^-}, \quad i = 1,\dots,n.$$

(7.131)

It follows from (7.124) and (7.131) that the maximum expected total profit, which can be realized, is given by

$$\sum_{i=1}^{n} E_{\lambda_i}\{Q(Y_i|u_i^*)\} = \sum_{i=1}^{n}\Bigg((c_{y_i} + c_{y_i}^- - c_{u_i}^-) \int_{0}^{u_i^*} \overline{F}_{\beta_i}(y_i)dy_i - c_{y_i}^- E_{\lambda_i}\{Y_i\}$$

$$- (c_{u_i} - c_{u_i}^-)u_i^* \Bigg).$$

(7.132)

Similarly, it can be considered the *Generalized Single-Period Decision-Making Model* under parametric uncertainty. In this case we have the following:

$$u_i = s_{k_i}\left[\left(\frac{c_{y_i} + c_{y_i}^- - c_{u_i}^-}{(\lambda_\bullet + 1)c_{u_i} - c_{u_i}^-}\right)^{1/k_i} - 1\right].$$

(7.133)

Substituting (7.133) into (7.130), we have

$$\lambda_\bullet^* = \underset{\lambda_\bullet \in (-\infty,\infty)}{\arg}\left(\sum_{i=1}^{n} c_{u_i}s_{k_i}\left[\left(\frac{c_{y_i} + c_{y_i}^- - c_{u_i}^-}{(\lambda_\bullet + 1)c_{u_i} - c_{u_i}^-}\right)^{1/k_i} - 1\right] = C\right)$$

or

$$\lambda_\bullet^* = \underset{\lambda_\bullet \in (-\infty,\infty)}{\arg\ \min}\left(\sum_{i=1}^{n} c_{u_i}s_{k_i}\left[\left(\frac{c_{y_i} + c_{y_i}^- - c_{u_i}^-}{(\lambda_\bullet + 1)c_{u_i} - c_{u_i}^-}\right)^{1/k_i} - 1\right] - C\right)^2.$$

(7.134)

Substituting (7.134) into (7.133), we determine the optimal order quantity for each from n different items to maximize the expected total profit under parametric uncertainty,

$$u_i^\bullet = s_{k_i}\left[\left(\frac{c_{y_i} + c_{y_i}^- - c_{u_i}^-}{(\lambda_\bullet + 1)c_{u_i} - c_{u_i}^-}\right)^{1/k_i} - 1\right], \quad i = 1,\dots,n.$$

(7.135)

It follows from (7.124) and (7.135) that the maximum expected total profit, which can be realized under parametric uncertainty, is given by

$$\sum_{i=1}^{n} E_{s_{k_i}}\{Q(Y_i|u_i^\bullet)\} = \sum_{i=1}^{n}\Bigg((c_{y_i} + c_{y_i}^- - c_{u_i}^-) \int_{0}^{u_i^*} \overline{F}_{s_{k_i}}(y_i)dy_i - c_{y_i}^- E_{s_{k_i}}\{Y_i\}$$

$$- (c_{u_i} - c_{u_i}^-)u_i^\bullet \Bigg).$$

(7.136)

7.7 Advanced techniques of signal processing in terms of hypotheses testing and misclassification probability

The foundation of the discipline of signal processing is a rigorous mathematical representation, which is crucial for a thorough comprehension of the material. The use of these mathematical ideas and techniques is made possible by the ongoing development of numerous technologically sophisticated fields that call for signal processing to extract crucial information. To extract specific information that is difficult to access in the time domain for practical applications, the signal might be processed in a variety of ways. Numerous applications, including as analysis, synthesis, filtering, and characterization or modeling, modulation, detection, estimation, classification, suppression, cancellation, equalization, coding, and synchronization all depend on the processing of such signals. If the result of testing the H_0 and H_1 hypotheses is the acceptance of the H_0 (or H_1) hypothesis, this means that the anticipated info signal, which is necessary to make an adequate decision, has been detected [1–3].

7.7.1 Optimizing the product acceptance process in terms of misclassification probability

Assume that many products are sent in for inspection and that each product's lifespan follow an exponential distribution with the probability density function (7.1). In this instance, we take into consideration a few scenarios of optimizing the design parameters of the item acceptance process, like the sample size m to check the mean time to failure (MTTF) λ and the separation threshold h to meet the risks to both the manufacturer and the consumer. The producer's risk, shown by α_0, is the likelihood of rejecting a good lot. The consumer's risk, shown by α_1, is the likelihood of accepting a subpar lot. We must use the test method that meets the following requirements in order to decide whether or not to approve a product lot:

$$\Pr(\text{reject a product lot}|\text{mean time to failure} = \lambda_0) \leqslant \alpha_0 (\text{Manufacturer's risk}) \quad (7.137)$$

$$\Pr(\text{accept a product lot}|\text{mean time to failure} = \lambda_1) \leqslant \alpha_1 (\text{Consumer's risk}) \quad (7.138)$$

Let's assume a situation, a product lot misclassification probability (total) is not determined in advance, but the size m of a random sample $(Y_1, \ldots, Y_m)$ of product items to test the MTTF λ is fixed. Now, finding the separation threshold h that reduces the possibility of a product batch being misclassified is the difficult task (7.139) as shown below:

$$\alpha(h|m) = \alpha_0(h|m) + \alpha_1(h|m), \quad (7.139)$$

Example 7.1. Let m items be selected at random using the probability density function from a valid lifespan distribution (7.1). Then, based on the random sample

$(Y_1, \ldots, Y_m)$ of size m from (7.1), a maximum likelihood estimator (MLE) of λ, is provided by

$$\hat{\lambda} = \sum_{i=1}^{m} Y_i \Big/ m. \tag{7.140}$$

It can be shown that

$$\hat{\lambda} \sim g_\lambda(\hat{\lambda}|m) = \frac{1}{\Gamma(m)(\lambda/m)^m}(\hat{\lambda})^{m-1}\exp\left(-\frac{\hat{\lambda}}{\lambda/m}\right), \; \hat{\lambda} \geqslant 0. \tag{7.141}$$

It follows from (7.139), (7.140), and (7.141) that the product lot misclassification probability (see figure 7.1) is given by

$$\alpha(h|m) = \alpha_0(h|m) + \alpha_1(h|m) = \int_0^h g_{\lambda_0}(\hat{\lambda}|m)d\hat{\lambda} + \int_h^\infty g_\lambda(\hat{\lambda}|m)d\hat{\lambda}. \tag{7.142}$$

To minimize the product lot misclassification probability, the optimal separation threshold h is determined via (7.142) as follows:

$$h = \underset{h}{\mathrm{argmin}}\, \alpha(h|m) = \underset{h}{\mathrm{argmin}}\left(\int_0^h g_{\lambda_0}(\hat{\lambda}|m)d\hat{\lambda} + \int_h^\infty g_{\lambda_1}(\hat{\lambda}|m)d\hat{\lambda}\right)$$
$$= \underset{h}{\mathrm{argmin}}\left(g_{\lambda_0}(h|m) - g_{\lambda_1}(h|m)\right)^2. \tag{7.143}$$

It follows from the above (see figure 7.1) that classification of the product lot is carried out through $\hat{\lambda}$ and h as follows:

$$\hat{\lambda} \in \begin{cases} \{\hat{\lambda}: \hat{\lambda} \leqslant h\} \rightarrow \text{reject the product lot,} \\ \{\hat{\lambda}: \hat{\lambda} > h\} \rightarrow \text{accept the product lot.} \end{cases} \tag{7.144}$$

It should be noted that the following situations:

$$\underset{h}{\mathrm{minimize}}[c_0 \times \alpha_0(h|m) + c_1 \times \alpha_1(h|m)], \tag{7.145}$$

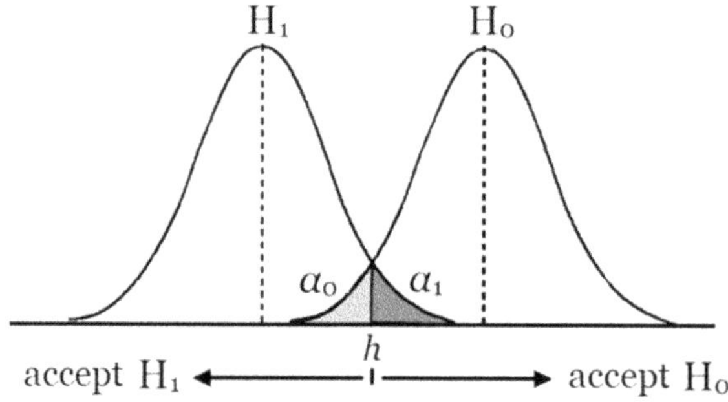

Figure 7.1. Product lot misclassification probability $\alpha(h|n)$.

$$\operatorname*{minimize}_{m,h}[cm + c_0 \times \alpha_0(h|m) + c_1 \times \alpha_1(h|m)], \tag{7.146}$$

$$\operatorname*{minimize}_{m,h}[\alpha_0(h|m) + \alpha_1(h|m) - \alpha^\bullet]^2, \tag{7.147}$$

$$\operatorname*{minimize}_{m,h}[(\alpha_0(h|m) - \alpha_0^\bullet)^2 + (\alpha_1(h|m) - \alpha_1^\bullet)^2], \tag{7.148}$$

where (c, c_0, c_1) represent the cost factors, $(\alpha^\bullet, \alpha_0^\bullet, \alpha_1^\bullet)$ represent the given values, can also be considered in a similar way.

7.7.2 Model of signal detection process in terms of hypotheses testing

This model is given in figure 7.2.

Consider the problem of testing the hypothesis H_0: $\mu = \mu_0$ against the alternative: H_1: $\mu > \mu_0$ in sampling $\mathbf{Y}=(Y_1, \ldots, Y_m)$ from $N(\mu, \sigma^2)$, where both parameters μ and σ^2 are unknown. The probability density function of $N(\mu, \sigma^2)$ is given by

$$f_{\mu,\sigma^2}(y) = \frac{1}{\sqrt{2\pi}\sigma} \exp\left(-\frac{(y-\mu)^2}{2\sigma^2}\right), \; |\mu| < \infty, \sigma > 0, \; |y| < \infty, \tag{7.149}$$

where

$$\frac{Y-\mu}{\sigma} = v \sim N(0,1) \tag{7.150}$$

is a pivotal quantity. A complete sufficient statistic for the parameter μ is given by

$$\bar{Y} = \sum_{i=1}^{m} Y_i \bigg/ m \sim \varphi_{\mu,\sigma^2}(\bar{y}) = \frac{\sqrt{m}}{\sqrt{2\pi}\sigma} \exp\left(-\frac{m(\bar{y}-\mu)^2}{2\sigma^2}\right), \; |\bar{y}| < \infty, \tag{7.151}$$

if μ is unknown, σ^2 is known. In this case,

$$\frac{\sqrt{m}(\bar{Y}-\mu)}{\sigma} = v_1 \sim N(0,1) \tag{7.152}$$

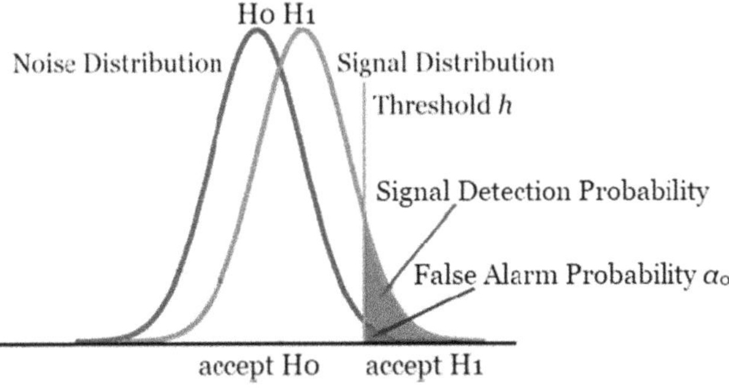

Figure 7.2. Model of signal detection.

is a pivotal quantity. A complete sufficient statistic for the parameter σ^2 is given by

$$S^2 = \frac{1}{m-1}\sum_{i=1}^{m}(Y_i - \bar{Y})^2 \sim \varphi_{\sigma^2}(s^2) = \frac{\left(\dfrac{m-1}{2\sigma^2}\right)^{(m-1)/2}}{\Gamma\left(\dfrac{m-1}{2}\right)}(s^2)^{\frac{m-1}{2}-1}\exp\left(-\frac{(m-1)s^2}{2\sigma^2}\right), \quad s^2 \geq \tag{7.153}$$

$$0,$$

if μ and σ^2 are unknown. In this case,

$$\frac{(m-1)S^2}{\sigma^2} = v_2 \sim \chi^2(n-1) \tag{7.154}$$

is a pivotal quantity. It will be noted that the statistics $\bar{Y}$ and S^2 are independent. Now the unknown parameter σ^2 can be eliminated from the problem as follows. From (7.151) and (7.154) it follows that

$$\begin{aligned}
\varphi_{\mu,\,\sigma^2}(\bar{y})d\bar{y} &= \frac{\sqrt{m}}{\sqrt{2\pi}\sigma}\exp\left(-\frac{m(\bar{y}-\mu)^2}{(m-1)s^2} \times \frac{(m-1)s^2}{2\sigma^2}\right)d\bar{y}\,\frac{s}{s} \\
&= \frac{s}{\sqrt{2\pi}\sigma}\exp\left(-\frac{m(\bar{y}-\mu)^2}{(m-1)s^2} \times \frac{(m-1)s^2}{2\sigma^2}\right)d\left(\frac{\sqrt{m}(\bar{y}-\mu)}{s}\right) \\
&= \frac{s}{\sqrt{2\pi}\sigma}\exp\left(-\frac{t^2}{m-1} \times \frac{(m-1)s^2}{2\sigma^2}\right)d(t),
\end{aligned} \tag{7.155}$$

where

$$T = \frac{\sqrt{m}(\bar{Y}-\mu)}{S}. \tag{7.156}$$

From (7.155) and (7.153) it follows that the probability density function of T is given by

$$\varphi_{m-1}(t) = \int_0^\infty \varphi_{\mu,\,\sigma^2}(\bar{y})\varphi_{\sigma^2}(s^2)ds^2$$

$$= \int_0^\infty \frac{s}{\sqrt{2\pi}\sigma}\exp\left(-\frac{t^2}{m-1} \times \frac{(m-1)s^2}{2\sigma^2}\right)\frac{\left(\dfrac{m-1}{2\sigma^2}\right)^{(m-1)/2}}{\Gamma\left(\dfrac{m-1}{2}\right)}(s^2)^{\frac{m-1}{2}-1}\exp\left(-\frac{(m-1)s^2}{2\sigma^2}\right)ds^2$$

$$= \int_0^\infty \frac{1}{\sqrt{(m-1)\pi}}\frac{\left(\dfrac{m-1}{2\sigma^2}\right)^{m/2}}{\Gamma\left(\dfrac{m-1}{2}\right)}(s^2)^{\frac{m}{2}-1}\exp\left(-\frac{(m-1)s^2}{2\sigma^2}\left[1+\frac{t^2}{m-1}\right]\right)ds^2$$

$$= \frac{\Gamma(m/2)}{\Gamma((m-1)/2)\sqrt{(m-1)\pi}}\left[1+\frac{t^2}{m-1}\right]^{-m/2}, \quad -\infty < t < \infty.$$

(7.157)

In this case,

$$T = \frac{\sqrt{m}(\bar{Y}-\mu)}{S} \sim t(m-1) \tag{7.158}$$

$$(t\text{–distribution with } m-1 \text{ degrees of freedom})$$

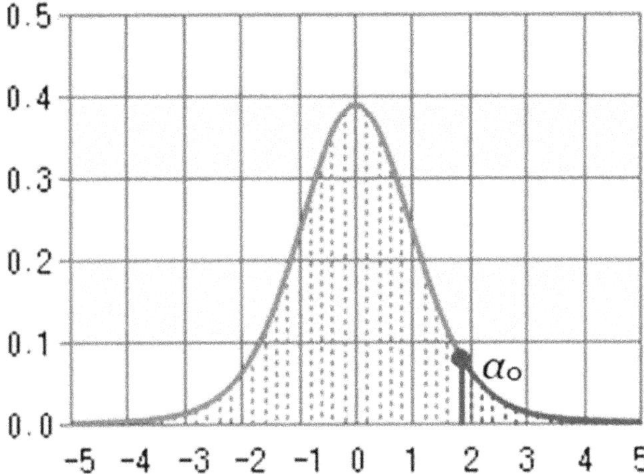

Figure 7.3. Student's t-distribution (percentile).

is a pivotal quantity. We choose the threshold h in accordance with the distribution of $T(\mathbf{Y})$ under H_0:

$$h = \arg\left(\int_h^\infty \varphi_{m-1}(t)dt = \alpha_0\right), \qquad (7.159)$$

where $\mu = \mu_0$, α_0 is the false alarm probability. It follows from the above (see figure 7.2) that detection of the expected signal is carried out through T and h as follows:

$$T \in \begin{cases} \{t: t \leqslant h\} \rightarrow \text{accept the hypothesis } H_0(\text{noise}), \\ \{t: t > h\} \rightarrow \text{accept the hypothesis } H_1(\text{signal}). \end{cases} \qquad (7.160)$$

If, say, the false alarm probability $\alpha_0=0.05$, T (random variable) is with $m-1=9$ degrees of freedom, then the threshold $h =1.833$ (see figure 7.3).

7.7.3 Statistical hypotheses testing whether two samples come from the same distribution under parametric uncertainty

Suppose that the first sample $\mathbf{X}=(X_1, \ldots, X_m)$ has size m with the cumulative distribution function

$$F_{\lambda_1}(x) = 1 - \exp(-x/\lambda_1), \qquad (7.161)$$

where the parameter λ_1 is unknown, and that the second sample $\mathbf{Y}=(Y_1, \ldots, Y_n)$ has size n with the cumulative distribution function

$$F_{\lambda_2}(y) = 1 - \exp(-y/\lambda_2), \qquad (7.162)$$

where the parameter λ_2 is unknown. Thus, the null hypothesis is H_0: both samples ($\mathbf{X}$ and $\mathbf{Y}$) come from a population with the same distribution (i.e., $\lambda_1=\lambda_2$). The alternative hypothesis is H_1: $\lambda_2 < \lambda_1$.

Consider the problem of testing the null hypothesis H_0: $\lambda_1=\lambda_2$ against the alternative hypothesis H_1: $\lambda_2 < \lambda_1$, where both parameters λ_1 and λ_2 are unknown. In this case, the unknown parameters (λ_1 and λ_2) can be eliminated from the problem as follows. It is known that

$$S_m = \sum_{j=1}^{m} X_j \sim f_{\lambda_1}(s_m) = \frac{s_m^{m-1}}{\Gamma(m)\lambda_1^m} \exp\left(-\frac{s_m}{\lambda_1}\right), \quad s_m \geqslant 0 \tag{7.163}$$

is the complete sufficient statistic for λ_1,

$$S_n = \sum_{j=1}^{n} X_j \sim f_{\lambda_2}(s_n) = \frac{s_n^{n-1}}{\Gamma(n)\lambda_2^n} \exp\left(-\frac{s_n}{\lambda_2}\right), \quad s_n \geqslant 0 \tag{7.164}$$

is the complete sufficient statistic for λ_2. It follows from (7.163) and (7.164) that

$$
\begin{aligned}
f_{\lambda_1}(s_m)ds_m &= \frac{s_m^{m-1}}{\Gamma(m)\lambda_1^m} \exp\left(-\frac{s_m}{\lambda_1}\right)ds_m = \frac{(s_m/\lambda_1)^{m-1}}{\Gamma(m)} \exp\left(-s_m/\lambda_1\right)d(s_m/\lambda_1) \\
&= \frac{(s_m/\lambda_1)^{m-1}/(s_n/\lambda_2)^{m-1}}{\Gamma(m)}\left(\frac{s_n}{\lambda_2}\right)^m \exp\left(-\left(\frac{s_m/\lambda_1}{s_n/\lambda_2}\right)\frac{s_n}{\lambda_2}\right)d\left(\frac{s_m/\lambda_1}{s_n/\lambda_2}\right) \\
&= \frac{[(m/n)f]^{m-1}}{\Gamma(m)}\left(\frac{s_n}{\lambda_2}\right)^m \exp\left(-\frac{m}{n}f\frac{s_n}{\lambda_2}\right)d\left(\frac{m}{n}f\right) \\
&= \frac{[(m/n)f]^{m-1}}{\Gamma(m)}\left(\frac{s_n}{\lambda_2}\right)^m \exp\left(-\frac{m}{n}f\frac{s_n}{\lambda_2}\right)\frac{m}{n}d(f),
\end{aligned}
\tag{7.165}
$$

where

$$\frac{S_m/\lambda_1}{S_n/\lambda_2} = \frac{S_m/S_n}{\lambda_1/\lambda_2} = \frac{m\hat{\lambda}_1/n\hat{\lambda}_2}{\lambda_1/\lambda_2} = \frac{m}{n}\frac{\hat{\lambda}_1/\hat{\lambda}_2}{\lambda_1/\lambda_2} = \frac{m}{n}F. \tag{7.166}$$

From (7.165) and (7.166) it follows that the probability density function of F is given by

$$
\begin{aligned}
g_{2m,\,2n}(f) &= \int_0^\infty \frac{[(m/n)f]^{m-1}}{\Gamma(m)}\left(\frac{s_n}{\lambda_2}\right)^m \exp\left(-\frac{m}{n}f\frac{s_n}{\lambda_2}\right)\frac{m}{n}f_{\lambda_2}(s_n)ds_n \\
&= \int_0^\infty \frac{[(m/n)f]^{m-1}}{\Gamma(m)}\left(\frac{s_n}{\lambda_2}\right)^m \exp\left(-\frac{m}{n}f\frac{s_n}{\lambda_2}\right)\frac{m}{n}\frac{s_n^{n-1}}{\Gamma(n)\lambda_2^n}\exp\left(-\frac{s_n}{\lambda_2}\right)ds_n \\
&= \int_0^\infty \frac{[(m/n)f]^{m-1}}{\Gamma(m)\Gamma(n)}\frac{m}{n}\left(\frac{s_n}{\lambda_2}\right)^{m+n-1}\exp\left(-\frac{s_n}{\lambda_2}\left(1+\frac{m}{n}f\right)\right)d\left(\frac{s_n}{\lambda_2}\right) \\
&= \frac{\Gamma(m+n)}{\Gamma(m)\Gamma(n)}\frac{m}{n}\left(\frac{m}{n}f\right)^{m-1}\left(1+\frac{m}{n}f\right)^{-(m+n)} \\
&= \frac{\Gamma\left(\dfrac{2m+2n}{2}\right)}{\Gamma\left(\dfrac{2m}{2}\right)\Gamma\left(\dfrac{2n}{2}\right)}\left(\frac{2m}{2n}\right)\left(\frac{2m}{2n}f\right)^{\frac{2m}{2}-1}\left(1+\frac{2m}{2n}f\right)^{-\frac{2m+2n}{2}}, \quad f > 0.
\end{aligned}
\tag{7.167}
$$

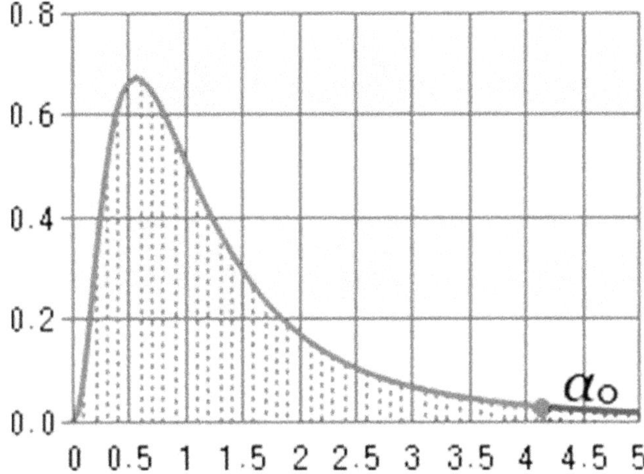

Figure 7.4. $F(2m, 2n)$-distribution (percentile).

In this case,

$$F = \frac{\hat{\lambda}_1/\hat{\lambda}_2}{\lambda_1/\lambda_2} = \frac{\hat{Q}}{Q} \sim F(2m, 2n) \quad (F\text{-distribution with } (2m, 2n) \tag{7.168}$$

degrees of freedom).

We choose the threshold h in accordance with the distribution of F under H_0:

$$h = \arg\left(\int_h^\infty g_{2m,\,2n}(f)df = \alpha_0\right), \tag{7.169}$$

where α_0 is the false alarm probability. It follows from the above (see figure 7.2) that detection of the expected signal is carried out through F and h as follows:

$$\hat{Q} \in \begin{cases} \{\hat{Q}: \hat{Q} \leq h\} & \to \quad \text{accept the hypothesis } H_0(\lambda_1 = \lambda_2), \\ \{\hat{Q}: \hat{Q} > h\} & \to \quad \text{accept the hypothesis } H_1(\lambda_2 < \lambda_1). \end{cases} \tag{7.170}$$

where

$$\hat{Q} = \hat{\lambda}_1/\hat{\lambda}_2. \tag{7.171}$$

If, say, the false alarm probability $\alpha_0 = 0.05$, F (random variable) is with ($2m=8$, $2n=6$) degrees of freedom, then the threshold h is equal to 4.146 804 (see figure 7.4).

7.7.4 Parametric estimation via shortest-length confidence intervals

The shortest-length confidence intervals for Q can be obtained as follows. The random variable (pivotal quantity) $F = \hat{Q}/Q$ has the probability density function (7.167). Using F, we can construct a $100(1-\alpha)\%$ confidence interval for Q from

$$P\left(q_1 \leqslant F \leqslant q_2\right) = P\left(q_1 \leqslant \frac{\hat{Q}}{Q} \leqslant q_2\right) = P\left(\frac{\hat{Q}}{q_2} \leqslant Q \leqslant \frac{\hat{Q}}{q_1}\right) \tag{7.172}$$

$$= \int_0^{q_2} g_{2m,\,2n}(f)\,df - \int_0^{q_1} g_{2m,\,2n}(f)\,df = 1 - \alpha + p - p = 1 - \alpha \tag{7.173}$$

by suitably choosing the decision variables q_1 and q_2. Hence, the confidence interval for Q is given by

$$\left[\frac{1}{q_2}, \frac{1}{q_1}\right]. \tag{7.174}$$

The length of the statistical confidence interval is given by

$$L\left(q_1,\, q_2 \big| \hat{Q}\right) = \hat{Q}\left(\frac{1}{q_1} - \frac{1}{q_2}\right) \propto L(q_1,\, q_2) = \left(\frac{1}{q_1} - \frac{1}{q_2}\right), \tag{7.175}$$

In order to find the shortest-length confidence interval $L(q_1,\, q_2)$ we should find a pair of decision variables q_1 and q_2 such that $L(q_1,\, q_2)$ is minimum.

The problem statement is expressed in terms of the choice variable p (probability) and the quantile functions q_1 and q_2 in the new, basic computational technique (version 1).

Minimize:

$$L^2(q_1,\, q_2) = \left(\frac{1}{q_1} - \frac{1}{q_2}\right)^2 = \left(\frac{1}{F_p^{-1}(2m,\, 2n)} - \frac{1}{F_{1-\alpha+p}^{-1}(2m,\, 2n)}\right)^2, \tag{7.176}$$

where the p–quantile function q_1 of $F(2m,2n)$ is given by

$$q_1 = F_p^{-1}(2m,\, 2n) \tag{7.177}$$

and the $(1 - \alpha + p)$-quantile function q_2 of $F(2m,2n)$ is given by

$$q_2 = F_{1-\alpha+p}^{-1}(2m,\, 2n), \tag{7.178}$$

subject to

$$0 \leqslant p \leqslant \alpha, \tag{7.179}$$

Numerical solutions. Using the computer program 'Solver,' the exact numerical solutions minimizing $L(q_1,q_2)$ can be found. The exact numerical solutions are given (if, for instance, $m=4$, $n=3$, and $\alpha= 0.05$) by

$$p = 0.049\ 497,\ q_1 = 0.278\ 166,\ q_2 = 24.278\ 13 \tag{7.180}$$

with the $100(1-\alpha)\%$ shortest-length confidence interval

$$L_{1-\alpha=0.95}(q_1,\, q_2) = 3.553\ 787. \tag{7.181}$$

If $\alpha = 0.95$, then the $100(1-\alpha)\%$ equal tails confidence interval is given by

$$L_{1-\alpha=0.05}(q_1, q_2) = 0.073\ 286. \tag{7.182}$$

with

$$p = 0.7445\ 65,\ q_1 = 1.751\ 146,\ q_2 = 2.008\ 966. \tag{7.183}$$

Inference. The proposed new simple computational method correctly recognized the adequate version of possible decision making and gave accurate numerical results:

$$Q \in \left[\frac{\hat{Q}}{q_2}, \frac{\hat{Q}}{q_1}\right]_{\alpha=0.95} = [0.497\ 768\hat{Q},\ 0.571\ 054\hat{Q}]$$

$$\in \left[\frac{\hat{Q}}{q_2}, \frac{\hat{Q}}{q_1}\right]_{\alpha=0.05} = [0.278\ 166\hat{Q},\ 24.278\ 13\hat{Q}]. \tag{7.184}$$

New simple computational technique (version 2): the problem statement in terms of the decision variables q_1, q_2, and p.

Minimize:

$$\left[F_{2m,\ 2n}(q_1) - p\right]^2 + \left[F_{2m,\ 2n}(q_2) - (1 - \alpha + p)\right]^2 \tag{7.185}$$

subject to

$$q_1,\ q_2 \geqslant 0. \tag{7.186}$$

In this case, if $\alpha = 0.05$ and $p = 0.049\ 497$, the optimal numerical solution of (7.185) is given by (7.180) and (7.181). If $\alpha = 0.95$ and $p = 0.744\ 565$, the optimal numerical solution of (7.143) is given by (7.182) and (7.183).

It is possible to build automated procedures for evaluating the quickest detection of a modification in quality control using the methodology previously suggested. To put it another way, this section deals with the 'quickest detection problem,' or the issue of choosing the best moment to issue warning signals when a fault occurs in a production system that is continually monitored. A model that is significantly different from those previously examined and is inspired by what is really done in practice can be developed to explain the random occurrence of errors in a production system.

7.8 Conclusion

The unique unified strategy of computational intelligence described in this study is the conceptually straightforward, effective, and practical approach for developing precise, optimal, or enhanced statistical decision rules in the presence of parametric model uncertainty. This method has potential uses in strategy, business, services, goods, procedures, and equipment, among other things. The method is based on the useful use of the invariance principle in statistical mathematics. We mostly used the exponential distribution to explain the method. Applications to other distributions at the log-location scale could come right after. The methods outlined in this chapter can be expanded in a number of ways to address various issues that may arise in Industry 4.0.

References

[1] Nechval N A 1988 A general method of constructing automated procedures for testing quickest detection of a change in quality control *Comput. Ind.* **10** 177–83

[2] Nechval N A 1997 Adaptive CFAR tests for detection of a signal in noise and deflection criterion ed T Wysocki, H Razavi and B Honary *Digital Signal Processing for Communication Systems. The Springer International Series in Engineering and Computer Science* 403 (Boston, MA: Springer), 177–86

[3] Nechval K N and Nechval N A 2014 Adaptive CFAR tests for detection and recognition of target signals in radar clutter *Adv. Syst. Sci. Appl.* **14** 129–43

[4] Nechval N A, Berzins G and Nechval K N *et al* 2021 Intelligent constructing efficient statistical decisions via pivot-based elimination of unknown (nuisance) parameters from underlying models *Autom. Control Comput. Sci.* **55** 469–89

Chapter 8

Role of artificial intelligence in industries for advanced applications

Resham Raj Shivwanshi and Neelamshobha Nirala

The rate of technological evolution is increasing rapidly, not just to solve the problem of humanity but also to explore what problems can occur to find the most significant issue that needs to be resolved first on a priority basis. An appropriate evolution requires a lot of data and computing power to make it possible to understand esoteric details quickly. Higher computing power is not enough to produce the most desired outcome. It requires a concise and advanced algorithm to generate satisfactory results. The development of artificial intelligence has been gaining additional weight over the last two decades because the availability of data and computing power is increasing rapidly. The field where the study of algorithms and information is performed to train the machine to get the response more likely to an intelligent human being is now a part of study for numerous science groups across the globe. It has changed everyone's life and consistently impacted almost every industry on a large scale. The development of industrial sectors such as IT, automobile, communication, energy, food, and healthcare is remarkable as these industries greatly grow in their domain. This chapter aims to highlight some of those important impressions of AI on industries responsible for bringing a significant shift in their timeline and consistently altering the future with unprecedented potential efforts.

8.1 Introduction

When steam engines were first discovered, they changed the mindsets of people towards transportation. Almost every industrial development has been significantly improved due to the impact of the newly developed transportation system, and this was the time when the first industrial revolution took place during the 18th century. Industrial development faced a second large-scale and rapid development in the late 19th and early 20th centuries with the advent of electricity.

Simultaneously, internal combustion engines, chemical industries, and various electrical communication technologies contributed significantly to the uplift of the second industrial revolution [1]. These two historical phases of time put enormous potential products and services in the market for human evolution that became a major cause of industrial impact on almost every person's life on the planet. The world felt the next widespread influence in the 20th century during the 70s when electronics and information technologies came into existence and showed huge acceptance among various industries [2]. This was the time when electronic chips, microcontrollers, and software industries were growing at a rapid pace. In the 21st century, newly developed algorithms, the Internet of Things (IoT), artificial intelligence, and data science are impacting almost every science and technology domain, leading the world towards the entirely new era of the fourth industrial revolution, also known as Industry 4.0. Recent advancements and evolution can be seen all over the place due to increasingly changing environments and human-made activities [3]. Rapidly fluctuating natural and human-originated movements often create disastrous situations at the cost of ever-increasing human greed and broader expectations from abundant resources. Today, technological advancement can address most issues related to natural and human-originated problems. Computers, for this purpose, are essential tools that efficiently execute specific programs and algorithms to analyze information about given situations properly. For the last two decades, the most demanding technology for information analysis has been artificial intelligence, a machine's ability to mimic human behavior. This revolutionary information processing technique is equipped with advanced algorithms that are usually efficient enough to perform specific tasks. AI algorithms are also gaining massive attention from researchers as they are good at extracting valuable insights from a large amount of data, such as statistics of population density, growth in GDP, weather forecasting, and studying critical global challenges, for which, in general, an individual human can't process and draw some valuable conclusions out of a large dataset [4]. These algorithms can now alter themselves to some extent to provide more efficient results by attaining improved performance to meet industrial and individual demands. Although improved techniques can come up with efficient solutions to various critical problems, their systematic exploration is necessary as it has huge potential to produce groundbreaking outcomes that may create devastating implications if guided in the wrong direction. The information and its meaningful utilization are increasing in the public domain, which makes it necessary to observe the actual impacts of the technology to lead the overall outcomes toward a better future. Artificial intelligence-based automation supports the rapid advancement of any known domain of human exploration, but it needs attentive improvements for balanced performance of various applications to get more sustainable outcomes. The role of AI for further development is very significant and requires consistent efforts of researchers, technologists, and developers to work collaboratively to implement reliable, high-performance algorithms [5]. Machine learning (ML) is the subfield of AI that deals with a small number of datasets to perform the outcome prediction-based task at a lower scale of computation cost. On the other hand, a deep learning algorithm consists of many neurons and neural networks connected for performing learning and information extraction tasks from the huge dataset to provide the most favorable outcomes in the form of desirable possibilities. These algorithms play an essential role in finding the solutions to various problems associated with the various domains, such as image recognition and analysis, audio signal analysis, temperature and weather analysis, astrological image and radio analysis, natural catastrophe analysis, the study of population and human evolution, the study of animals and plants, the study of ecological and cosmological analysis, and oil, natural gas, economical, sociological and public health analysis, clean water, farming, and soil-based analysis, and petroleum and energy-based analysis. Among these most prominent fields of AI interventions, healthcare, food, energy, clean water, recycling, and sensitization are the domains that need greater attention from the researchers to address the current social and economic issues [6]. Healthcare industries face enormous challenges nowadays due to increasing newly discovered diseases among humans, animals, and plants [7]. Also, diseases are spreading due to untimely diagnosis and critical behavior. Many patients suffer from different types of cancers that are initially hard to diagnose but predictive with their subtle initial signs. Detection and diagnosis through expert doctors are sometimes affected due to human error and excessive patient load [8]. In such cases, AI techniques may be a better option for improved detection and diagnosis.

8.2 Seven top technologies of AI that are responsible for profoundly influencing the fourth industrial revolution

Artificial intelligence (AI) has been studied for more than 60 years since it was introduced in computer science. With various exploration and innovations, the field of AI is improved a lot in the past sixty years, but it is still a fascinating topic for researchers today. Technology leading companies such as Alphabet, Microsoft, Facebook, and Apple are showing great interest in AI to meet the growing demands of consumers. Incorporating other technologies such as cloud computing, edge computing, sensor, or sensor networks, AIOT, deep learning, data mining, and data science, AI is becoming more reliable for decision-making tasks, leading the world toward a fast-growing industrial revolution that is named Industry 4.0. In the following subsections, some of the most prominent fields of AI are explained to comprehend recent technological trends, their pros and cons, and to analyze how to navigate these technologies for sustainable growth systematically.

8.2.1 Artificial intelligence of things (AIOT)

As mentioned, the AIOT is an enhanced version of the IoT that involves a number of sensor nodes connected and interacting to collect and process their surrounding information. According to the information collected at each node, various decisions can be made for advanced industrial or domestic operations [9]. IoT technology is a connected network of dedicated devices that can collect specific information to make optimized decisions at the level of local or globally distributed network ends. When AI combines with the IoT, it becomes possible to implement more efficient systems in terms of energy consumption, precision, reliability, robustness, and up-gradation shortly [10]. An intelligent predictive maintenance system (IPMS) is one of the applications of this technology which incorporates a mutually connected sensor network through a vast manufacturing plant to assess failure detections and corrections. The sensor nodes in this system take real-time data to process the information at a centralized computing system and allow the overall facility to give an alarm for any upcoming or existing failures. This system can also be implemented to correct the failure on its own by following specific rules and protocols the system administrator provides at the time of its deployment [11].

AIOT is now also popular in the agricultural domain, where sensor-based soil, crop fertility, and crop disease monitoring are currently being carried out for improved production with better quality. Cloud-based, mutually connected sensor nodes perform better in analyzing weather and pest conditions using mobile phones. Smartphone-based systems connected through a centralized computing facility to process the real-time IoT data are accessible for farmers and professionals to use. Incorporating advanced deep learning technology with AIoT makes a system that precisely helps to obtain valuable insights from the vast amount of data available. Obtained insights are then used to train the system for making useful decisions to govern error correction and quality management tasks related to precision agriculture (PA). As PA is responsible for monitoring crops and soil more precisely by observing inter- and intra-crop areas, it receives wide acceptance in the local and global crop research societies [12]. The basic principle behind PA is to measure the quality and performance of small sections of crops and farmland since there might be customized requirements for different areas regarding irrigation, fertilizers, and disease detection facilities. AIoT, in that case, shows its best useability and defines the best possible steps for better performance [13].

Figure 8.1 shows the basic block diagram of an AIoT system that comprises six basic blocks. The first few blocks represent sensors responsible for collecting environmental or surrounding data from various physical sources in the form of temperature, pressure, humidity, speed, light intensity, audio signals, radio signals, or different types of forces. These data are then converted from physical to electrical

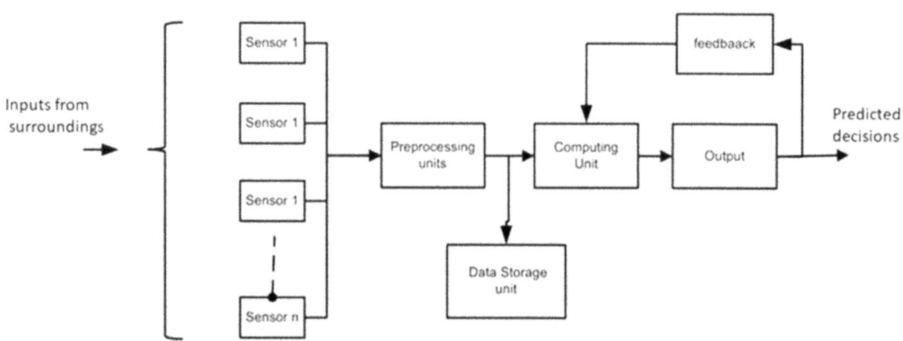

Figure 8.1. A block diagram of fundamental AIoT systems.

quantities, then analog to digital form with the help of preprocessing units. The preprocessing unit is also responsible for filtering out irrelevant noise in the incoming input signal and adding some amplification parameters if the strength is low. The input data can be stored in a data storage unit for offline and future analysis to improve the system's performance. Preprocessed information is then fed further into the computing unit. This unit holds the required functionality in algorithms, programs, and decision-making models to compute the desired outcome by analyzing input data. The output block is again rechecked for error and losses by retaking a small amount of information by computing units through a feedback loop. The computing device and feedback loop then work collaboratively to provide the most reliable and error-free output.

The AIoT system works very well for distributed sensor network-based decision-making systems. After coming up with sensor-based decisive outcomes, the system may connect through various switches and actuators to perform different automotive tasks. One of the best examples of this kind is an automated home system. With growing cellular and 5G technology, several sensors such as pressure, light, and audio are now connected with lightning speed, allowing the automation system to act quickly for any given task. A 5G technology-based advanced home automation system can turn on the lights, music, and temperature controlling system according to the mood of the person living inside the house. The AIoT devices are now making these types of systems in such a way that they may provide a better user experience with timely updates to newer functionality for any upcoming future up-gradation [14].

8.2.2 Additive manufacturing

Additive manufacturing is the new hope for design, manufacturing, and construction companies as the technology provides multiple benefits with the finest precision using advanced hardware, machinery, and software systems. By the name additive, it signifies design and development through addition, where the addition of raw material systematically produces a high-quality product. The additive manufacturing process is called 3D printing, starting from a three-dimensional computer-generated model. After completing the designing process of the required object in the computer system, the model file must be rendered to a 3D printer. This computer-generated model, also called the computer-aided design (CAD) model comprises of machine instructions that command the machine controller to develop a 3D physical model [15]. Its primary advantage is the minimum wastage of manufacturing during the process of product development [16]. Artificial intelligence-based devices, algorithms, and tools play a major role in governing the functionality of 3D printing machines. A sensor-based feedback system, along with a precisely designed computer program, decides how much material should be taken at a time to bind together to create different layers of the product. How much temperature and speed can rise and how to proceed further in case of system error, without impacting the production, can be managed by an intelligent controller or computing device fixed along with the 3D printing machine. As AI-based technological evolution is growing faster, new ways of designing and developing may be available in the near future. Different types of manufacturing materials and a mixture of materials are explored and continuously tested by various research groups, as their studies are now becoming more approachable due to the presence of AI devices. Deep learning technologies are also essential for detecting real-time manufacturing defects for error corrections and reducing human interventions in the production area [17]. There are mainly three types of additive manufacturing followed by industrial manufacturers: liquid-based, solid-based, and powder-based. These methods of manufacturing can be further classified into a different subcategory, which is shown in figure 8.2. According to the figure, liquid-based technology that incorporates liquid raw material as input is divided into two subcategories; one is molten, and another is polymerized. Polymerizable technology involves liquid resin and laser or UV light to cure resin so that the required area becomes solid and shaped to create the desired model. The molten technology is based on melting the raw material through a high-temperature nozzle and adding layer by layer of molten material onto the base plate so that the required object may take the shape of the desired product. The solid bonded-object-based 3D printing technique involves the addition of solid material plates or sheets layer by layer and sticking them together by glue, welding, high pressure, or heat treatment.

The process of layer addition continues until the required shape of the modeled object is achieved. Another form of 3D printing uses a powder base of raw material, further divided into melting and binding. In powder melting-based additive

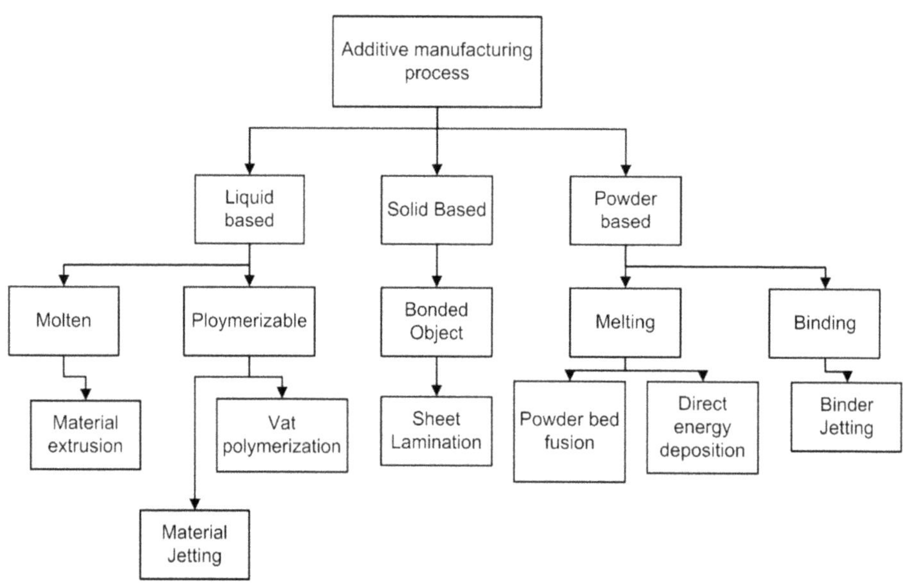

Figure 8.2. Types of additive manufacturing technology.

manufacturing, the raw material can be taken in the form of a polymer substance or metal substance [18]. This process requires a bed of metal or polymer. Then with the help of a laser-based particle melting technique, powder substance binds together. This process repeats itself systematically till the formation of a final product. Selective laser sintering is also one of the techniques to produce a metallic product from a metallic powder bed [19].

As the technology allows for the implementation of precise melting, binding, cooling, and error-correcting systems, advanced additive manufacturing machines will continue to rise. AI is fueling the rapid expansion of these techniques by accepting a consistent up-gradation in terms of larger storage capacity, higher computation power, and an optimized algorithm.

8.2.3 Data science

Data science is the interdisciplinary branch that utilizes science, technology, algorithms, and systems to extract valuable insights from many structured or unstructured datasets and apply those insights to make conclusions for any given input set. Machine learning and artificial neural network play an essential role in the flourishing field of data science. With the increasing number of storage and daily online and offline digital activities, an enormous amount of data is created. The generated data type is sometimes structured, which is easy to understand, but often non-structured and complex to compute to obtain valuable insights. Structured data is often available with proper labels and systematic patterns, whereas unstructured data does not contain appropriate labels. Although, both types of datasets can be categorized based on repetitive patterns and analogies inhibited by the data.

Data science is the discipline that solves significant problems related to various datasets and data challenges. It is tailored to obtain strategies, techniques, and valuable algorithmic patterns from the input datasets. With the idea of Industry 4.0 in 2011, all the available and upcoming industries are intended to achieve a higher quality of production, reduced errors, and less production time [20].

To enable this, experts from various branches must come together to obtain meaningful outcomes using advanced data science strategies. Figure 8.3 shows the outline of different departments, such as mathematics, computer science, and domain expertise. These are further categorized as statistical analysis, data processing, and machine learning, collaboratively creating a specialized domain such as data science [21].

During the early industrial revolution, machines were not complicated, but now demands are increasing, and devices are becoming more complex. This complexity makes it harder to investigate the breakdown and real-time errors at the level of the human workforce. Automated technologies for real-time maintenance and predicting upcoming analyses play an essential role. They generate real-time data and statistics, which can be analyzed by employing data science strategies to overcome real-time problems and upgrade the machines further to meet consistently growing consumer demands.

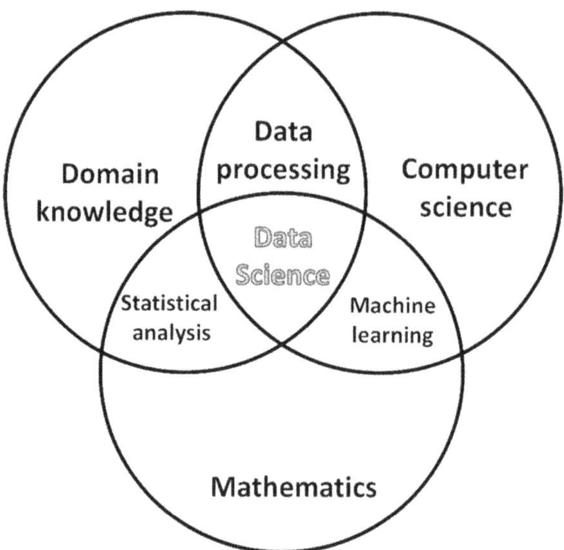

Figure 8.3. Collaboration of data science with other scientific streams.

8.2.4 Cloud computing

Cloud computing is the technology responsible for providing centralized computing, data storage, virtual systems, and application software platforms to several connected users through a personal computer, mobile, laptop, or printers via the Internet [22]. The system is helpful in many cases where the user does not want to engage in active system management. A cloud management system can manage several tasks independently or remotely with administrator instructions. This facility allows the use of on-demand computing power for any required period to complete remote operations precisely. Various industries are nowadays acquiring this facility for data analysis and predictive tasks. Automobiles, aviation, transportation, space exploration, satellite communications, power sectors, food production, and health-care industries are more common and have exclusive users of cloud computing technology [23]. The advancement in the healthcare industry, particularly in rural areas, required more accessibility to cloud computing devices; however, with consistent efforts, many rural and urban locations are becoming the prominent beneficiaries of this technology. Nowadays, on-demand computing and application management systems for healthcare vital parameter analysis are available widely and render valuable outcomes for the user's satisfaction.

Cloud computing systems can be made up of big clouds by connecting many computing facilities and servers available in different geographical locations. It helps maintain data coherence and prevents the loss of crucial data over time. The distributed network of servers and online application management system also helps to prevent essential data from theft or being destroyed by known or unknown cyber attacks.

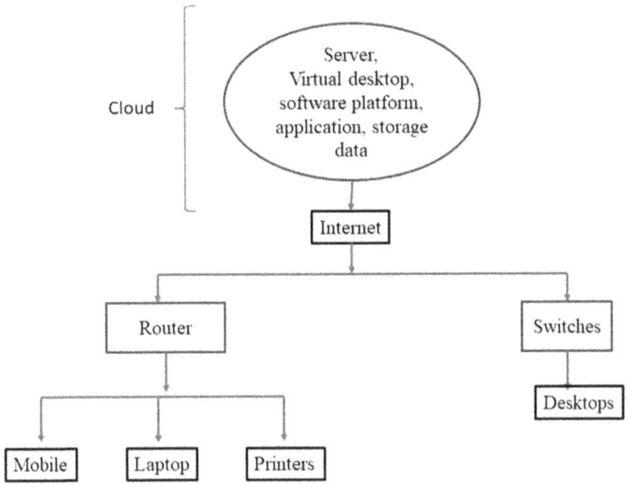

Figure 8.4. Basic block diagram of a cloud computing system.

Figure 8.4 shows a cloud system connected through mobile, laptops, printers, and desktops; however, several other real-time sensor-based devices can be simultaneously connected with this system. The widespread examples of cloud-connected real-time sensors are weather data-collecting devices, natural catastrophe detection systems, forest fire alarm systems, and various wireless sensor networks (WSNs) such as motion sensors, sound sensors, pressure sensors, and air pollution sensors. Expanding cloud-based applications are becoming more popular, which may reveal better solutions to humanity's more significant problems by incorporating other trending technologies such as WSNs and AIoT. The advent of virtual reality and cloud-based games, along with augmented reality, profoundly impacts societal psychology [24]. Along with these periodic upgrades, the technological growth of present-day devices with a cloud-based system may cover a considerable part of the industrial revolution.

8.2.5 Computer vision

Computer vision is the domain of multiple scientific branches that aim to provide computers with the ability to see and comprehend images to make valuable predictions [25]. Computer vision also includes different image processing tools such as image enhancement, image filtering, and image denoising, significantly improving the computer vision experience.

Figure 8.5 shows the fundamental flow chart of computer vision operation, which starts from the image acquisition. Images are made up of pixel intensity variations directly proportional to the light intensity variation of the object captured in the image. However, it is not always feasible to capture all the available information of the image; thus, most applications require preprocessing of the raw input image. Image capturing devices are vastly improved as advanced microprocessors, sensors, and algorithms are becoming available. However, many applications require specific

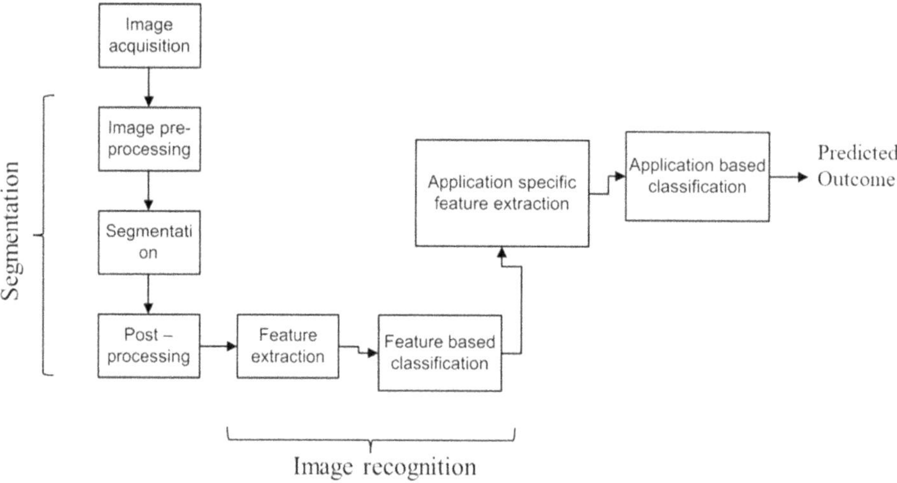

Figure 8.5. An explainable block diagram of the computer vision task.

preprocessing steps to get desired input data to properly govern the image identification process. It involves filtering to remove unnecessary noise, an enhancement to improve the pixel details, and segmentation to select or extract the region of interest (RoI). After this stage, valuable insights from the region of interest can be obtained by applying efficient feature extraction techniques. Application of suitable feature extraction techniques renders essential parameters to understand the useful information in the piece of an input image. These necessary parameters are further forwarded to the feature-based classification system, which classifies various features in a certain number of groups. This group can be the outcome of some applications where classification is the system's aim. Some applications required further identification of groups to understand the overall region of interest to fulfill the demands of the image recognition system. These arrangements for the computer vision system can be made to create image identification tasks, such as face detection or object detection techniques.

The computer vision system is widely applicable in several industrial applications, such as satellite imagery, geo imaging, transportation, construction, agriculture, manufacturing, autonomous vehicles, traffic management systems, and healthcare technologies [26]. Medical image analysis for the detection and diagnosis of acute diseases is the most significant and globally appreciable application of computer vision. This technology is becoming an immensely explored topic among researchers and industries to understand the nature of deadly diseases. The images for this purpose can be obtained through different imaging modalities such as computed tomography (CT) imaging, magnetic resonance imaging (MRI), positron emission technology (PET) scanning, x-ray imaging, and ultrasound imaging. Depending upon the suitable modalities, human body parts and various organs can be scanned to obtain a detailed image for the profound analysis of any abnormality inside.

Similar technology is becoming more efficient in the field of oncology for the detection of lung cancer at its early stage. The whole procedure can also be called computer-aided detection (CAD) of lung cancer. As lung cancer cases are increasing and reaching an alarming stage, with 2.38 lakh (238,000) cases registered in the year 2022 in the USA, the research and development of CAD technologies are becoming part of numerous research groups [27]. CAD technology is consistently improving and will soon be accessible for most parts of the world with much better accuracy than traditional techniques (table 8.1). A better understanding of lung cancer CAD can be taken by going through the essential experimental blocks explained below in figure 8.6.

The automated lung cancer detection technique requires multiple steps synced to achieve desired outcomes with improved performance and greater accuracy. The

Table 8.1. Algorithmic flow of lung cancer detection technique.

Sl. No.	Step	Description
1	Dataset selection	(a) Lung image database consortium (LIDC) online repository (b) Type of modality (CT image) (c) Format of data is digital imaging and communications in medicine (DICOM) in '.dcm'
2	Selection of programming environment	a. Python (a) Python (or any suitable environment such as MatLab)
3	Extraction of visual information or morphological information	(a) Using the pydicom library (for python)
4	Image preprocessing	(a) Image enhancement (b) Noise reduction
5	Lung parenchyma or boundary detection	(a) Image segmentation by applying the appropriate technique
6	Extraction of RoI	(a) Removing background from the original image
7	Nodule segmentation and feature extraction	(a) Applying a deep learning technique or CNN-based automated nodule segmentation technique
8	Classification	(a) Severity detection by applying a customized CNN model. (Designed and optimized for better outcomes.)

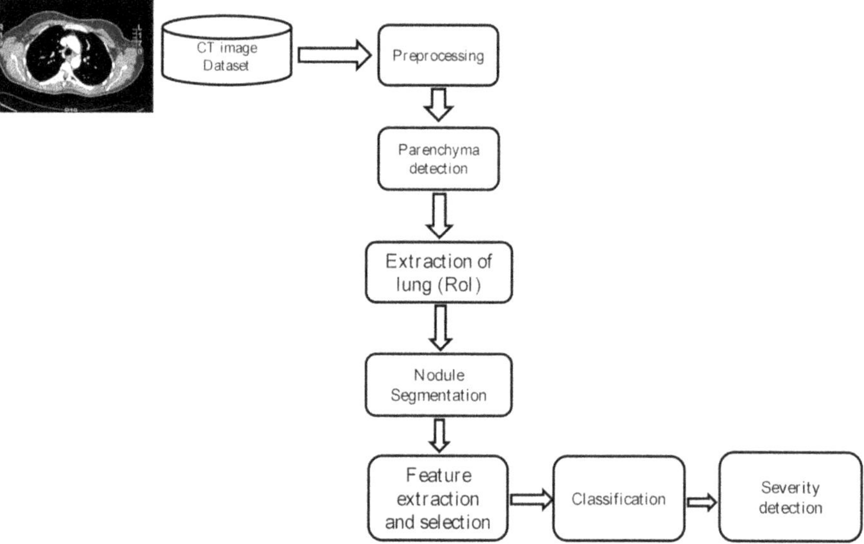

Figure 8.6. Different stages of lung cancer CAD system.

CAD algorithm starts with selecting the dataset and opts for the required integrated development environment (IDE) to perform various algorithmic operations. The procedure consists of several small functions and necessary executable blocks, which are explained further in this section.

8.2.5.1 Dataset selection

Datasets for lung cancer analysis are available online as a public repository in varied form that allows an individual to choose one of them as per the comfortability and significance of research outcomes. LIDC is one of the online repositories containing many lung cancer CT images (1018 scans), along with the annotations given by expert radiologists. The whole dataset contains 2669 lessons marked as a nodule greater than 3 mm. Among these nodules, 928 (34.7%) were checked and identified by four expert radiologists, along with their characteristics annotated in the given files along with the datasets [28]. Our study shows the experimental analysis of the mentioned dataset along with the significant result obtained during the execution of specified algorithms.

8.2.5.2 Selection of programming environment

The programming environment for the proposed algorithm's execution and experimental analysis is responsible for implementing, altering, and improving the algorithm for an automated disease detection system. Therefore, it is necessary to choose it wisely so that maximum performance of the production in terms of reliability, ease of access, program reproducibility, reduced time, and cost of execution can be achieved. As a research group or an individual researcher, choosing one environment with a better support system and valuable online programming libraries is good. However, it sometimes also depends upon one's comfortability in the IDE. A python IDE is chosen for this research as it provides better performance and ease of access for the study and experiments.

8.2.5.3 Extraction of visual, structural, or morphological information

The LIDC dataset contains CT scan images in the DICOM format that carries '.dcm' files. Each scan has information on the internal structure of the human body in the form of radiation intensity variation. These intensity variations are given in the Hounsfield unit (HU), a measure of radio density for interpreting CT images. These images need to be converted into IDE-compatible formats, which is '.npy' in the case of python IDE. '.npy' files are composed of numeric values in the 16-bit configuration. Every scan can be visualized into a 3D or 2D image by plotting it through the matplotlib library. A library named 'pydicom' can also be used to read and extract information from DICOM files, among the multiple other functions available inside the IDE. In the proposed technique, the 'pydicom.dcmread' function is used to read the 'DICOM' files. Figure 8.7 shows the extracted image slice from the DICOM file. After obtaining the intensity values, the scanned files are further provided to the image preprocessing blocks to apply image enhancement and noise reduction operations.

8.2.5.4 Image preprocessing

Image preprocessing is not a mandatory step in the image recognition or computer vision domain as it may distort the valuable information inside the input image. However, it is sometimes essential when the object inside the input image is not

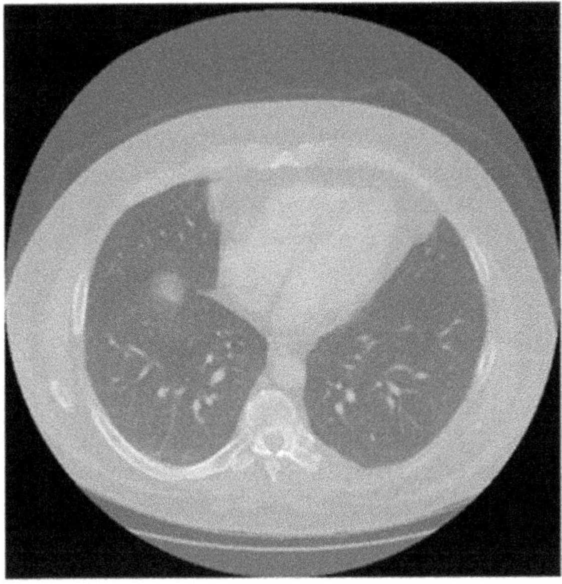

Figure 8.7. Extracted image slice from DICOM image (dataset ID, LIDC-IDRI-0031, slice number-66).

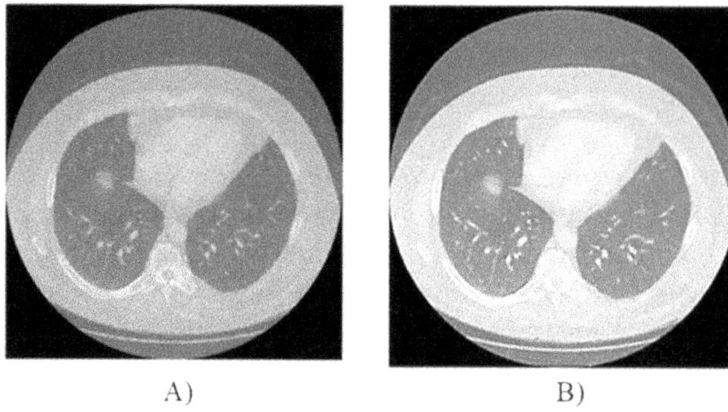

A)   B)

Figure 8.8. (A) Input image, (B) preprocessed by contrast enhancement technique.

visible, or there is a presence of unwanted noise. Image enhancement or image filtering techniques provide better visibility and the ability to comprehend the required information through various image recognition algorithms.

Figure 8.6 shows the advantage of applying the image preprocessing technique. The algorithm used to improve the image's contrast is contrast limited adaptive histogram equalization (CLAHE). Figure 8.8(A) is the input image, and figure 8.8(B) is the contrast-enhanced image. This algorithm improves the contrast so that amplifying noise content inside the image may not have occurred during the enhancement operation. Image filtering through a canny edge-detecting algorithm can also be applied for detecting the edges and boundaries of the lung, which may further help to detect lung parenchyma. In this study, filters for edge detection are not applied; instead, the input image is given to a deep learning model for the parenchyma detection, which shows better outcomes for the lung segmentation operation.

8.2.5.5 Parenchyma detection

Parenchyma detection is the process of selecting the outer boundary of the lung and separating it through the rest of the region of the CT image. Multiple algorithms are available for this process, such as the neighboring anatomy guided method, random walk method, atlas-based and shape-based method, region growing, and graph cut method [29]. Among the various other ways, the deep learning-based automated parenchyma detection and segmentation method are becoming more popular as it shows better accuracy and ease of operation during experimental tasks.

8.2.5.6 Extraction of RoI

The extraction of RoI involves extracting the essential information of the lung by multiplying the lung mask obtained through the parenchyma detection technique. This operation provides the original lung image that contains all the information and intensity variations of the inner side of the lung boundaries. In contrast, the outer region is removed by considering it as a background object. The background object involves ribs, bones, the heart, and the non-essential part that does not come inside the lung boundary.

8.2.5.7 Nodule segmentation and feature extraction

Nodule segmentation and extraction of features are the most critical tasks in the entire lung cancer CAD system. Among the various algorithms, such as fuzzy connectivity and region growing, local adaptive segmentation, morphological and threshold-based techniques, and support vector machines, image segmentation through an optimized CNN model can be more appropriate as it reduces human error by reducing the manual intervention and allows further rapid optimization with growing automated technologies. Deep learning and advanced computation technologies can also be helpful in the future advancement of automated nodule segmentation techniques. As the ability of CNN-based models to learn useful features from the input data is increasing, the rapid growth of reliable nodule segmentation and feature extraction techniques will become more feasible shortly. The LIDC dataset comes with annotated information along with structural and categorical features of annotated nodules. These features are available in their rank values, and each rank value signifies a specific property related to nodules or lesions. A CNN-based deep learning model can be trained by providing the feature information and the severity level of cancer malignancy.

On the other hand, in the multiclass classification system, various images of nodules and their malignancy level can be provided as input for training a deep learning model, which may further detect the malignancy state of a new unknown test image. The feature information provided with the dataset are available in the form of subtlety, internal structure, calcification, sphericity, margin, lobulation, spiculation, texture, and malignancy. All these features are provided with scores ranging from 1 to 6 that describe different levels of properties in the form of nodule features.

The plot shown in figure 8.9 is obtained using the pylidc tool and annotated information available in the LIDC dataset [30].

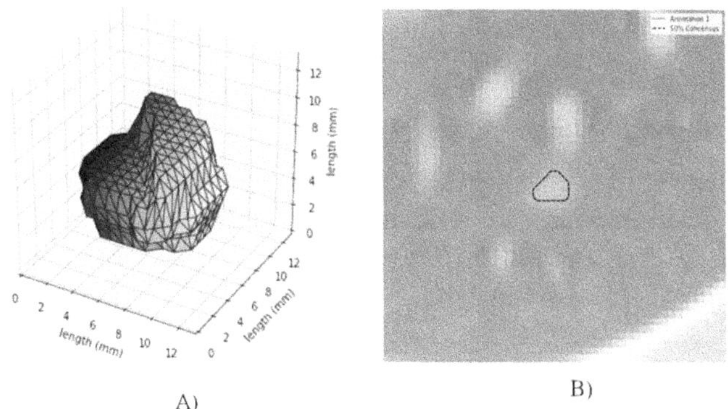

Figure 8.9. Annotated nodule, LIDC dataset, ID—LIDC-IDRI-0031 (A) 3D-plot, (B) 2D-plot.

8.2.5.8 Classification

Lung nodules can be classified based on their various features and properties. Its classification becomes significant when the classifier renders the malignancy score with more excellent reliability and accuracy. However, other forms of the classification algorithm, in terms of nodule characteristics, may help to achieve better diagnostic efficiency using the lung cancer CAD system. The reliability and robustness of classification also depend upon a significant feature selection technique. As per the existing studies, it has been found that subtlety and spiculation are the two most essential characteristics of nodules among all, as they showed better results in terms of malignancy detection [30] (figure 8.10).

The presence of feature scores and their percentage values among the dataset can be comprehended by plotting the feature distribution for the overall dataset. Characteristic features of nodules are divided into six different types of scores from 1 to 6, although not for all characteristics because some of them are given '5' as the maximum score. Each score for the different characteristics carries different properties of the nodule. Tables 8.2 and 8.3 are provided to describe the significance of scores of nodule characteristics [31]. The information above for the nodule characteristics is given along with the dataset, which can be combined with other relevant image features to achieve better outcomes. Other relevant image features such as circularity, perimeter, roughness, elongation, minimum intensity, maximum intensity, mean intensity, intensity standard deviation, and eccentricity can also be considered for improved training and testing accuracy of the newly implemented CAD algorithm.

The above procedure of the lung cancer CAD system is explained under the computer vision section, providing a detailed description of a standard automated lung cancer detection system and its experimental analysis. After carefully comprehending the detailed explanation above, it may be possible to implement an early cancer disease detection and diagnostic system.

These analyses are presently recommended to be performed by expert doctors or by reliable technologies under the supervision and proper concern of the domain experts. However, with the consistent efforts of researchers and scientists, it may soon become possible to get an efficient diagnosis or detection of acute disease through an individual computer vision system.

8.2.6 Natural language processing (NLP)

Natural language processing (NLP) is the technology that aims to provide interaction between humans and computers via human natural languages and linguistics. The functionality of NLP lies between AI, DL, ML, and linguistics, as depicted in figure 8.11. Linguistics is the study of human vocal and scripting languages, which are the primary methods of communication between humans. Humans can communicate via languages due to their ability to understand the meaning of specific pieces of information. Languages and their deep understanding are stored as a memory in the brain that helps to make out the valuable sense of sentences by comparing the received and available information in the human

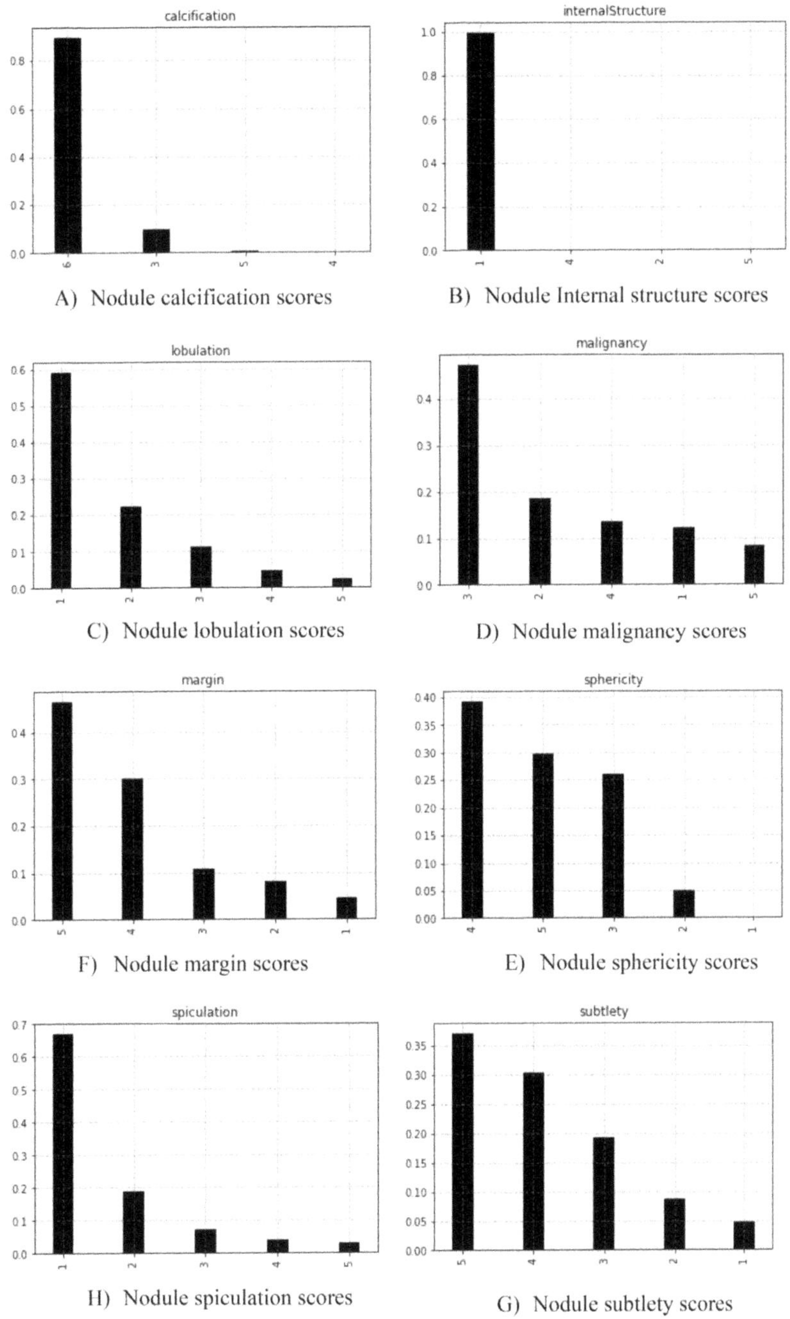

Figure 8.10. Characteristics of nodules and their corresponding scores available in the entire percentage of datasets.

Table 8.2. Significance of scores for different characteristics of nodule-I.

	Calcification	Internal structure	Lobulation	Malignancy
1	Popcorn	Soft tissue	Marked	Highly unlikely
2	Laminated	Fluid	—	Moderately unlikely
3	Solid	Fat	—	Indeterminate
4	Non-central	Air	—	Moderately suspicious
5	Central	—	None	Highly suspicious
6	Absent	—	—	—

Table 8.3. Significance of scores for different characteristics of nodule-II.

	Margin	Sphericity	Spiculation	Subtlety
1	Poorly defined	Linear	Marked	Extremely subtle
2	—	—	—	Moderately subtle
3	—	Ovoid	—	Fairly subtle
4	—	—	—	Moderately obvious
5	Sharp	Round	None	Obvious
6	—	—	—	—

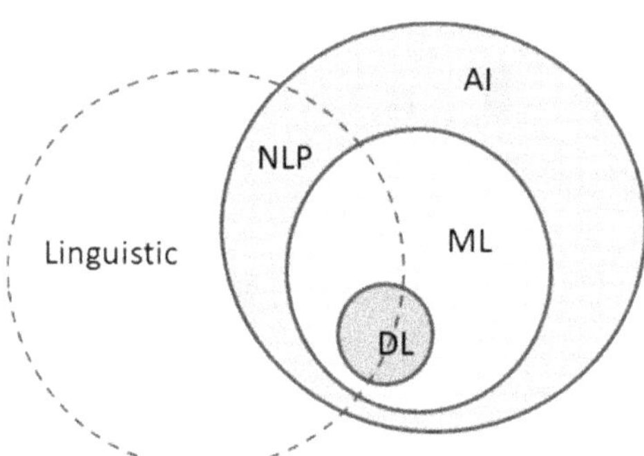

Figure 8.11. How AI, ML, DL, and linguistics contribute to NLP.

memory. In a similar pattern, algorithms of NLP are designed first to comprehend the meaning of linguistic inputs using AI and then store the insights in the form of weights of neurons inside the artificial neural network (ANN) [32].

Storing and updating the weights of neurons for some specific insights refers to learning. When these networks of connected neural networks (NN) are too deep and multiple layers are available between the output and input layers of the neural net, it is called a deep learning network. NLP is mainly focused on linguistic processes that require a good amount of learning capability, which are made up of an optimum number of NN layers. A long-short term memory (LSTM) neural network is widely accepted for the natural language processing that is made by a recurrent neural network (RNN) [33]. The RNN is the type of deep learning network made by a specific feedforward network. These feedforward networks are different from other traditional feedforward networks. The neuron of layers can store the input from the previous step for an indefinite duration. The hold duration depends upon the input data and insights of upcoming data. NLP can extract useful insights from speech data due to the efficient performance of the LSTM network that renders better outcomes for time series and sequential data such as speech-to-text or text-to-speech conversion. Natural language processing involves both natural language understanding (NLU) and natural language generation (NLG), which are the two essential types of this system. These two components of NLP can also be divided into three different segments.

The NLU component comprises lexical ambiguity, syntactical ambiguity, and referential ambiguity [34]. In contrast, the NLG component includes text planning, sentence planning, and text realization. Natural language understanding may face serious issues sometimes. These issues are considered ambiguous (ambiguity) in the computing process. Lexical ambiguity occurs during the process of NLU when the machine cannot sense the actual meaning of a word due to the availability of multiple definitions for the same word. Another form of NLU ambiguity is syntactical ambiguity, which occurs due to the availability of various meanings of one sequence of words. The third cause of NLU ambiguity is referential ambiguity which arises due to unwanted pronouns, and some specific sentence is intended to provide more than one meaning of a given sentence (figure 8.12).

Natural language generation is also one of the essential procedures of NLP that can be governed by taking care of three different components, text planning, sentence planning, and text realization. During the phase of text planning, the NLG algorithm is intended to retrieve essential insights from the stored database [35]. The next step is sentence planning, which intends to produce reasonable words, and relevant sentences

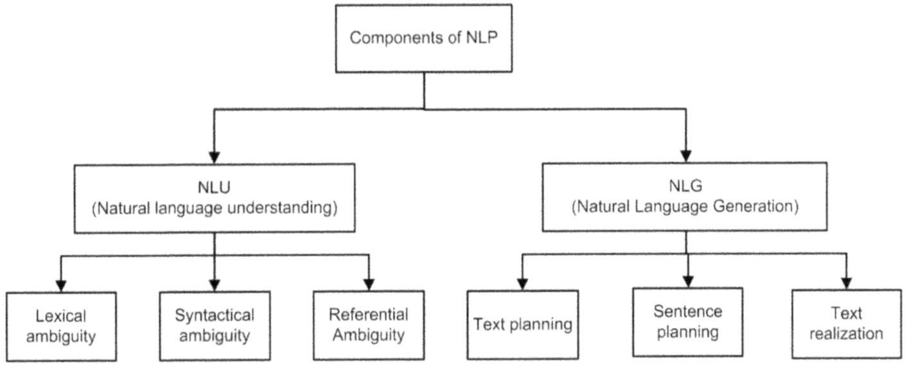

Figure 8.12. Fundamental components of natural language processing.

and set up the tone of sentences. The final stage is text realization, which is responsible for connecting different sentences to the actual and meaningful structure of the sentence.

NLP technology is in its early phase, but as AI is a building block, its rapid advancement is more likely, and its widespread acceptance is inevitable. The modern digital era is full of intelligent devices which are closely in touch with humans to meet their demands in daily life. Interaction between smart devices and humans is now becoming more accessible, but there are still some gaps between what is desired and what is available. Advancements in the NLP technique may open new ways to establish more manageable and reliable communication between humans and machines.

8.2.7 Robotics

Increasing demand for high-quality products and services is always creating a scope to take the assistance of the technology for performing human-level tasks without making significant errors and simultaneously managing the quality of functions. Automating most human-level tasks also enables systematic management and quality control of any given operations. There are several industries where automated technologies perform much better than humans, such as food packaging, assembly lines and machine manufacturing, healthcare, tools and equipment, and textile industries [36]. Automated technologies and robotics reduce human efforts in implementing many products and services. Robots are capable enough to execute any given task multiple times with high precision. There is no specific time for their working hours. Thus they can work continuously for many days without taking rest. Due to its ability for timely updates, the more competitive and challenging tasks can be performed by the manufacturers and developers. Timely observing the errors and correcting working principles leads the manufacturing industries towards less production cost, increased performance, higher quality, and increased production rate. Robots and automation are not limited to the physical world and are now available virtually in chatbots and automated text messengers. Nowadays, Alexa, Siri, and Google Assistant are playing a vital role in automating many digital tasks, from home automation to calling a number and sending a text to someone [37]. AI and internet technologies are making it feasible to monitor robotic and automation functionality from any point of the globe. That makes this system equipped with global security and quality protocols. Digital manufacturing and consumer platforms help to put real-time feedback in the complete system directly, which enables the manufacturers to see the overall performance. This allows them to make necessary upgrades to create more satisfactory outcomes at the user's and manufacturer's end. Some industries deal with hazardous and life-threatening manufacturing processes. Robotics and AI technology are essential in developing better solutions to these problems.

Apart from these, defense industries are also implementing robotic accessories, equipment, and vehicles that work efficiently for various defense tasks. Bomb-defusing robots and unmanned aerial vehicles (UAV), drones, and advanced weapon systems are essential to defend the border and protect the people from internal and external threats, which are being developed in the field of robotics. There are widespread applications of robotics, from space exploration to precious mining metals on Earth, improving with time. Advanced technology makes it feasible to achieve significant progress at the fastest possible rate but managing the direction of progress is also essential so that it always moves towards betterment instead of creating severe issues.

8.3 Conclusion

AI and digital data can be considered a new fuel for industrial growth that consistently pushes human capabilities' boundaries. The feasibility of solutions to tackle any existing and upcoming challenges can now be calculated with the help of advanced scientific tools. This makes it possible to visualize the process and outcomes before making any physical attempt. By going through the research and experiments iteratively, a better result in the virtual platform can be obtained, which may be considered valuable insight to perform the actual task. In medical science, a number of studies can be performed in a simulation first, and then they come into existence. As technology is getting more advanced, there is a need for more precise simulation software to simulate the actual process precisely. Increasing data centers and computing resources may greatly help address this situation. Apart from that, researchers, experts, and industrialists from different domains must work collaboratively to design safer, reliable, and high-performance protocols for industrial upgrades. Industrial up-gradation has several merits on its own, but it also has a darker side for humanity in the form of unwanted consequences. Rapid technological growth impacts the mental and physical health of humans. Hence it is recommended to create a good balance between technologies and human-level experiences to follow sustainable development. Technological growth is an inevitability that certainly takes the world towards an entirely new era of technology-depended society. Therefore, it is up to us how we lead this technological evolution by making a good balance between technologies and human ethics. Industrial expansion is impacting not only the health of humans but also the overall health of the planet and disrupting ecological balance. As in the present scenario, it is possible to visualize the problem and its solutions profoundly. It can be assumed that advanced technologies may undoubtedly help us overcome these situations soon.

References

[1] Popkova E G, Ragulina Y V and Bogoviz A V 2019 Fundamental differences of transition to Industry 4.0 from previous industrial revolutions *Studies in Systems, Decision and Control* 169 (Berlin: Springer) pp 21–9

[2] Nancu D, Sima V and Gheorghe I G 2020 Influences of the Industry 4.0 revolution on the human capital development and consumer behavior: a systematic review *Sustainability* **2020** 4035

[3] Xu M, David J M and Kim S H 2018 The fourth industrial revolution: opportunities and challenges *Int. J. Financ. Res.* **9** 90

[4] McLennan S *et al* 2020 An embedded ethics approach for AI development *Nat. Mach. Intell.* **2** 488–90

[5] Dwivedi Y K *et al* 2021 Artificial intelligence (AI): multidisciplinary perspectives on emerging challenges, opportunities, and agenda for research, practice and policy *Int. J. Inf. Manage.* **57** 101994

[6] Kumain S C, Kumain K and Chaudhary P 2020 AI impact on various domain: an overview *Int. J. Manag.* **11** 1433–9

[7] Pérez Santín E *et al* 2021 Toxicity prediction based on artificial intelligence: a multidisciplinary overview *WIREs Comput. Mol. Sci.* **11** 1–32

[8] Shen J *et al* 2019 Artificial intelligence versus clinicians in disease diagnosis: systematic review *JMIR Med. Informatics* **7** e10010

[9] Dong B, Shi Q, Yang Y, Wen F, Zhang Z and Lee C 2021 Technology evolution from self-powered sensors to AIoT enabled smart homes *Nano Energy* **79** 105414

[10] Zhang T, Zhao Y, Jia W and Chen M Y 2021 Collaborative algorithms that combine AI with IoT towards monitoring and control system *Futur. Gener. Comput. Syst.* **125** 677–86

[11] Barton M, Budjac R, Tanuska P and Gaspar G 2022 Identification overview of Industry 4.0 essential attributes and resource-limited embedded artificial-intelligence-of-things devices for small and medium-sized enterprises *Appl. Sci.* **12** 5672

[12] Khanna A and Kaur S 2019 Evolution of Internet of Things (IoT) and its significant impact in the field of precision agriculture *Comput. Electron. Agric.* **157** 218–31

[13] Chen C, Huang Y, Li Y, Chang C and Huang Y 2020 An AIoT based smart agricultural system for pests detection *IEEE Access* **8** 180750–61

[14] Dong B, Shi Q, Yang Y, Wen F and Zhang Z 2021 Nano energy technology evolution from self-powered sensors to AIoT enabled smart homes *Nano Energy* **79** 105414

[15] Tay Y W D, Panda B, Paul S C, Noor Mohamed N A, Tan M J and Leong K F 2017 3D printing trends in building and construction industry: a review *Virtual Phys. Prototyp.* **12** 261–76

[16] Pratheesh Kumar S, Elangovan S, Mohanraj R and Ramakrishna J R 2021 Review on the evolution and technology of state-of-the-art metal additive manufacturing processes *Mater. Today Proc.* **46** 7907–20

[17] Farhan Khan M *et al* 2020 Real-time defect detection in 3D printing using machine learning *Mater. Today Proc.* **42** 521–8

[18] Chin S Y, Dikshit V and Priyadarshini B M 2020 Powder-based 3D printing for the fabrication of device with micro and mesoscale features *Micromachines* **11** 29–40

[19] Kafle A, Luis E, Silwal R, Pan H M, Shrestha P L and Bastola A K 2021 3D/4D printing of polymers: fused deposition modelling (FDM), selective laser sintering (SLS), and stereo-lithography (SLA) *Polymers* **13** 3101

[20] Nasution M K M 2021 Industry 4.0: data science perspective *IOP Conf. Ser.: Mater. Sci. Eng.* **1122** 012037

[21] Sajid S, Haleem A, Bahl S, Javaid M, Goyal T and Mittal M 2021 Data science applications for predictive maintenance and materials science in context to Industry 4.0 *Mater. Today Proc.* **45** 4898–905

[22] Alouffi B, Hasnain M, Alharbi A, Alosaimi W, Alyami H and Ayaz M 2021 A systematic literature review on cloud computing security: threats and mitigation strategies *IEEE Access* **9** 57792–7807

[23] Askary Z and Kumar R Cloud computing in industries: a review *Lect. Notes Mech. Eng.* **20** 107–16

[24] Longan M, Dimita G, Michels J D and Millard C 2021 Cloud gaming demystified: an introduction to the legal implications of cloud-based video games *Paper No. 369*

[25] De Ketelaere B *et al* 2022 A fresh look at computer vision for industrial quality control *Qual. Eng.* **34** 152–8

[26] Brunetti A, Buongiorno D, Trotta G F and Bevilacqua V 2018 Computer vision and deep learning techniques for pedestrian detection and tracking: a survey *Neurocomputing* **300** 17–33

[27] Xia C *et al* 2022 Cancer statistics in China and United States, 2022: profiles, trends, and determinants *Chin. Med. J. (Engl).* **135** 584–90

[28] Pehrson L, Nielsen M and Ammitzbøl Lauridsen C 2019 Automatic pulmonary nodule detection applying deep learning or machine learning algorithms to the LIDC-IDRI database: a systematic review *Diagnostics* **9** 29

[29] Xu M *et al* 2019 Segmentation of lung parenchyma in CT images using CNN trained with the clustering algorithm generated dataset *Biomed. Eng. Online* **18** 2

[30] Hancock M C and Magnan J F 2016 Lung nodule malignancy classification using only radiologist-quantified image features as inputs to statistical learning algorithms: probing the Lung Image Database Consortium dataset with two statistical learning methods *J. Med. Imaging* **3** 044504

[31] Opulencia P, Channin D S, Raicu D S and Furst J D 2011 Mapping LIDC, RadLex™, and lung nodule image features *J. Digit. Imaging* **24** 256–70

[32] Lauriola I, Lavelli A and Aiolli F 2022 An introduction to deep learning in natural language processing: models, techniques, and tools *Neurocomputing* **470** 443–56

[33] Zhang S 2021 Language processing model construction and simulation based on hybrid CNN and LSTM *Comput. Intell. Neurosci.* **2021** 1–11

[34] Gnanasekaran R K, Chakraborty S, Dehlinger J and Deng L 2021 Using recurrent neural networks for classification of natural language-based non-functional requirements *CEUR Workshop Proc.* **2857**

[35] Dong C *et al* 2021 A survey of natural language generation *ACM Comput. Surv.* **1** 1–36

[36] Javaid M, Haleem A, Singh R P and Suman R 2021 Substantial capabilities of robotics in enhancing Industry 4.0 implementation *Cogn. Robot.* **1** 58–75

[37] Ali B, Ravi V, Bhushan C, Santhosh M G and Shiva Shankar O 2021 Chatbot via machine learning and deep learning hybrid *Modern Approaches in Machine Learning and Cognitive Science: A Walkthrough. Studies in Computational Intelligence* 956 (Berlin: Springer) 255–65

IOP Publishing

Advanced Signal Processing for Industry 4.0, Volume 1
Evolution, communication protocols, and applications in manufacturing systems
Irshad Ahmad Ansari and Varun Bajaj

Chapter 9

Artificial intelligence based flexible manufacturing system (FMS)

Ashutosh Mishra

Automation of manufacturing processes has been an essential and necessary aspect in the industrial set up since its very inception. Development in artificial intelligence (AI) methodology and robotics has made it possible to deploy AI in general and machine learning (ML) methods in particular to enhance the automation in production management, process planning, and intelligent scheduling. AI methodologies have also been used in preventive maintenance, process automation, supply chain management, product development, warehouses, enterprise resource planning, and product design. This chapter covers the use of some of the expert systems (ESs) in various manufacturing processes in flexible manufacturing systems. A rule-based system has been described in the process planning. The components and procedure of a knowledge-based scheduling strategy has been described in detail. An outline of ML methods in some processes has also been discussed in this work.

9.1 Introduction

Industrial automation is obtained by an autonomous system, which has the several features. The system should be able to model the environment at a level commensurate with its expected output. It should be able to make knowledgeable decisions based on its current model, a set of constraints and inference rules. It should predict what is going to happen next and also reduce the future contention for resources. The autonomous system should react to the situation and also should be able to communicate with the external part or machine.

The automation requires *planning*, *scheduling*, and *processing* the parts in particular or an item as a whole. These three problems of automation are described in three steps: (i) problem description, (ii) knowledge-based system approach, and (iii) description in brief with some of the salient features of the knowledge-based system in the problem domain.

doi:10.1088/978-0-7503-5247-5ch9   9-1

Henry Ford was perhaps the first figure to perceive the concepts of automation many decades ago and implemented hardware in the production of automobiles, keeping in view that each American family should possess at least one to drive in. The philosophy of automation, i.e., less and less human intervention and the principle of mechanization, still persists in every industry and organization where human beings work. The implementation strategy changes with steps of technological development. Equipment, instruments and tools, and the hardware modules are being controlled by the computer, maintaining a trade-off between its hardware and software module. Manufacturing is divided into fixed and flexible methodology, corresponding to predominantly hardware and software, respectively. Automation is essential because it lowers labor costs, boosts product quality, speeds up production, cuts down on wasteful inventory, saves money on material handling, and gives manufacturers more say over the manufacturing process.

In an analysis of how artificial intelligence (AI) can improve flexible manufacturing systems (FMSs), the author examines the central aspects of the field. In addition, the benefits of using a knowledge-based approach in FMS design and operational control are discussed [1].

An essential principle of production is to maximize output while minimizing expenses and maximizing quality. The emerging 'Intelligent Production' paradigm is capable of meeting these needs. The intelligent manufacturing system is, in and of itself, a flexible manufacturing system, able to adapt to shifting production needs and environmental conditions. These adaptable responses allow for more efficient use of space, lower overall production and investment costs, and higher output [2].

The challenge of running a versatile manufacturing profitably on a microgrid depends on the electricity variable with high electricity rate. The reinforcement-learning algorithm is taught to perform online optimization of the production machine in conjunction with an estimate of future energy prices to cut down on cumulative energy expenditures. The idea is tested in practice on a production machine with variable speed capabilities [3].

Expert systems have been used in the textile and garment industries to help with quality control and the identification of manufacturing flaws. The use of AI in fields such as engineering, manufacturing, management, and others has been reported alongside textile and garment industry training programs [9]. Research into AI has been ongoing since the 1970s, and its discoveries and contributions have been a major step toward improving the efficiency and effectiveness of human labour. These days, it's hard to go anywhere without encountering at least one intelligent machine, such as a gate scanner, a ticket counter for a movie, a vending machine, an automated teller machine, a washing machine, etc. Numerous fields and industries have made extensive use of expert systems built on foundations of rule-based expert systems, frame-based expert systems, neural networks, genetic algorithms, etc [10]. Increased use of AI, the IoT, and robotics in smart manufacturing and digital transformation for transition to 4.0 industrial set up [11] is just some of the evidence of this trend. Data, assumptions, methodologies, and relevant conditions are just some of the procedural issues raised by the widespread adoption of ML and data science (DS) methods. The industry and FMS both stand to benefit from a clearer understanding of the pitfalls inherent in this data-driven approach, as well as some concrete recommendations for how to best put this science to use. An intelligent manufacturing analytics framework with five stages and suggested protocols for implementing DS in FMS techniques has been created. When it comes to the digitization of manufacturing and the ability to combine data from manufacturing processes and quality measurements, machine learning and deep learning techniques are both essential and potentially useful [12]. The use of predictive quality in this industry allows manufacturers to make informed predictions about the quality of their products using information gleaned from the production process. Research papers published in the scientific literature between 2012 and 2021 that address predictive quality in production were analyzed in a thorough and methodical review study. The manufacturing industry has benefited greatly from the use of computers in the form of automated test facilities, material requirements planning, computer-aided design, computer-aided manufacturing, computer-aided process planning, and computer-aided quality control. Expert systems, has emerged as a tool to improve the quality of information provided to the responsible experts and users for operating modern automated and integrated systems, and help them in reliable operation of their systems, but knowledge that is based on past experience and cannot easily be cast into mathematical formulas for conventional algorithmic programs, has yet to be fully exploited [13]. AI landing for sheet metal-based drawer box defect detection with deep learning aims to solve the problem of multiple defect detection through the design and implementation of industrial process integration with AI via automated optical inspection (AOI) (ALDB-DL) [14]. Defect classification in area of interest (AOI) images using a convolution filter and a global average pooling (GAP) in a convolutional neural network (CNN) algorithm based on computer vision and deep learning. CNNs and long short-term memory networks (LSTMs) with a skip connection (PdM) [15] have been used for multivariate time-series forecasting. Prediction of machine failures was tested using CNNs, LSTM models, and a CNN-LSTM hybrid. Data from sensors like voltage, pressure, vibration, and rotation are included in the dataset, along with details about machine failures, maintenance, error conditions, and telemetry. Predictions made with the CNN-LSTM were the most accurate and trustworthy. Industrial automation makes use of deep learning techniques in areas like process planning, production management, and processes a scheduling [16]. Production systems for industrial applications [17] have been the subject of research highlighting the difficulties and exploring the potential of machine learning.

Software-based systems such as computer-aided manufacturing (CAM), computer-aided design (CAD), computer-integrated manufacturing (CIM), computer-aided process planning (CAPP), automated computing-aided planning system (ACAPS), etc., automate various stages of the manufacturing process.

AI in general and expert system in particular have been fully deployed in various manufacturing dimensions such as design, planning, and control. AI with its problem-solving paradigms such as constraint-based search, heuristic search and production rules are used for ACAPS planning and scheduling. In the next section, process planning and scheduling have been discussed.

9.2 Background of AI in automation and FMS

Deployment of AI methodology and ML methods would enhance the following aspects of FMS.

AI can be used to generate quick, data-based decisions, improve production outcomes, push process effectiveness forward, reduce operational costs, and develop products with superior scalability. In addition, AI performs admirably when tasked with translating and comprehending human speech. Workers and supervisors will find it less complicated to convey their needs to computer programs in this way. Software users, for instance, would rather conduct a search for a specific feature than wade through a maze of submenus. Thanks to AI, programs can understand the user's goals and act more intuitively, resulting in better results with fewer mistakes. The impact of AI in production make valuable sense in term of adopting new technologies in manufacturing sector. For as long as there have been new technologies, the manufacturing sector has been eager to adopt them. There have been drones and industrial robots used in production since the 1960s. The dawn of the next wave of automated advancement is on the horizon. If businesses use AI to reduce wasteful stockpiling and cut costs, this could be a promising breakthrough for the manufacturing sector. The manufacturing sector, it has been said, must be ready for organized manufacturing plants in which the supply chain, design team, production line, and quality control are all coordinated to form an intelligent engine that provides essential knowledge insights. Advances in Artificial Intelligence and Their Impact on the Manufacturing Sector: Improvements in computer graphics have long been used in quality control to spot manufacturing flaws as they occur. However, with the advent of ever-increasing informational demands in manufacturing and the fact that plant managers no longer wish to compensate workers for data entry, this practice has become increasingly untenable.

The following are some of the many benefits of AI: artificial intelligence equipped with computer vision can explain how data is taken in. With the help of a camera, a stock transaction can be created automatically whenever a factory worker takes pre-ordered raw materials from a shelf. The natural user interface will consist of performing the action itself, rather than entering data or scanning items into a computer. The IoT will be affected by AI. Through the IoT, we can reach consumers who may not realize they need a specific product or service, but who nonetheless could benefit greatly from having it delivered to them.

In addition, the IoT can relay detailed telemetry to manufacturers and retailers so they can assess product quality and identify potential causes of failure. In a nutshell, the IoT is an internal tsunami of data that can be used by AI for reasoning and learning. This will pave the way for more sophisticated generative design processes, where products are rethought in ways that mimic evolution.

9.3 Application areas of AI

There are various areas of industrial origin where AI methodologies have been deployed today. Some of them are discussed as follows:

Logistics robots, autonomous vehicles factory automation, IT operations, design and manufacturing, IoT, warehouse management, process automation, predictive maintenance, AI-based product development, AI-based connected factories, AI-based visual inspections and quality control, AI for purchasing price variance, AI order management, and AI for cyber-security.

Short descriptions about some of the important AI applications are given below.

9.3.1 Supply chain management

Manufacturers can benefit from AI-enabled systems that help them evaluate alternative scenarios for last-mile delivery performance (in terms of time, cost, and revenue). Accurate delivery time predictions can be made with the help of AI thanks to its capacity to plan the most efficient routes, monitor driver activity in real time, and factor in environmental factors like weather and traffic reports. AI can also help factories improve their inventory management and capacity planning. By establishing a real-time and predictive supplier assessment and monitoring model, businesses can be alerted the moment a failure occurs at a supplier and the extent of the disruption to the supply chain can be determined instantly. When operating automated machinery, operators must rely on their experience and intuition to keep track of multiple signals on multiple screens and make fine-tuned adjustments as needed. Aside from the already heavy workload, this system places additional burdens on the operators in the form of troubleshooting and testing. This leads some business owners to ignore the importance of maximizing profits, prioritize the wrong tasks, and cut corners wherever possible. There are two issues with this method: human-intensive systems increase the likelihood of human error, the breakdown of vital factory machinery, and a general decline in production output, and their reliance on personnel with extensive experience makes it more difficult to replace factory workers. When an experienced operator leaves a company, they take valuable factory knowledge with them.

Robotics and AI have the potential to greatly increase manufacturing plant output with minimal increase in labor costs. Other uses include automating a number of labor-intensive processes within manufacturing facilities. It can reduce the number of resources needed to run a factory by keeping a close eye on everything all the time and immediately notifying technicians of any problems they might encounter. This can be done by creating a central repository for all operational data along with context, which will make employee transitions much simpler. Output can be modified in response to changes in demand and manufacturing tactics.

Siemens is a leader in industrial automation. Through the use of AI algorithms, cloud-based analytics, and computer vision, the company has partnered with Google to increase output in the factory.

9.3.2 AI in design and manufacturing

Using software enhanced with AI, multiple optimal designs for a single product can be generated. The software, which is also known as generative design software, necessitates specific input from engineers, including things like raw materials, size and weight, manufacturing methods, cost, and other resource constraints. Changing these settings allows the algorithm to produce new potential layouts.

9.3.3 AI in warehouse management

Numerous warehousing tasks can be automated with the help of AI. Manufacturers can keep a constant eye on their stockrooms and improve their logistics thanks to the data they collect in real time. Moreover, with the aid of demand forecasting, factories can anticipate customer needs and stock up on inventory in a timely manner, all without incurring excessive transportation costs.

Robots in warehouses can be used for a variety of tasks, including inventory tracking, lifting, moving, and sorting, freeing up humans for higher-level strategic work and reducing the likelihood of injuries.

Costs associated with managing a warehouse can be lowered, productivity can be increased, and fewer workers will be needed if quality control and inventorying are automated. In turn, this means manufacturers can boost both their revenue and their bottom line.

9.3.4 AI process automation

Automatic bottleneck detection and elimination is now possible with AI-powered process mining tools. Manufacturers can now easily evaluate regional performance benchmarks, streamline processes, and standardize procedures with the help of these tools.

Robotic process automation (RPA) is another example of how robots can be put to use; in this scenario, robots operate autonomously to carry out routine tasks in the manufacturing setting. The only time a human has to step in is when the robots encounter something completely out of the ordinary. It is also possible for robots to inspect and screen processes automatically using computer vision.

Improvements in workplace safety, employee morale, and output are just a few additional benefits that can result from automating processes.

McKinsey found that by automating processes with AI in the semiconductor industry, yield could increase by as much as 30% while simultaneously decreasing scrap rates and testing costs.

9.3.5 AI for predictive maintenance

Value from AI in manufacturing is estimated to be between $0.5 and $0.7 trillion globally due to predictive maintenance, according to a report by McKinsey. The first step in implementing Industry 4.0, according to BCG, is implementing predictive maintenance.

Capturing and processing big data (including audio, video, and GPS) from sensors on the shop floor can aid in spotting anomalies or inefficient equipment and preventing unscheduled downtime. Any number of things, from a faulty part on the production line to a strange noise coming from the engine, could be to blame. Fixing problems before they happen will save money and increase productivity in the factory. In addition, it is more cost-effective to repair faulty parts than to buy a whole new machine.

9.3.6 AI-based product development

Manufacturers can use AI-based product development to create multiple simulations and test them using augmented reality (AR) and virtual reality (VR) before beginning production (virtual reality). Manufacturers can cut down on the time and money spent on trial and error and get their products to market faster. Assist the company's engineers in identifying potential issues with the final product and devising solutions to those issues before they reach the consumer. Maintenance and bug fixing should be streamlined.

In order to stay ahead of the competition and create marketable products, manufacturers are increasingly turning to AI-based product development.

The AI algorithms also help in product development because they get better with each iteration thanks to their feedback loop.

Consider General Electric's 'Brilliant Factory' as an illustration. GE established such a facility in Pune, India, to lessen the frequency of maintenance and maximize output. They saw an increase in OEE of 45%–60% in their networked machinery.

9.3.7 AI-based visual inspections and quality control

AI-driven flaw detection makes use of computer vision, which employs high-resolution cameras to keep tabs on every stage of manufacturing. An automated system of this kind could detect problems that a human observer might miss and initiate maintenance immediately. This contributes to fewer product recalls and less waste.

Immediate detection of abnormalities, such as toxic gas emissions, also aids in the prevention of workplace hazards and boosts factory worker safety.

Augmented reality overlays are another AI-based system that helps find quality issues by comparing actual assembly parts to those provided by suppliers. With the help of AR, remote training and support is possible, allowing technicians from anywhere to connect with those at a facility and provide guidance.

This also makes it easier to manage data for all procurements in one place and keep track of where and when parts were purchased from various vendors.

9.3.8 AI order management

Methods for processing orders must be flexible, economical, and responsive to changes in supply, demand, consumer preferences, and production tactics. Robots and AI-based systems can help factories streamline order entry and other 'copy-paste' tasks, while sensors can keep tabs on stock and generate purchase orders on their own, handling the complication posed by various order types coming in from various sources, thus improving the order and planning of inventory.

9.4 Some of the real industrial set up and AI application

Larsen & Toubro (L&T) Construction employs AI and image processing methods to guarantee welding quality. Welds in the plant can be analyzed in a controlled environment, but L&T faces a greater challenge when attempting to analyze welds in pipelines and other 'one-off' applications. Despite the fact that the images are captured locally at the edge, the analysis is performed in the cloud in order to train and make use of more computational resources. Siemens' Amber factory in Germany is an example of a facility that uses an alternate method to product testing. Because of the high cost and time required to test each individual component in this fast-paced production setting, AI is being used to predict quality based on production data and previous quality test results.

9.4.1 Maintenance applications

According to the Arcis Resources Corporation's estimates, the maintenance of machinery is the most common area where AI applications are being put to use because of the significant cost savings they provide. The assumption here is that the operational and maintenance costs will account for 80% of the total cost of ownership for automation tools. The solution for cabinet inspection presented by Brain Creators enables proactive, real-time asset management via asset maps. Another hassle is the time spent after an inspection on the paperwork required by authorities. AI can free up maintenance workers to perform higher-value tasks.

9.4.2 Warehouse logistics, MES, and ERP applications

At Toyota Material Handing India Pvt. Ltd forum, the forklift application developed by Toyota Material Handling generated a lot of interest. The application chronicles the history of the evolution of the forklift from a manually operated device to one that is guided by wires to one that operates completely independently. This has many uses, including improved security, better teamwork, and more efficient operations. The digital twin is used for training, verification, and optimization, all of which take place in the cloud. Swarm intelligence and distributed learning are made possible by on-board intelligence and cloud-based communication.

This failure analysis, preventative maintenance, and optimization opportunity made possible by cloud-based learning with a digital twin saves time and money for Toyota over the course of the entire product lifecycle. Fully automated logistics that can handle smaller batch sizes are made possible by deeper enterprise resource planning (ERP) integration.

From manufacturing execution systems (MES) to ERP and beyond, AI is becoming increasingly popular for enhancing system usability. Synthesized responses to questions like 'how much material is used?' complement pre-built user interfaces and reports. Here, users like L&T, as well as AI experts, must work closely together, as context is essential. AI can be used to perform quality control, as well as fill out and generate reports, relieving employees of administrative duties.

9.4.3 Operational simulation and optimization

In spite of the fact that shoppers anticipate uniform quality and flavor, food and drink items are subject to the inherent variations in their ingredients. For international brands, the urgency of this issue cannot be overstated. Siemens's AI-based offering for the brewery industry was developed in collaboration with a brewery customer and aims to ensure quality and flavor consistency. As an added bonus, this project helped prevent the loss of specialized knowledge by documenting the steps involved.

9.5 Process planning

9.5.1 Problem description

In the manufacturing domain, the process planning sets up a correlation among processes, parameters, and machine or machine tool to convert from the design state (drawing) to the product state (an item) by machining a part or assembling the parts. The major steps for planning are process selection, elaboration, and sequencing. There are many factors such as shape tolerance, surface finish, size, material, type, etc., that affects the selection and sequencing in the planning.

9.5.2 Selection

The selection process identifies the auxiliary manufacturing steps required to make the component. The selection steps cover material process, machine, tool fixture, and end effector selector. For example, a spherical metal part with holes and cuts is fabricated using the following procedures, chosen based on the part's geometry, size, and precision demands. A metal bullet's approximate shape can be achieved through forging, while the milling and drilling processes can be used to fine-tune the bullet.

9.5.3 Elaboration

Elaboration is defined as the steps which are concerned with generation of parameters for the selected subprocess, which is a cutter path generation module. For example, tools must be chosen such as numerical control, the program must be written for the computer-aided numerical control, the program must be written for the computer-aided numerical control, and cutting speeds must be determined. This step is also turned as cutter path generation module.

9.5.4 Sequencing

Sequencing determines how to order the subprocesses.

9.6 Design strategies

The design strategy in FMS requires the following facts:

1. Determination of part type family.
2. Selection of machine tools for manufacturing each part type.
3. Selection of machine tools for each part type.
4. Selection of material handling.
5. Design of fixtures to be used for each part.
6. Determine the layout of the FMS.
7. Process planning.
8. Specifying the information processing paradigm.

9.7 Rule-based model

A rule-based model can be developed to perform the above stated functions to meet the requirement.

(i) **Selection of part-type**

Let there be several part types such as P_1, P_2, P_3, ..., P_{16} for a two-family spectrum. The FMS contains the knowledge of different levels and dimensions such as work piece geometry, precision and lot sizes, and job dimensions, etc. For example, a rule is processed like this:

Rule 1: IF the shape is circular and IF material is aluminum and IF dimension accuracy is less than 0.6 mm THEN select part type (3).

(ii) **Selection of machine tools**

For each type of job specification and part type, a particular machine is assigned. If the job specification does not match with the function of a machine tool, then a combination of machine tools are considered. For example,

Rule 2: IF processing requires some lathe AND selection material handling system THEN use combined machine tools.

The selection of the material handling system is based on the cost, maintenance, efficient utilization of equipment, and improved facility layout. For the selection of manual, automatic crane and robot-based processes, the following rule is generated.

Rule 3: IF material or job is heavy and IF environment is hazardous and IF cost of equipment does not matter and IF utilization rate is high THEN deploy on industrial robot.

9.8 Structure of FMS

Loading/unloading station: This part of the system is responsible for loading and unloading of the material or finished parts manually or automatically by robot.

Work transportation device: Carts are used to transport jobs from the loading area to the powered machine tools, and the same is true for the machines themselves.

The structure of an FMS is shown in figure 9.1.

Material handling equipment: The material handling equipment or device is used to transport work in process or tools to assigned positions. Industrial robots or overhead cranes are deployed for the purpose.

General purpose machine tools: The general-purpose machine tools such as horizontal or vertical machining centers which can perform milling and hole-making processes. Computer numerical control (CNC) lathes are used to handle rotational parts.

Auxiliary equipment: FMS includes work piece clearing, on-line inspection, automated measurement, and gauging equipment to achieve better quality of products and detection of wear.

System controller: The system controller supervises the entire functional process and components of FMS using system status data reported by sensors. Special

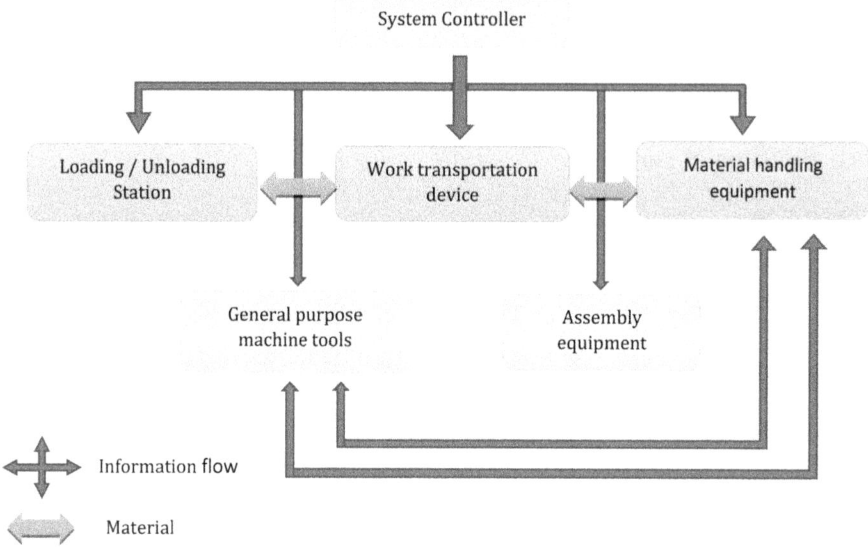

Figure 9.1. Structures of an FMS.

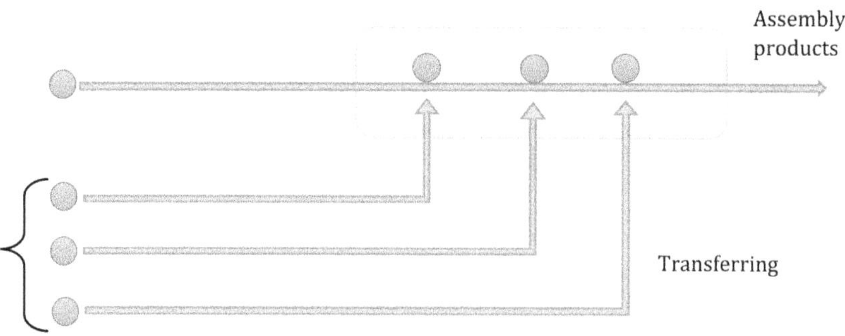

Figure 9.2. Basic functions of automation.

purpose-dedicated computing together with measuring instruments are required for the purpose.

Tool room and storage: These are located off-line. In the tool room, all the tools used in the system are stored. For the temporary storage of work pieces/finished parts a storage place is required.

Automated assembly

The automation in the present-day technology is not restricted to FMS but extended to simple automatic assembly processes. The assembly process is to bring two or more parts together and fix them in their specified relationship with either permanent or temporary fastening. Three basic functions are: feeding, transferring, and fastening as shown in figure 9.2. In automatic assembly, parts or components are supplied in bulk from a feeder, and these are oriented, fed to, and loaded on the specified work stations.

AI-based manufacturing: a comparative view

A comparative view of AI-based manufacturing ESs with their maker, functions, and features are given in table 9.1.

Table 9.1. ESs in FMS.

Name and maker	Functions	Features
*GAR Descotte and Latcombe University of Grenoble, France	*Macro level planning activities	*A rule-based model
		*50 rules and general-purpose problem solver *Implement in HACUSP.
*CMPP	*To define manufacturing practice for families of parts.	*Implement in COPPL an English-like problem-oriented language.
*Computer Managed Process Planning *United Technology Research Center with US Army, USA	*Interfaced to CAD and CAM in aviation industries.	*Computer process planning.
*KAPP: Knowhow and Knowledge Assisted Production Planning System	*It consists of CAD and user input, decision making know-how data-based and know-how acquisition subsystems	*Written in common LISP and partially Small Talk-80
*Iwata and Fukuda, Japan subsystems	*On the basis of machine surface recognition, it selects the machine cutting tools and fixture delaminating the cutting conditions.	*It runs on TEK 4405 systems
*SIPP: Semi-intelligent Process Planner. *Nau and Chang, University of Maryland, USA	*It is for generative process planning of machined parts. *Integrated with automated manufacturing research facility (AMRF). It produces plans for an automated machine shop.	*It uses frame-based model with 60 frames. *Implemented in Prolog.
*XPLAN: Export Process Planning *Lenau and Alting Technical University of Denmark	*It works for past specifications, process selection, process sequencing and cutting tool and time selection.	*It is based on decision tree logging processing system.
*TURBO–CAP *Wang and Wysk Pennsylvania State	*It works for machine surface identification, process planning and sequencing, NC code generation, knowledge acquisition and database management.	*Rule-based and combination of rule and frame model is used for knowledge representation. *Rule- and frame-based methodology.

*SAPT: System for Automatic Programming Technology *Miacic V.R.	*It is computer-aided process planning for families of parts prismatic and rotational.	*It is based on LISP and PROLOG.
*ROBEX Bowden	*Assembly planning in robot-based flexible assembly systems. *Prototype of generative planning system.	*Rule-based model.
*Zhang and Hou	*Process planning for gear manufacture. *Outer and lower KB, one to store rules and manufacturing environment and other knowledge about manufacturing independent of factories.	*It uses a rule-based model.
*Husbands, P. and Mill, F. G. Edinburgh, USA	*It aims at developing a process planning shell. *It is used for component representation.	*It is implemented in C-Prolog on Unix-based Sun Work stations.
*XPS-1 utrc, USA SACK JR, C.F	*It is a generative planning system. *It has three major areas: data director decision, modeling, and relational database in its methodology	*It is implemented in Fortran 77.
*Beckendroff University of Hanover *AVOPLAN	*Automatically generates a process sequence and it is integrated to a computer-aided generative process planning system.	*A hybrid rule- and frame-based system.
* Liu Y.S. and Allen, R	An embedded coding and classification mechanism handles the switching between variant and generative approaches automatically.	*It uses comm. and language of an MPP.

9.8.1 Process planning in FMS

The process planning requires the detection of surface, process step sequencing, solution of machine tool, jig fixture, and machining parameters. The following rule depicts the combination of some of the steps.

Rule: IF material is mild steel and IF length is 108 mm and IF major diameter is 48 mm and IF the number of surfaces is six (cube) and location of surface is external, THEN select machine MI.

In this section, we deal with the interaction of computer science in general and AI in particular with mechanical and production engineering as sub areas. One of the most versatile and widely used interactive paradigms is the knowledge-based (KB) automated assembly with robot. The other problems are collision-free path planning and robot-based vision systems have potential to use of AI methodology to resolve the several related issues. The three specific problems are described in detail.

9.8.2 Process control and scheduling

This is the continuous control over the flexible system connected with the execution of choices made utilizing control methodology and checking and adjusting framework conduct. There is an assortment of scientific methods to control the issue of FMS [8]. These include the following methods:

- Numerical programming
- Measurable methodologies

For example, mean esteem utilization and varieties of place/transition net like e-net, hierarchical models, and different types of other control ideas.

As discussed, formation-based approaches give apparent advantages for it makes it exceptionally simple to supplant or add separate 'parcels' of information instead of complete enhancement of the existing calculation-based line program. Buller [4] was one of the first to examine the major approaches, which are:

An example-directed assessment procedure utilized capture the unique idea of booking environment.

A nonlinear planning technique for coordinating assembling interaction and asset tasks.

A search algorithm to accelerate the hunt interaction. The booking issue is settled utilizing deterioration approaches to conquer the high intricacy experienced if the administrator observing algorithm is utilized straightforwardly.

9.8.3 Scheduling

The scheduling approach in FMSs is like utilizing dispatch rules, network stream analysis, multilevel or hierarchical methodology, and whole number projects. Simulated intelligence can be utilized effectively as subordinate to the methodologies, decreasing the difficulty in modeling a mathematically complex, powerful framework.

An accurate systems approach to scheduling the production framework was examined by two projects in the manufacturing scheduling realm presenting various stages of adaptability. It was shown that numerous qualities of master frameworks exist in the heuristic methodology. It also upholds the applicability of information-based procedures to supplement application operations research techniques and heuristic strategies are applied to this issue. Utilizing artificial innovation to look at and solve the issues of everyday booking of a programmed part programming techniques for CNC machines, selection of quality control charts, and the robot's way of arranging in the work area is confused with odds.

For manufacturing products, it is required that a number of grades and conditions are to be met. The resources may be machines, a specific machine, or the amount and quality of products that contain certain variables of the scheduling process.

Basically, there are two types of scheduling: predictive scheduling, which is concerned with creating a schedule of production, and reactive scheduling, which deals with adaptation and self-organizing to accommodate the changes in the environment, consisting of completion order, resources, time, and the process equipment methods.

The scheduling process has the following components:

- A process in which a product is to be manufactured within the due date of the scheduled time.
- *The routine:* A set of products to be processed and the respective methodology, i.e., a product material (raw) and the process methodology combination.
- The respective machine or machine tools correspond to the product–material–process method combination.
- *Soft constraints:* The due date or schedule time span, which can be relaxed or may be compromised.
- *Hard constraints:* Such as quality of the product, the optimization strategy and certain marketing delivery policy.

Although conventional scheduling methods using the Gantt chart still have their value and due importance, but in the real-time scheduling strategy, algorithmic or mathematical optimization methods for computation are more applicable. As the environment changes due to information exchange, interaction with external agencies' schedule expertise requires a complex infrastructure. The impact of information technology has compelled us to adopt a different methodology of problem-solving, i.e., a knowledge-based system (KBS) altogether. Many KBSs or expert systems for manufacturing, process planning, and scheduling have been developed as listed in figure 9.1, but in the following paragraphs, different architectures of KBS for scheduling are described.

9.8.4 Intelligent scheduling

This module is a cell-level scheduling strategy that is a catalyst for proper resource utilization and timely response to its objectives and goals together with the constraint. It has three essential components: knowledge base, inference engine, and a learning mode.

9.8.5 Knowledge base

Various machine/part configurations and scheduling environments supporting varying computational complexity are included. The knowledge is organized in a hierarchical manner. The knowledge is composed of procedural and declarative knowledge. The procedural expertise stands in for the data and regulations associated with the various working conditions. These criteria and principles form the foundation for making scheduling decisions. Schedules are made in accordance with the most important principles. The dominant principles are facts and rules that are dominant under specific conditions and reduce the size of the scheduling problem. For example,

Rule: IF several machines have the same tools attached and IF these are capable of the same functions THEN pool them in same machine group.

The above rule can be represented in **prolog** as:

IF have-same (machine, tools) AND same (function-capacity) THEN pool (same machine)

9.8.6 Declarative knowledge

It is made up of data collected from the factory floor, along with dynamic dispatching rules and scheduling heuristics.

Shop-floor status information: The current status of all machines and material handlers is supplied by sensory equipment attached to the PC or workstation-based controllers. The sensory output is converted to facts as assertions in the sense that the aspersion can be transformed to a rule. For example, a simple fact can be represented as:

- Machine (Mach, 11)
- Current job (Job, 21)
- Grinding (Mach 21, Job 21)
- Processing-time (11 min)
- Waiting (Job 29, Mach 11)

These assertions (facts) are: Mach (Machine) 11 is machine, Job 21 is current job, processing time is 11 min, Mach 21 is grinding Job 21, and Job 29 is waiting on Mach 11. Such types of knowledge bits are generated in the shop-floor status.

Dynamic dispatching: In real-time scheduling processes, dynamic dispatching rules are required to model the decisions process. Job rules have local and global references. Rules connecting a particular machine job queue and rules connecting functions and job queues of several machines are termed local and global, respectively. The work in the next queue (WINQ) rule is with reference to the example of the global dispatch rule, which states that the job with the least work in the next machine queue has the highest priority in the current machine queue. In dynamic dispatching, the environment changes are modeled with different assumptions and hypotheses in the various sets of rules. These are called *reactive rules*.

Scheduling heuristics: Heuristics are employed in the scheduling process rather than algorithmic or mathematically formulated relations of the parameters which describe the dynamics of dispatching rules. In real-time scheduling, in general, heuristics may or may not be favorable to the working conditions and constraints, i.e., environment.

9.8.7 Knowledge-based scheduling system

Production management requires the monitoring and controlling of two factors: time and cost for efficient production through automation methodology. As described in this section, Buckley and Murthy (1997) developed a production planning and scheduling system for a metal industry producing 10 000 different pipeline fittings [18].

- It supports long-term and short-term scheduling.
- The long-term scheduling is for the year, which ensures the schedule by week, considering the capacity of the machines.
- The information requirements for scheduling are a set of order information about the product, the amount delivered in a week, and priority.
- The product fittings for pipelines and processes are cutting, heating, or forming pipes.
- The hard constraint is that all orders must be scheduled.
- Schedule the operations in order, the preference constraints between operations must be maintained.
- Work-in-process and flow times should be minimized, and orders should be scheduled for the given week.

The short-term schedule consists of around 100 orders, each with 10 operations, and necessitates a detailed weekly schedule showing production hours.

Based on this information, guidelines are formulated for the selection of orders based on three factors: normal operations, available resources, and out-of-the-ordinary times. The variables are the time of day the process is initiated, the length of time it takes to complete, the number of options available, the frequency with which supplies are replenished, and the user's relative importance.

Routing: It consists of a given order with self-first inverse order, i.e., last in and first out (LIFO), critical routing first, and simple routing.

Operations: The sequences to be maintained are increasing operation number of first in and first out (FIFO), decreasing operation number of LIFO, an increasing number of alternatives and resources (critical operation first), and decreasing number of alternative resources (simple operation first).

Resources: There is a specific order in which tasks must be completed, such as first, the most basic resource, and then the most important.

Internals: Forward from the specified release date and backward to the specified due date are the options for choosing intervals.

AI-based architecture: The AI-based system architecture consists of a knowledge base, an interface component, users and developers, and a knowledge-based algorithm.

Knowledge base: The knowledge base comprises the facts and figures and their correlation with the application domain, i.e., production management.

The facts and figures are orders, products, the required quality of the product, resources, and the schedule strategy. These facts and figures are represented in correlated form through heuristic rules as hard and soft constants. The heuristic knowledge of the domain is an expert requirement and constraint of the application domain.

The interface: Two types of interface components have been developed for the problem: user interface and developer interface. Graphics play a central role in the user interface, which also allows for interactive scheduling, is window-oriented sensitive, displays alternate plans and machines, resource capacity load, and a Gantt chart.

- The user is able to view the impact of scheduling decisions on the problem-solving process as it currently stands.
- Schedules can be created or fixed in real time with the help of this feature by selecting elements to add or remove, rearranging their order, or swapping out resources entirely.
- The long-term interactive scheduling process is shown in two windows to the user; one shows operation instructions, and the other weekly load distribution and their effects, and the overview of the weekly capacity diagram of one machine. It provides a shift of the order to another week if the schedule violates the given capacity limit.
- The user has the freedom to choose from three distinct short-term scheduling strategies that vary in terms of operations, resources, and time intervals. PSK uses a Gantt chart to display its schedules. Users have the option of switching things up.
- The developer interface provides links to commercial data-based systems and networking. It also provides a shop-floor data-acquisition system and other programming languages.
- **The KB algorithm**: It is basically the inference and search mechanism. The algorithm for predictive and reactive scheduling, the predictive algorithm is based on the expert reasoning process, which also involves heuristic search production to find the good or desired solution, i.e., creating step-by-step scheduling.
- **The reactive feature**: The changes in the environment, external or internal, require the scheduling strategy to be adaptive enough to accommodate the external event, i.e., short-term acceptance of higher priority or internal events such as machine breakdown. These events may require the scheduling of the production process.

9.9 Conclusion

AI-based FMS automation requires *planning*, *scheduling*, and *processing* for any industrial process automation. The architecture of FMS, a model, is described elaborately. The rule-based and knowledge-based approaches are discussed for industrial automation. AI-based manufacturing processes are discussed with many aspects in a comparative view manner—dynamic dispatching and knowledge-base scheduling in one of the novel approaches in this context. Explainable AI and other modern approaches will be considered for FMSs with practical implementation through the industrial process.

References

[1] Jain P K and Mosier C T 1992 Artificial intelligence in flexible manufacturing systems *Int. J. Computer Integr. Manuf.* **5** 378–84

[2] Peter Košťál A M and Michal D 2019 Possibilities of intelligent flexible manufacturing systems *IOP Conf. Ser.: Mater. Sci. Eng.* **659** 012035

[3] SchirinBär R J B, Bauer J and Franke J An artificial intelligence approach for online optimization of flexible manufacturing systems *Appl. Mech. Mater.* **882** 96–108

[4] Holubek R and Kostal P 2013 The intelligent manufacturing systems *Adv. Sci. Lett.* **19** 972–5

[5] Danisova N, Ruzarovsky R and Velisek K 2011 Design alternatives of intelligent camera systemfor check parts at the intelligent manufacturing-assembly cell *Inf. Technol. Manuf. Syst. II, PTS 13* **58** 2262–6

[6] Mishra R B 2010 *Artificial Intelligence* (Delhi: PHI Learning)

[7] Danišová N, Ružarovskỳ R and Velíšek K 2013 Designing of intelligent manufacturingassembly cell by moduls of system catia and e-learning module creation *Adv. Mater. Res.* **628** 283–6

[8] Dhruv Lal A K 2022 Madan role of artificial intelligence *FMS: A Recent Trends in Production Engineering Review* 5 (Ghaziabad: HBRP Publication), 1–11

[9] Dawit Teklu Weldeslasie M A 2019 The application of expert systems in manufacturing sector *Int. J. Innov. Technol. Explor. Eng.* **9** 4444–9

[10] Tan C F, Wahidin L S, Khalil S N, Tamaldin N, Hu J and Rauterberg G W M 2016 The application of expert system: a review of research *ARPN J. Eng. Appl. Sci.* **11** 2448–53

[11] Chia-Yen L and Chien C-F 2022 Pitfalls and protocols of data science in manufacturing practice *J. Intell. Manuf.* **33** 1189–207

[12] Hasan T 1 and Meisen1 T 2022 Machine learning and deep learning based predictive quality in manufacturing: a systematic review *J. Intell. Manuf.* **33** 1879–905

[13] Kuo T, Mital A and Anand S 1994 An introduction to expert systems in production and manufacturing engineering: the structure, development process and applications ed A Mital and S Anand *Handbook of Expert Systems Applications in Manufacturing Structures and Rules (Intelligent Manufacturing Series)* (Dordrecht: Springer)

[14] Ruey-Kai S, Chen L-C, Pardeshi M S, Pai K-C and Chen C-Y 2021 AI landing for sheet metal-based drawer box defect detection using deep learning (ALDB-DL) *Processes* **9** 768

[15] Abdul Wahid John G B and Intizar M A 2022 Prediction of machine failure in Industry 4.0: a hybrid CNN-LSTM framework *Appl. Sci.* **12** 4221

[16] Maschler B and Weyrich M 2021 Deep transfer learning for industrial automation *IEEE Ind. Electron. Mag.* **15** 65–75

[17] Mayr A *et al* 2019 Machine learning in production—potentials, challenges and exemplary applications *Procedia CIRP* **86** 49–54

[18] Buckley S J and Murthy S S 1997 Guest Editors' Introduction: AI in Manufacturing *IEEE Expert* **12** 22–3

IOP Publishing

Advanced Signal Processing for Industry 4.0, Volume 1
Evolution, communication protocols, and applications in manufacturing systems
Irshad Ahmad Ansari and Varun Bajaj

Chapter 10

Applications of deep learning in revolutionizing industrial sectors

Rohit Roy Chowdhury, Tapomoy Koley and Kausik Basak

With the advancement in modern deep learning strategies, various industrial sectors have witnessed a paradigm shift in their work management and technological outcomes through incorporating complex mathematical reasonings to understand and analyze multi-hierarchical datasets. Contextually, deep learning-based applications are also evolving day-by-day to deal with the arising challenges in real-life scenarios. As Industry 4.0 inspires accumulation of multi-sensor data with smart devices, connected over the Internet, artificial intelligence with deep learning enables harvesting such information to cater to data-driven decisions with significant throughput and efficiency. In this Industry 4.0 paradigm, the majority of industries are moving towards automation for various tasks. This allows companies to have a locus through which data flows and can be collected to understand the consumption patterns for the consumers and take decisions that support these behaviors. Since industries are heavily backboned with a huge number of databases, deep learning can thrive almost in every sector, paving the road of its growth in a smarter way. This chapter showcases some of the present-day applications of deep learning-based algorithms in the domains of the stock market, marketing industries, bioinformatics, and cybersecurity with an emphasis on how modern techniques are revolutionizing these fields.

10.1 Introduction

We are living in an age where machine learning, more precisely deep learning, is playing a paramount role in automating applications around us—whether it is a smart home, a driverless car, or automatic control of various electronic goods. The advent of deep learning has entirely transformed our lives into a different paradigm, paving its way in multiple day-to-day aspects, with or without our choice to embrace it. With continuous evolution of various frameworks and algorithms in this field, we can observe a sharp rise in its importance and demand across numerous fields and technologies. In recent times, in almost all domains of technological revolutions including the fields of healthcare, agriculture, transportation, automobiles, cyberse-curity, economy, mobile technology, etc., one can find applications of artificial intelligence (AI), including deep learning, to reduce human intervention and effort. There is a significant play of deep learning in data science as high-value predictions can pilot to better and smarter decisions in real-time. Also, as a technology, deep learning assists in analyzing huge databases with significant accuracy and perform-ance, easing the tasks of data scientists through different automated processes. With the advent of automatic sets of generic methods, deep learning has paved a new way of data extraction and interpretation in industries. With the enormous computa-tional power in modern processors and clusters, various industries are replacing conventional methodologies with deep learning architectures to achieve data-driven decisions, creating more accurate and precise outcomes.

Deep learning, a subset of AI, helps in providing inferences mostly based on complex mathematical reasoning to drive a large dataset, imitating the functionality of a human brain. It learns the hidden patterns present in a data bank and utilizes the learning for complex decision making. The disciplines, which are anchoring-machine and deep-learning domains, include mathematical modeling, probability and statistics, linear and vector algebra, data structure and algorithms, AI, image and video analytics, computer vision, augmented reality, etc., which are tangled with each other from foundation to application stages, making machine learning an inevitable tool in almost all application fields in modern age. While data structure and mathematics form the basis for various deep learning algorithms, computer vision and augmented reality, on the other hand, they create an application environment where many deep learning-based algorithms can play an important role in segmenta-tion, classification, clustering, prediction, and much more. Also, with the advent of ultra-fast computational processors and enormous storage capabilities of recent computer systems, it becomes facile to implement complex and sophisticated learning algorithms on huge databases, enabling us to explore every form of data around us to produce fast insights and create fact-based decisions.

Apart from analyzing the data for various decision-making processes, deep learning approaches also allow the option to make multiple decisions based on changing trends. It can be seen in the financial sector where the stock market is based on sentiments and therefore analyzing the posts being made on social media and articles being written, can be channelized as a tool to comprehend market sentiment. Likewise, modern techniques in deep learning have reinforced different dimensions in the education sector to properly articulate student-specific program strategies based on the preferences of subjects and behavior, making a breakthrough in improving study patterns. Moreover, other sectors like automobiles, cybersecurity, finance, etc. have also been fortified with the inclusion of deep learning computations. This chapter mainly focuses on the application areas of deep learning in various industrial sectors, such as the stock market, marketing industries, bioinformatics, and cybersecurity. In the following sections, this chapter provides a wide landscape of deep learning aspects, starting from its relevance in Industry 4.0, to its learning strategy, to various application areas in different industries. In the future, AI and deep learning will see more improvements with developments in quantum computing and big data due to enormous computational speed and abundance of data.

10.2 Relevance in Industry 4.0

From an industry perspective, the modern age requires data-driven technological solutions to decipher problems in almost all areas of human civilization. Standing at the Industry 4.0 generation, where it focuses on continuous accumulation of data from multiple smart sensors and devices connected over networks, deep learning can leverage analysis of such large amounts of data in a fruitful and efficient way. In parallel, with big data and the Internet of Things, the whole paradigm is creating an interdisciplinary arena that ascertains hidden information from large datasets using knowledge-driven computational algorithms. These mathematical reasonings help to learn complex patterns from the data and create actionable intelligence, easing the pathway for obtaining meaningful insights about the process, followed by channelizing these decisions in remote places to perform specialized tasks. Such modern decision support mechanisms are not only capable of delivering optimal decisions, but also enable continuous monitoring, forecasting, optimizations, quality control, and management. While the conventional learning algorithms are quite successful in a variety of applications, ranging from manufacturing industries, data storage, compression-to-edge computing and cybersecurity, a lot of research is still happening to provide more efficient analysis and high throughput in terms of productivity of any organization.

In this era of deep learning, industries are executing algorithms in various conceivable areas, including in-house and outdoor applications, to discover underlying patterns and trends. Recent advancements in the analytical domain have nurtured the revolution in data analytics to a supreme extent. With advancements in hardware for increased storage capacity, GPU technology for enhanced micro-calculations, and frameworks like TensorFlow, PyTorch, etc., the usefulness of deep learning has become more mainstream. Almost every organization, irrespective of its size, depends on data, related analytics and scientific mechanisms to make critical business decisions—in which case deep learning has become a backbone to all such comprehensions. In this Industry 4.0 revolution, digital ecosystems are generating digital businesses which are gaining value through enhanced interactions between business, people and things. The need-of-the-hour technologies and knowledge-based computations are greatly motivating new age innovations that can help organizations make more precise, apt data-shabby decisions.

Deep learning and AI-based solutions can now be seen in real-life usage, from face unlocks in smartphones, automated cars, and even fraud detection in banks. In research, too, the introduction of complicated new layers and the optimization of frameworks are lowering both physical and data requirements. Along with data availability, a lot of progress is made in data engineering. For example, in the field of the stock market, the concept of utilizing bots to make stock selections and trading has increased. This has allowed investment banks to check and execute trades much faster and with better accuracy. Deep neural networks are enormously researched and implemented to try and accurately make predictions on market scenarios. Contextually, user segmentation and targeted marketing are becoming core areas in which deep learning is hugely explored. Several existing methodologies are also improved so as to include deep learning in the computational design architectures. Additionally, in the biological domain, with huge datasets and tasks, conventional regression and classification algorithms very often face bottlenecks, in which case, deep learning algorithms can pave the way to explore different dimensions with more accurate inferences. The data and the deep neural networks both need to be deployed in a certain way in order to gain a significant foothold in various industrial works.

10.3 How does a network learn?

A generalized process schematic of any learning system is depicted in figure 10.1. Given a problem environment, the foremost task is to collect a sufficient amount of data that represents the problem well. Such a dataset comprises various features that are significant in describing the problem; several feature extraction or selection strategies are used to figure out the important attributes. Since the learning of any deep neural network requires a significant amount of data, it is very crucial to accumulate as much data as possible from the problem domain for the development of the resident knowledge. The entire dataset, after pre-processing, is divided into training and testing groups—usually 70%–80% of the processed data is kept for training or learning, whereas the remaining data is reserved for testing the algorithm. The deep neural network is trained on the training dataset to accurately learn the hidden features, targeting to identify the function that appropriately maps the input data to output. The number of input layer neurons of such a deep architecture match with the number of features selected. Several hidden layers (numbers may vary from problem to problem) are formed, in which several activation functions are selected, such as convolution, ReLu, sigmoid, etc., to track the linear and non-linear relationships in the dataset. Along with this, number of neurons is finalized in different hidden layers. Different optimization techniques are executed to reduce the error and optimize network weights, bias, etc., leading to creating a trained model enriched with the knowledge of the problem domain. After the training phase,

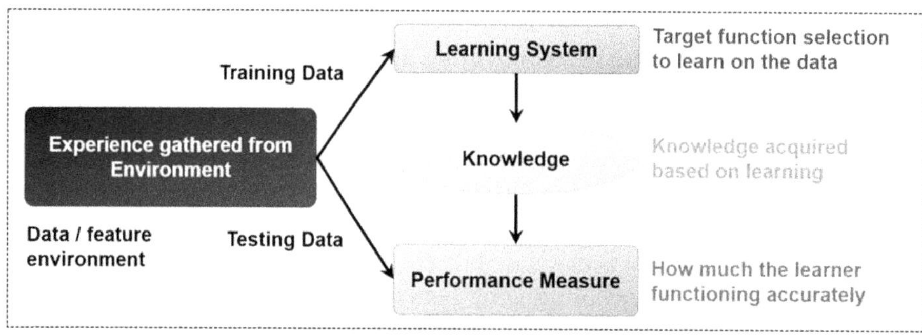

Figure 10.1. Process schematic of a learning system.

performance of the computational framework is examined using the testing dataset, in terms of various measurement metrics, e.g., accuracy, precision, recall, sensitivity, specificity, f1-score, etc. Without much elaborating on the technical side of various deep learning techniques, let us focus on its applications in various industrial sectors.

10.4 Application in stock market

10.4.1 A brief about the stock market

Let us start with some general introduction! The stock market allows people to trade in equity of publicly traded companies. In the market itself, there are two subdivisions of equity—the future or derivatives market and current market. Although, the derivatives market has higher volatility, it allows an advantage over the current market [1, 2]. The advantage is that a large bulk of equity can be purchased by paying only 10% of the bulk value, which is called the margin. The full and final payment is usually made at the end of the contract period—the last Thursday of every month. In case the last Thursday is a holiday, then the last day is moved to the next working day. The derivatives market also provides a flexibility in which it is possible to sell off the shares that were pre-purchased with the margin paid, but allows the full profits to be held. This is accomplished by a system called margin maintenance [2]. This means the margin for the amount purchased has to be maintained as the equity price changes over a daily basis. Hence, if the price increases, more margin has to be added and if it decreases, then the seller will also lose on price. In the Indian stock market, equity is monitored depending on the variations of a few significant points on an everyday basis. These are OPEN, HIGH, LOW, CLOSE, and OPEN INTEREST. OPEN signifies the opening price of an equity at 9:15 am on a working day. CLOSE is the price at which the equity is closed on a working day. It is taken as the average of the OPEN price from 3 pm to 3:30 pm. HIGH indicates the highest valuation that the equity has reached on that specific day and LOW determines the lowest valuation that it has reached on a day. OPEN INTEREST signifies the number of people with an active order [3]. Due to the highly random and chaotic behavior of the market, a forecast made is usually accurate when done over a smaller time frame, instead of a larger time frame. The root cause depends on the fact that the changes in values appear on a more-or-less daily basis. Importantly, there are also external factors that need to be considered. These include factors such as public sentiment, government decisions in policies, foreign financing impact, international trade law changes, and many more.

10.4.2 A generalized approach

A generalized block diagram of a stock market prediction process is presented in figure 10.2. At the beginning, stock-specific market data are acquired from various online historical archives such as the National Stock Exchange (NSE), Bombay Stock Exchange (BSE), etc., depending on the application domain and relevance of the product in that domain. Usually, these datasets carry noises and artifacts, which eventually hampers the learning process of the deep neural framework, leading to an overfitting issue. Therefore, pre-processing is crucial in such a context to denoise the

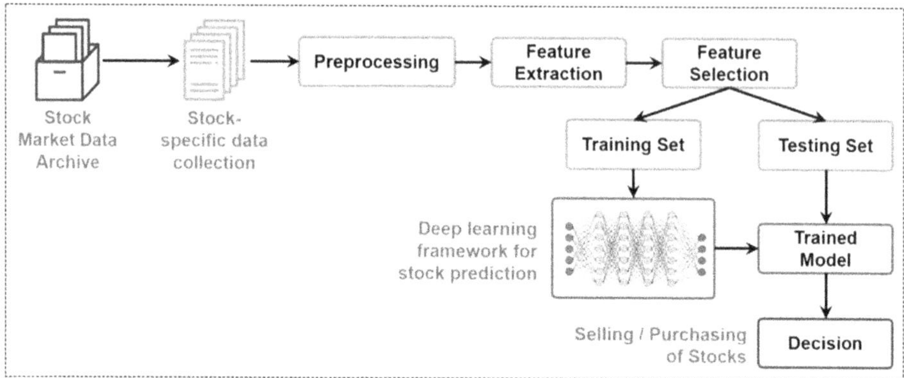

Figure 10.2. Block diagram of a generalized stock market analysis/prediction framework.

dataset. Once denoised, the dataset is further curated for selection of features which are relevant to the application domain, discarding other attributes. From the machine learning viewpoint, it is also desirable to find out the significant feature-list which can further assist in learning of the network and provide more accurate results. Any presence of insignificant features may lead to drastic reduction in the accuracy score. Thus, feature extraction and selection are required to figure out significant features. Various algorithms are present to perform feature significance tests— principal component analysis, forward and backward sequential selection, and random forest are some of them. After selecting significant features, the entire dataset is partitioned into training and testing groups; usually 80% of the overall data is utilized for learning, whereas the remaining 20% is reserved for testing purposes. The deep learning-based framework is trained on a dataset, randomly nominated from the original data matrix. In this process, the network tries to learn the hidden pattern, representing the class-specific discriminating properties, through tuning of the synaptic weights and other parameters. Once trained, model validation is examined against the test data with respect to various measurement metrics, e.g., accuracy, sensitivity, specificity, f1-score, etc., to accurately pursue the analysis of a final decision.

In the case of market analysis, considering the past movements to predict the next movement, deep learning can be considered as a powerful tool to analyze the data and cater a data-driven decision. Furthermore, deep learning concepts are not only significant in price forecasting of the equity market, they can deliver significant insights on various other markets, entangled with the stock market directly or indirectly. Examples include: commodities and foreign exchange, which are also highly traded markets. Apart from forecasting, the application can be extended to hedging as well, which is a complicated statistical procedure for saving investments [4].

10.4.3 The RNN and LSTM models

Contextually, recurrent neural networks (RNNs) have a notable amount of advantages [5]. RNNs have multiple deep-learning models in the same pipeline, where the models may be similar or different types. The idea is to feed the output of one neural network as an input to another. This allows a comprehensive system to be developed with more flexibility in network topology to solve various complex problems. Hence, RNNs have gained a lot of success in the field of stock market forecasting. However, considering the RNN framework, since contextual information is passed through several layers, it will cause the various weights and gradients to be a victim of decay or exponential blow up. Both of these can cause issues in the actual training of the network. In such a scenario, the long short-term memory (LSTM) can mitigate these problems. LSTM, a special variant of RNN, resembles a type of node that was developed to fit well in these kind of problems [5, 6]. The node diagram of a LSTM module is depicted in figure 10.3(a).

The LSTM node can take a set of multiple values as an input. As the name suggests, the objective of this node is to store important data or functions in its short-term memory for a long-term impact analysis. LSTM modules have three gates, the forget gate, output gate, and input gate. The input gate allows inputs of various sizes. LSTM has two sections in the node—a hidden state and a cell state of the previous timestep [6]. Both of these are short-term memory components and can

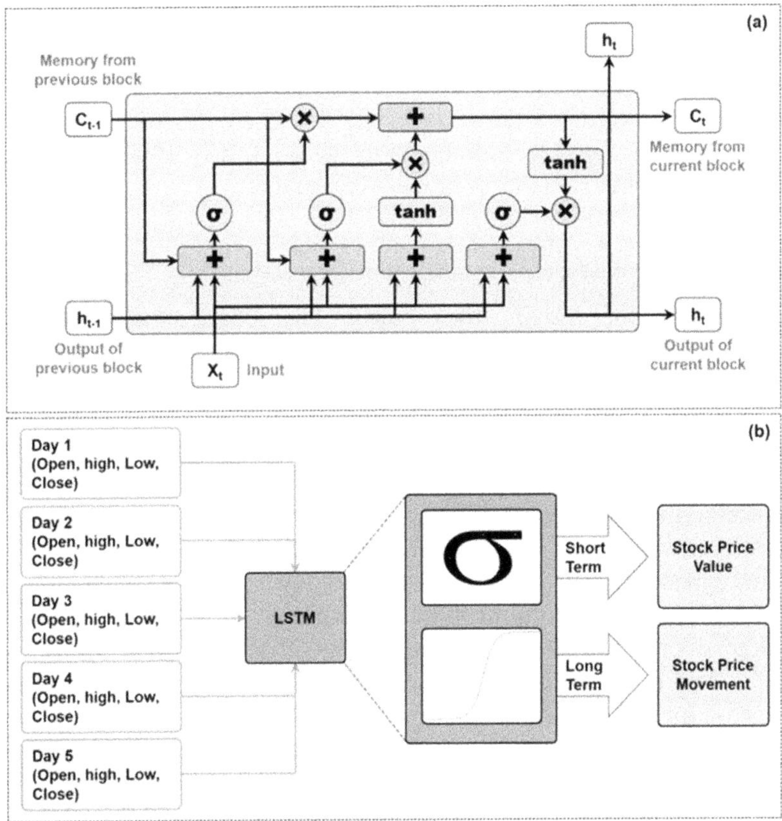

Figure 10.3. (a) Node diagram of a LSTM module. (b) A sample LSTM-based framework, used in stock market prediction.

withstand important attributes from the previous timestep's information. Simultaneously, the architecture also includes a forget gate in order to figure out which information should be discarded. Mostly, the forget gate implements a binary sigmoid function—for 0, it discards all the information, whereas it stores data for 1. An LSTM node also has a facility of providing flexibility to cardinality of the input and output. It can be one to one, many to one, one to many, and many to many as required. Importantly, in the majority of cases, the model accuracy for such a structure is significantly higher than the conventional machine learning methods [5].

Let us consider one example to understand its operation! In stock market analysis, the LSTM node takes an input as Open, High, Low and Close of 5 days, as depicted in figure 10.3(b). The 5 days are provided as an input and the LSTM node will predict for the next day or a gap of certain days. The long-term memory aspect helps to track the direction of movement, whereas the short-term aspect assists in quantifying the value of movement. If we need to predict only the next day, then the value will look at the five input days and provide result for the sixth day. However, if the prediction is to be made for the next three days, then in that scenario, it will make the sixth day prediction and omit the first day of the input. By adding the predicted score to the input, it will run the model again and therefore try to predict for the seventh day and so on. The sigmoid function plays the main function in the regression, whereas the tanh function is used if the change is positive or negative. However, the main issue lies in the factor that any error at one level of prediction carry forwards to the next level. This decreases the accuracy as days progress. Accounting for the randomness in the data, it is usually stated that it is better to predict for the short term in the stock market.

10.4.4 Deep learning researches in stock market analysis

While diving into the Indian stock market, the contribution of the NSE is noteworthy— it can be portrayed as the world's largest derivative exchange depending on the count of trading contracts (as statistically established by the Futures Industry Association in 2021). In the NSE, NIFTY 50 is a potential index, consisting of 50 industries from various sectors that have the highest valuation in free market capitalization [7, 8]. The periodic movement of the NIFTY dictates the performance and overall sentiment to gauge the performance of the stock market. To analyze such time-series data and comprehend the heteroskedasticity, a generalized auto-regressive conditional heteroskedasticity (GARCH) model is most often used in many financial institutions [8, 9]. It points to the irregular variations of the error term in any regression-based problem. The GARCH system has three steps to it: (a) fit the data in an auto-regressive model, (b) calculate the autocorrelations in the error term, and (c) perform a test for significance [8]. One of the key aspects of creating a successful investment portfolio is risk management—understanding and managing the volatility of a stock. A stock with a high beta value signifies that it will move a lot. This allows for larger gains with respect to return of investment. However, this can also fail as it means that there is a chance that a stock can also move into the opposite direction, resulting in heavy loss. The GARCH model can be implemented to understand the volatility of the market depending on its movement. It can have multiple different variations. These include models such as exponential GARCH (EGARCH) and the threshold GARCH (TARCH) model [8, 9]. In the stock market, low moving trends are usually seen as sudden and steep movements, whereas high moving trends are usually not that steep and possesses a stable slope. These kinds of movements are called asymmetric volatility phenomena. It was validated that the EGARCH and TARCH models are specifically built to understand these movements. A comparative study was carried out between EGARCH, TARCH, and a LSTM-based model on the NIFTY 50 in terms of understanding and showcasing volatility [9]. The analysis showed that the predicting capability of the EGARCH and TARCH models was better, whereas, LSTM showcased better responses at regression-based output and recognizing patterns in the data [9].

Pair trading is another popular trading strategy in the stock market. Pair trading has two elements—stock selection and trading. Stock selection is the part where it is important to correctly select two highly correlated stocks [10]. These are usually picked as two stocks in the same industry or from different industries where the impact on one can be perceived in another. The primary idea is to figure out two stocks with high beta values (which represents volatility of a stock). The baseline assumption in this trading is that the trend in one stock will be replicated by the other stock [11]. Hence, if the trend of one stock can be predicted and verified then the same strategy can be applied to the other one without requiring separate analysis. This usually ends up with better gain since two transactions are being made instead of one [10, 11]. The first step of choosing stocks can be considered to be an optimization problem. Choosing the correct pair of stocks would maximize the profits and this is something that a genetic algorithm (GA) is good at [12]. GAs can be used to pick the stocks and then an LSTM-based approach can be implemented to obtain final prediction. The performance of such a GA-LSTM network was evaluated, utilizing a distance method along with a cointegration approach, to select appropriate stocks [13]. A GA was applied to the whole stock pool to pick out two stocks, which were then used in an LSTM-based network to predict direction. A comparative analysis showed that the combined technique outperformed all other baseline networks. The network that was trained and tested on the Taiwanese Stock Market reflected a remarkable 20% better result than other options [13]. Another approach in pair trading is to make use of deep reinforcement learning [14]. It trains an agent on multiple scenarios that are both positive and negative based on a reward–penalty system. When the agent makes a favorable decision, it is awarded with some kind of a reward and also charged a penalty when a mistake is made. Maintaining risk and volatility are the prime factors for working with reinforcement learning. Similar to the risk-reward system, a new system was put forward called the reward volatility system [15]. Reinforcement learning also appends the added features where it can be automated, hence requiring minimum supervision compared to the rest.

From another perspective, the stock market is reliant on the economic decisions taken by larger corporations and governments [16]. The decisions reflect a high impact on the volatility of the price of a company-specific stock. Along with that, the stock market is also responsive to supply and demand. The most basic macro-economic principle is applicable here [16, 17]. As demand for a stock increases, it will also increase its price and vice versa. Therefore, as the majority of the public invests in the market, it becomes open to be affected by sentiments. Contextually, social media plays a large role in the pricing as influencers voice opinions while simultaneously public statements made by high-ranking officials in the company can lead to a negative impact [17]. Deep learning architectures that are trying to forecast the movement of a stock can benefit heavily from a sentiment value. However, because of the chaotic and varied nature in the way social media posts are made, understanding the sentiment manually becomes a taxing and time-consuming task [18]. Not to mention, consider the huge dataset that is generated by people's opinions, which can be difficult to label. In such a scenario, a deep learning model cannot only understand what kind of sentiment it is, but also quantitatively implement a score that can showcase how positive this is [15, 18]. The sentiment value is a good indicator towards understanding what kind of movement will actually happen on the following day. Depending on these sentiment scores a decision support framework was proposed in [18]. This is a suitable tool to be used by traders who are not looking to invest for a longer period of time but instead are looking forward to short- and medium-term gains. The public sentiment regarding an event or an equity is usually short lived and therefore it suits the short time duration of a swing trade.

While making a trade, a trader looks at several numerical factors along with candlesticks patterns. However, this information is usually historic. Predicting for the next day itself is a challenge and the error term becomes more and more proficient as we move forward. These issues can be resolved by means of an RNN pipeline while using the sentiment score as a factor to control the values and their spikes [18, 19]. Such a decision support framework was made to create a report which would be good for the next 30 days and would be looking at technical factors, such as money flow index (MFI), the support and resistance of the equity, Fibonacci retracement levels, etc. The root means square error, mean absolute error, and mean absolute percentage error were found to be 4.13%, 3.24%, and 1.21%, respectively, for such a model [19]. Hence, the trader can make a more educated prediction or analysis of a trade with the presence of more information.

10.5 Application in marketing industry

10.5.1 A brief overview

Standing in the twenty-first century, we won't be wrong if we claim that the education sector is one of the largest consumer bases. As the education systems move away from the traditional pen and paper-based framework, technology-based solutions, extra help, and tuitions in the form of app-based learning are becoming more prominent. This has become even more prominent in the post-COVID scenario. Due to COVID restrictions, people have shifted to an online-based curriculum, in which case, the acceptance and overall use of such edtech startups has multiplied [20]. With the abundance of larger userbases and more metadata, the companies are looking forward to better and informed decisions. Such a huge abundance of data opens up the field of implementing data science problems to resolve various issues. The data can be used to predict and monitor behavior of users and to perform targeted marketing along with helping in making hundreds of other decisions. Its impact can be massive in understanding which design changes to implement, which courses or features to roll out and also to understand market demand. Edtech companies usually have a paid and free version of their product— the free version with limited functionality for students to have a more rudimentary experience of their offerings, while at the same time also being significant enough that a free user also gets value out of the application. This is done for future prospects, as a satisfied free customer today may convert into a paid customer down the line. Hence, understanding the requirements of a customer and offering them curated experiences as per their need, and to have predictive classification and regression-based solutions, deep learning can play a paramount role. Interestingly, these factors can be extrapolated to other marketing industries as well.

10.5.2 Resolving issues through deep learning

The utilization of deep learning in education-based marketing tends towards two primary sections—either it's a forecasting problem or it is to provide marketing-based classification [20, 21]. Contextually, big data, being prevalent in industries, is a major pillar in successful implementation of deep learning. By making use of big-query and non-structured structured query language-query, in the form of Hadoop and MongoDB, it has become the more preferred option. In the case of analyzing and predicting continuous data, usually linear regression is implemented. For a non-linear trend in data, either multiple linear regression or support vectors is a preferred choice. Even though all of these tried and tested algorithms have a lot of flexibility, they still do make use of linear relationships and try to circumvent the problem by increasing the dimension or they may try to do it using multiple regression lines [21]. Deep learning-based models have higher flexibility. Deep learning models can train themselves on to data that may not be linear in nature [22, 23]. Compared to ensemble methods, e.g., random forests, deep learning models have the added advantage to efficiently tackle various problem statements in both regression and classification domains. In a study, deep learning and more traditional models were compared for a marketing strategy determination, and it was established that the deep learning model made more effective usage of the input features and was able to provide better decisions on which campaign will be made for a particular customer [23].

An important issue in real life data is that the data is rarely balanced. It is more often that the data will be skewed towards one end. This means that a deep learning-based approach can be better sampled because by using regression as a form of regularization, bias can be managed to some degree [24]. The concept of onsite customer profiling and hyper customization, which has been used in a lot of physical retail stores, also had a positive impact. By utilizing state-of-the-art technologies, deep learning was implemented to validate if the collected data could impact on strategic marketing. It included factors like facial expression recognition, age and gender recognition, and in-store customer tracking. This information could be used in different cases to receive a number of added benefits, such as how to position products and store and assemble data for future analysis [25]. In a survey performed in Europe, it was evidenced that such technology was something that the fashion stores were looking toward as this could be executed as an assisting tool to place different products [26]. Most retail stores experienced that the product placement plays a very major role in sales. The application of deep learning, in this case, was established to be an effective tool towards forecasting overall sales of retail stores [25, 26].

An important parameter companies are usually aware of is churn [27, 28]. It is a major concern and diminishing churn in a company is usually one of their major foci. Churn is defined as a customer who has utilized an app for a short time and will no longer return to the platform. Going by the usage styles of a person (can be used as a feature), a convolutional neural network (CNN)-based approach can be used which will try to figure out patterns in the usage and can predict whether the customer will turn into churn or not [28]. However, it was identified that simply applying convolution to the dataset was leading to a lot of data loss. Hence, to combat this, an extended convolutional decision tree (ECDT) was suggested [29]. Along with a grid search algorithm for optimization, the ECDT–GRID system proved to be the most effective. A comparative analysis between a few deep learning-based models with conventional machine-learning techniques was performed and it was observed that decision trees and random forests had better performance [29]. Due to the majority voting nature of random forests along with the branching decision advantage of decision trees, these algorithms showcased a better advantage in terms of classification. The random forests made use of multiple trees and reached the result using a majority voting. Even if one tree cannot capture the trend it is likely that others will do. Contextually, deep learning has the option of portraying results as probability scores, instead of only one outcome. Although, when it comes to finding out patterns in the data, CNNs cater higher advantage. Therefore, if the data from the convolution layer is merged with a random forest, the combined model can reach higher accuracy [29].

10.5.3 Recommender systems

The recommender systems are of a huge value in industries as well [30]. The implementation strategy of a generalized recommender system, used in such a marketing domain, is depicted in figure 10.4. The architecture is made based on a hypothesis that people usually have an affinity for a specific action. If that affinity can be recognized, then the person can be kept engaged for a longer period of time. Recommender systems analyze historical data of people's actions and preferences over a certain period of time and formulate correlations and connections between these activities. The idea is that if a user performs an action A1, action A2, and so on, then the model after a certain point will be able to accurately predict the next action. This can be used in targeted marketing and showing relevant ads. Netflix, Amazon, and YouTube are companies with excellent recommender systems. However, the data used

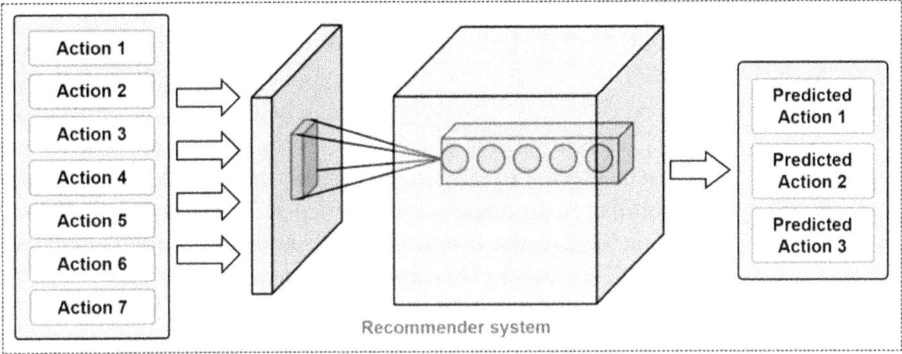

Figure 10.4. Implementation of a recommender system in predicting actions.

for training such models needs to have low variance—if variance is high then in that case it can become difficult to model correct relationships. In such cases the recommender system will provide random predictions till the point that a connection can be made. Designing a neural network comprising a recommender system can be achieved with very simple models, making use of forward propagation and a network of hidden layers, whereas more complex ones can make use of RNNs and other networks in order to make multiple predictions at the same time.

Usually, the buying and usage patterns of an individual are unique. Even though all of them can be stated to fall under multiple clusters over a longer time frame, in a short period the recommendations may be completely different based on a necessity or previous purchase history. For example, the same can be stated from the high success rates of Netflix and Amazon's recommendation systems. The content that is shown on the 'for you' page of Netflix has a much higher affinity to be watched than a customer actually searching for something. A good recommendation engine can cater the content based on preferences of the user [30]. Deep learning technologies with convolutional filters can be used, in this context, to oversee the recent and past purchases or actions and identify the patterns. Based on matching patterns with historical data, activities can be suggested to a user [31]. This has a lot of usage in applications that are trying to engage the customers for longer screen time. If the content being shown is something that the user prefers and it is there without the need for a user to actively search for it, the retention rates would be higher. The only downside of this is that the system requires the user to pursue significant actions on the application itself for it to start matching to proper content [30, 31]. Even though the system can look for patterns from the search, the actual output sample is observed to be smaller initially and grows in size as more topics and actions get included. Hence, an active user can see a more curated list of content than an inactive user. To address this problem, the latest model of recommender system tried to fix it by first looking at interest-specific genres of users [31]. In the initial phase the search is broader where the user is recommended with broader categories that they may find interesting. Once more data is collected, this can be modified to showcase specific elements.

With the utilization of collaborative filtering using online available data or historical data, RNN neural networks were designed to mitigate this challenge. The main focus in this network was to have two preferences, being provided as an output —short-term prediction representing a broader range, whereas the long-term aspects tend to be more specific [32]. Knowledge graphs are a very underrated tool that can be added along with these systems to obtain the predictions and a better comprehension of the output. Similar to autoencoders, the idea is to have two agents—(i) a knowledge representation (KR) of the training dataset and (ii) a knowledge learner (KL) that learns from the KR model. This signifies that instead of directly feeding the model features, it is fed a representation of the features which is an output of KR model. This allows the actual features to be turned into a more normalized and well represented set of values, for the KL model. The final output was computed in terms of Pearson's correlation similarity and an adjusted cosine. When compared with online course recommendation data, it achieved a reduced mean absolute error than traditional recommender systems [32]. Moreover, recommender systems play a critical role in maintaining customer retention by providing them suitable contents. As more accuracy is gained, smaller actions can be taken which can look to guide a user to get the optimal experience from a service. This will eventually increase the rate at which users use applications while at the same time can have a high impact on churn reduction.

10.6 Application in bioinformatics

10.6.1 Diverse areas and scope

Bioinformatics can be simply portrayed as a multidisciplinary field that encompasses computational analysis of biological data at the molecular level, enabling storing, organizing, and analyzing a large volume of biological data. Standing in the 21st century, where the different dimensions of the conventional healthcare framework are getting revolutionized with the manifestation of the digital ecosystem, the field of bioinformatics has also shown a paradigm leap. Over the past decade, rapid evolution in sequencing and other -omics technologies, such as genomics, transcriptomics, proteomics, and metabolomics, have generated an extensive quantity of biological data and paved the fast progression of the multi-omics field. As predicted by experts, only the field of genome sequencing will harvest around 40 exabytes of data per year by the next decade [33]. Integration and interpretation of these multi-omics data have huge potential in unraveling the hitherto unexplored biological phenomena, which side by side requires diverse applications of data science to bring out some meaningful conclusions. This kind of high-throughput profiling of biological systems is influencing the domain towards entering into an era of 'big data'—handling and analysis of such large data requires high-throughput and powerful computational infrastructure and algorithms [34]. Recent advances in computational programming, data organization, and visualization algorithms along with massive evolution in the AI field have paved the way to fruitfully conducting bioinformatics researches to resolve biological problems. In addition to offering extraordinary prospects towards new biological inventions, the accessibility of such big data also poses novel challenges from the viewpoint of data mining and analysis. Storage and interpretation of such enormous biological data necessitates data science tools, as they impose technical difficulties because of the diverse data sources and storage facilities. The present-day data science is efficient in tackling the intricacies of enormously huge datasets with constant expansions. Therefore, the biological and medical science society is showing much attention in data science with a hope that involvement of data science can generate meaningful biological insights by processing and filtering large datasets through the appliance of various machine learning models. Without putting much attention on the biological part, this section focuses on deep learning applications in bioinformatics, more precisely on the genomics segment; the knowledge can be further extended to other omics-based studies and applications.

10.6.2 Deep learning implementation in genomics

With huge genomic sequences, their sub-fragments and inter-relationships, this field is one of the complex fields in which deep learning can be implemented [33, 34]. Over the last decade, a lot of development has been carried out in this field. Prior to implementation of deep learning, usually we need to encode and normalize the signature nucleotides A, C, T, and G into a string of numbers that can be aptly comprehensible to the learning framework. With deep learning models and their complex mathematical reasonings, classification and prediction problems become easy to tackle. A generalized block diagram of such an implementation framework is shown in figure 10.5. Sequence data, acquired through various online repositories and laboratory archives, are preprocessed to reduce the noise and remove identical sequences, leading to creation of a curated dataset. After encoding and normalization of the sequence-reads, the sequences are usually fragmented into equal lengths of data using a sliding-window technique with a fixed stride size. The fragmented dataset is then divided into training and testing groups for learning and validation of the computational framework [34]. During the training phase, each fragment of data is fed into the input of the network. Over the process, network weights and parameters get optimized to map the input to output with reduced error. Once training is accomplished, the computational model is validated using testing data to showcase its performance in accurately predicting the unknown class of any given sequence fragments. Although the final output may depict different information, the strategy of counteracting the problem remains similar. Often, a soft-max

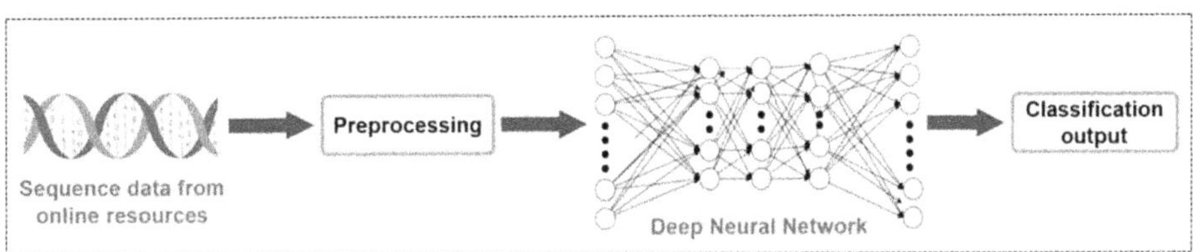

Figure 10.5. Block diagram of a deep learning-based implementation strategy.

layer may be incorporated to showcase a probabilistic score of different output classes. This eventually helps to identify the percentage associations of a single input with multiple outputs.

10.6.3 Challenges and strategies

The first issue arises due to distribution differences, as an effect of non-uniformity in various biological structures. As the sample specimen changes, so does the information and the patterns. For instance, proteins, which are elements of functional categories, have distinct expressions and physical interactions. Solving this issue is an open research topic as well, even though some industry-specific options can be used. These include quantile normalization and empirical Bayes adjustment [33]. Some recent solutions have come up in the form of domain adaption and transfer learning [35]. Another issue is due to dependent variables. A case of dependent examples can be represented well by a card draw from a 52-card deck. The action of drawing a card after a card has been drawn is independent even though probability of the draw itself has changed. Similarly, in the dataset, interactions which seem to be independent initially may not be so. Failure to adapt to this change can lead to bias in the model along with overly optimistic results. Protein-to-protein interaction can appear in both training and testing sets. Random cross validation is not the correct approach to fix this issue [35]. There can be masked variables which can change the result and this is known as confounding. It can be difficult to identify and resolve at the same time. There is also an issue of data imbalance. In other fields, data imbalance may also occur. However, it can be corrected without leading to a huge data loss. Interestingly, the same cannot be said for biological data. None of the real datasets have balanced data and it can be difficult to omit them because they are highly interdependent [35, 36].

Even with all these limitations, breakthroughs in model development were performed—a eukaryotic sequence classification system called Tiara was developed [36]. Eukaryotic genes are parts of DNA that represent a kind of template for RNA production. Eukaryotic genes are linear in nature and are compacted by histones with telomeres at each end. It was shown that the Tiara model could classify between six classes where three were prokaryotes, two were eukaryotes and one was unknown [36]. The model which was trained on the NCBI genome database, and used manual selection for the genome markers. This was required as the entire information taken together would add-up to a large amount of data, using which model fitting was difficult. The Tiara model had a two-step classification in which a user set a threshold of confidence (as an input). If that threshold was not met, then it classified an input as unknown. Compared to another tool with a similar function (EukRep), it managed to outscore it by 2.5% [36].

In the present age, shotgun sequencing has transformed the field of microbiology [37]. However, current systems for detecting bacterial species are following the conventional rule—aligning the sample with existing database. Besides being a time-consuming and resource-heavy task, this leads to high false positives, increasing scores for wrong classifications [38]. De novo binning strategies were developed in order to reconstruct the genome sequences that needed to be fully characterized. Although the developed methodology was not significant in terms of classical working systems, it opened the door for the implementation of machine learning approaches. However, considering the pitfalls, research in this field is still in the nascent stage. The main issue is that even though models can be fit well for specific datasets, the same robustness cannot be achieved for data from other datasets and environmental samples [38].

Antimicrobial resistance is becoming a global threat. As these pathogens continue to resist multi-drug categories due to the abundance of antibiotics and their usage, this is creating issues in treatment of patients [39]. If a pathogen becomes resistant then it will be difficult for drugs to neutralize them and can require a stronger dosage, which can come with attached side effects. Since such patterns can be observed in many antibiotic drug-resistant pathogens, it can be pointed out using machine learning or deep learning approaches that eventually identify the patterns [39]. In an attempt to classify genome sequences resistant to specific antibiotic classes using the Comprehensive Antibiotic Resistance Database (CARD), a model structure was attempted [40]. Rather than using the entire sequence, a fragment of the sequence of length 100 bp was utilized to train the model. A deep neural network-based computational framework (DeepARG) was developed, which could accurately characterize different antibiotic-resistant drug classes. The sequence data was encoded where each nucleotide of the sequence was assigned a specific numeric value to work with the deep learning system. Initially, the sequence length was extended using a hidden Markov model, and then an artificial neural network-based algorithm was employed to classify antibiotic-resistant classes. The overall outcome showcased an approximate 80% accuracy in prediction [40]. The DeepARG model was able to precisely classify with promising scores, better than most of the existing models.

10.6.4 Concept of federated learning

Over the past decade, deep learning-based computational models have been increasingly used in various applications related to the field of bioinformatics— ranging from gene sequence data, to enzyme classification, protein structure modeling, drug development, and more recently analyzing COVID-19 sequence data worldwide. This leading architecture family releases its variants, as well as independent research often coming out with a variant in the core architecture layer, trying to shoot more accurately and reduce computational complexity and parameter sets. These advanced architectures eventually get researched and established by scientists, focusing on various application domains. However, the path between building a deep learning-based model as a research or proof-of-concept to large industrialized solution with wide acceptability, has a critical challenge in training and testing the models with a larger set of data, which is acquired in different healthcare/laboratory frameworks across the world. Wider model training is indeed a very important pre-requisite as these data have variability based on demographic, condition, acquisition device used, or even disease variants [41, 42]. On the other hand, presence of strict privacy laws which guide storage and circulation of data often act as a hindrance for sharing the data across systems, which is essential for model building. Federated learning was introduced as a solution to solve the apparent paradox of training the deep learning-based models without the need of sharing the actual data [42].

Federated learning was first introduced in 2017, and was predominantly a centralized-server centric-learning approach [43]. Federated learning essentially eliminates the need for sharing the data outside the defined network boundary, which is already in line with privacy laws. Therefore, the data stays within the bound perimeters. The learning architecture enables training from the local data, followed by subsequent transmission of the learning or gained knowledge from one place to other. The gained knowledge is seemingly trapped within the parameter set of the local site level models. Thus, it enables building efficient deep learning-based models and improve their generalization capability [41–43]. A functional block diagram of a centralized federated learning architecture is depicted in figure 10.6. A central server orchestrates the entire federated learning process and derives the final model. Each local site trains with its local data parallelly and sends its model updates/attributes to the server. The central server then adjusts its own weight depending on the local parameters. The final parameter set is transmitted back to each local site for subsequent use. In subsequent times, research in this field led to the discovery of different types of federated learning architectures, such as decentralized federated learning architectures (e.g., IIL—institutional incremental learning, CIIL—cyclical institutional incremental learning) and heterogenous federated learning architecture. Following the immense interest and future potential, 'TensorFlow Federated' was launched to facilitate open research and experimentation with federated learning [44].

Often, data used in machine learning needs to be normalized and preprocessed prior to computational implementations. Since the data is usually acquired from different sources and may utilize different formats, merging various data trends from diverse sources increases the complexity. However, it can create a more generalized

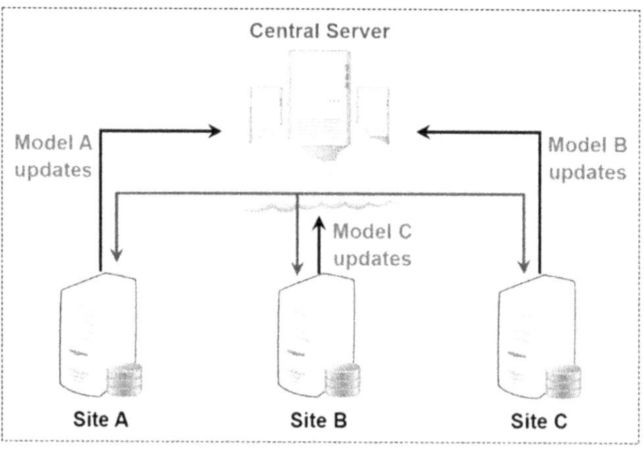

Figure 10.6. Functional block-diagram of centralized federated learning.

model, embedding a larger knowledge base in a learning framework. Federated learning makes use of this principle. By accumulating data from various resources into the model, it aims at creating one unified architecture that can cater a larger data-variety. For example, during the COVID pandemic the symptoms, constituent information, and factors in a COVID patient were varying geographically. This is because COVID was mutating in different manners, based on the location and several other demographical parameters. Hence, it was difficult to come up with a unified model that could accurately figure out a COVID patient. It means that the tests were losing effectiveness and countries had to bar people from migrating to other countries as they could not only carry the COVID pathogen but also go undetected. This is an issue that can be solved if a federated learning approach is taken. The algorithm SARS-COV-2 clinical decision support (CDS) was developed to encounter this [45]. It is made up of a combination of transfer learning and shared information. Where most AI models aim to explain their local dataset, federated learning allows a faster deployment architecture with reduced data issues. To protect the patient privacy, a generalized model is shared instead of having a central repository of data. Using transfer learning and then utilizing this model with added layers to fit a data would speed up the process of designing a model. A pre-trained model would also reduce the overall training time.

An improvement on the traditional approach, a federated averaging (FA) algorithm was developed, following a client–server architecture. Every data contributor acted as a client where they created and trained the model only on their local data, and then uploaded the network attributes to a centralized server. The values were adjusted on the central server based on a centralized aggregation with an aim to diminish the local loss. This, in turn, reduced the global loss as well. Post aggregation, the server resent the updated values back to the client, which were updated again. This process was repeated until both the global and the client losses converged on a single point. It was found that this architecture outperformed models that were trained locally and even models that were trained on a global dataset [45].

10.7 Application in cybersecurity

10.7.1 The need

Cyber threats have been increasing at an alarming pace over the past few years. Apart from an anticipated advancement in technology usage in our daily life, the unanticipated factor of the recent pandemic has played a major role for such increment. The changes in our social fabric triggered by the recent pandemic like remote working and almost a forced adoption of online routes both by consumers and service providers have opened up the floodgates. According to IBM's report, the average total cost of a data breach was accounted be to around \$4.24 million in 2021 with a yearly increment of around 10% (the highest in the 17-year history of the report) [46]. More importantly the report found that the cost was significantly higher for organizations with a lesser adoption of security AI and automation, zero trust security framework, and cloud security. More than ever these cyber-attacks have now become an established weapon in modern warfare, courtesy of the recent

Ukraine war. In response to all these, protecting an organization from cyber threats is a priority for the Chief Information Security Officers (CISOs). In a report, published by Accenture, it was found that 85% of CISOs strongly agreed that cybersecurity strategy was developed with business objectives, such as growth or market share, yet 81% of CISOs agreed with the existing challenges, to remain ahead of the cyber-attackers, is very crucial in this age [47].

10.7.2 Landscape and desired solutions

In the backdrop of increasing awareness about cyber threats, to address these, organizations were traditionally using monitoring systems/security information and event management (SIEM) solutions of different service providers. These solutions gave them a capability to collect system logs from different devices and report them in a single/unified pane of glass in real time. The solutions were coupled with rule-based systems and rules were configured leveraging past human experience. The rules, when failed, created alerts/events. These alerts/events were then analyzed by security operations center (SOC) engineers to detect actual threats. This mechanism gave them a first level of defense, but due to underlying demerits of the rule-based system which failed to adapt/learn, it led to generation of too many false positives, for which traditional solutions started facing challenges. On the other hand, AI started getting implemented within the database of the solution or outside by creating a separate data lake. In this context, deep learning algorithms started gaining popularity in addressing various problem domains with better outcomes than rule-based systems. Therefore, the requirement of a powerful and advanced model that can address the challenges of the cyber world became quite evident because of inherent challenges of the cyber world as mentioned below:

(a) Hardware infrastructure of the organizations were expanding and so did the volume of device logs, which were getting generated by monitoring the devices. This, in turn, led to the necessity of feeding large data to the models.

(b) Advancement of monitoring mechanisms and SIEM solutions led to increase in data dimensionality.

(c) Continuous change in nature, as well as complexity of cyber-attacks.

(d) Need of processing unstructured data.

Deep learning models, which are aptly equipped to encounter the above-mentioned challenges, started to get increasingly experimented with and utilized in the domain of cybersecurity for various different applications. While adoption of deep learning-based models in this domain is still in a very nascent stage, a lot of solutions face stringent obstacles before crossing the borderline of a proof-of-concept model to transform into an industry grade solution.

10.7.3 Deep learning researches

In a survey performed by Berman *et al* [48], around 75 research papers showcased various types of deep learning-based models that were predominantly being used in different aspects of cybersecurity. Percentage ratios of such models are depicted in figure 10.7(a). Interestingly, both unsupervised and supervised models were utilized in this area. Usage of unsupervised models can be correlated with a lack of labeled data collected/preserved from the past. Within the family of unsupervised learning algorithms, autoencoders appear to be a leading choice [48]. AI-based models in cybersecurity can largely be categorized into three different areas—(a) threat detection, (b) threat handling, and (c) threat preparedness. Deep learning-based algorithms are effectively applied in all three areas, producing valuable outcomes. Figure 10.7(b) depicts the interconnectivity of such a deep learning architecture model.

Threat detection: deep learning solutions are able to effectively process huge numbers of logs and other metrics. These logs and metrics are captured for tracing various threats by different SIEM solutions (like IBM QRadar, Splunk, Microsoft Azure Sentinel, etc.) or from devices directly (like FortiGate firewall logs, Windows/Linux server logs, etc). Cyber threats range in various areas and can be categorized in various different areas. Specifically for intrusions, the 'intrusion kill chain' framework (figure 10.8), defined by the researchers of the Lockheed-Martin corporation in 2011, gained wide acceptance [49]. Deep learning is gradually getting applied in different layers of the intrusion kill chain.

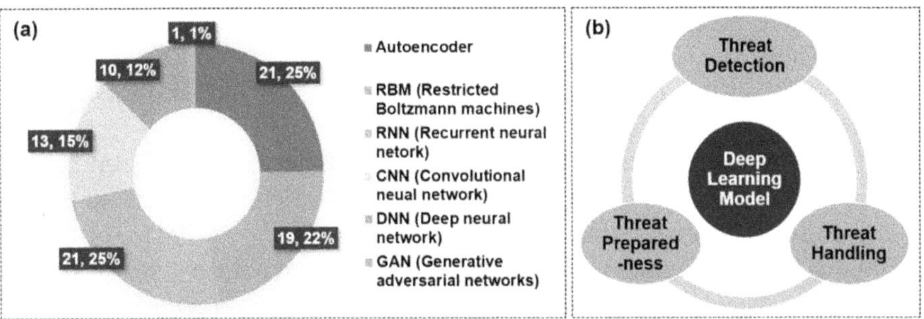

Figure 10.7. (a) Deep learning models in cybersecurity. (b) Application of DL-based models used in various research papers.

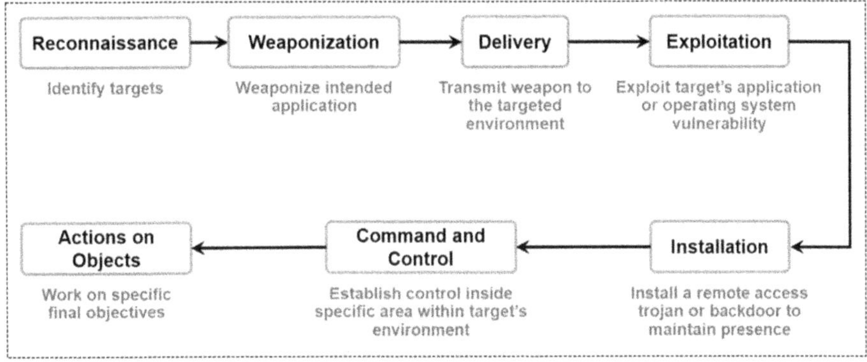

Figure 10.8. Intrusion kill chain.

Threat handling: deep learning methods are getting increasingly applied in the areas which aid the process of threat handling by SOC engineers after an event has occurred. The applications in this area typically are threat prioritization, grouping of alerts, pointing to the right playbook for resolution or even automatically applying the required action. Deep learning-based classifier models are getting used in this area to enable organizations in cutting down the required manual effort and providing a faster response after a threat has occurred. SOC engineering teams are being increasingly burdened with bigger volumes of threat; currently there is lot of focus on in this area.

Threat preparedness: preparing for future threats is another aspect where the organizations are focusing nowadays. Techniques like word embedding which uses a deep transfer learning approach at its core is helping to build effective NLP based models which can process a huge volume of threat articles which are getting published and eventually learn, understand, and prepare organizations before such attacks actually happen. The transfer learning embedding models like Glove and Word2Vec, which are being developed by Stanford University, help in building advanced models in this area.

10.7.4 Challenges of using deep learning models in industry

Although deep learning is opening a lot of opportunities, there are some associated challenges. These are:
(a) Problem with availability of labeled data, which is required for training the deep learning models.
(b) Problems with availability of historical data, as data storage in the monitoring tool database is costly. Unless a company has created a separate lake for the same storage, which is a challenge.
(c) Even if unsupervised algorithms are used, they would require tuning model thresholds, which often creates problems due to lack of past incident history.
(d) Adoption of deep learning models within an organization is also a challenge as most of the organizations are comfortable thinking in a rule-based way.
(e) Issues related with AI explainability of the complex models used often act as a deterrence for adoption.
(f) Model maintenance of these deep learning models once they are implemented in production requires considerable effort.

Having listed the above challenges, the benefits driven by the deep learning models is forcing organizations to break the barriers and work out the challenges to enable more and more adoption.

10.8 Case study: stock market prediction

10.8.1 About the dataset

The performance of an LSTM-based model was tested using a dataset of stock market information, available at Kaggle (https://www.kaggle.com/code/breenda/lstm-rnn-arimax-ar-models-and-analysis/data). The dataset contains stock market information of organizations which are part of the NIFTY 50, resembling the top 50 companies in regards to free-floating market valuation. Out of these 50 companies, we implemented an LSTM-based model to forecast the stocks for ICICI bank, and compared its performance through prediction capability, errors, and directional accuracy against other computational approaches. The dataset provides the following information on a day-to-day basis:

OPEN—Opening price of the equity on a particular day at 9:15 am.

HIGH—Highest value that the equity gained on a specific day.

LOW—Lowest value that the equity gained on a specific day.

CLOSE—Average of the traded prices between 3:15 and 3:30 pm on a specific day.

VWAP—Volume weighted average price is an average price of all of the trades which are done on a specific day.

Volume—Volume of shares traded on a specific day.

Turnover—The stock liquidity, calculated by dividing the total number of traded shares by the outstanding shares in a specific time frame.

10.8.2 Statistical analysis and preprocessing

In usual cases, the actual prices of shares are set by the buyer on a trading day, whereas selling prices are fixed by sellers. A trade is executed when a seller is selling at a price which is equal to or lower than the buying price of a buyer. Therefore, the four fundamental parameters—OPEN, HIGH, LOW, and CLOSE—may not be an accurate representation of the pricing, at which the trades occur on a specific day. Hence, for this analysis the volume-weighted average price (VWAP) is being used. The VWAP provides a weighted measurement and is a better representation of the stock price. Additionally, the volume, which showcases the number of shares being traded, is a significant metric to understand the movement. On a specific day, a movement in pricing of a share with low volume reflects the upward trend is not being supported by the market majority and may not be sustainable. Therefore, traders look at both volume and changes in pricing to support each other for making an educated decision.

Figure 10.9 represents movement of ICICI shares over the past few years with respect to volume and VWAP. As reflected from figure 10.9(a), there were several volumetric transitions during 2019–20, accounting for the dip in prices due to market movements because of the COVID-19 pandemic. As a result, a lot of people purchased stocks in order to make a profit when the market would be stabilized again. In the same period, figure 10.9(b) also depicted the downfall of VWAP as more people purchased ICICI shares in bulk. Ideally, analysis can be done better on uniform distributions. Therefore, observing the data-spread, a box–cox transformation was implemented on the data to make it more uniform [50]. The box–cox transformation modified all data based on an optimal lambda (λ) value. Usually, the λ value takes a range between −5 and 5. In our approach, the optimal value was chosen on the basis of which gives the smoothest curve. The λ value was found to be 0.168 186 while applying the box–cox transformation on our data. Figure 10.9(c)

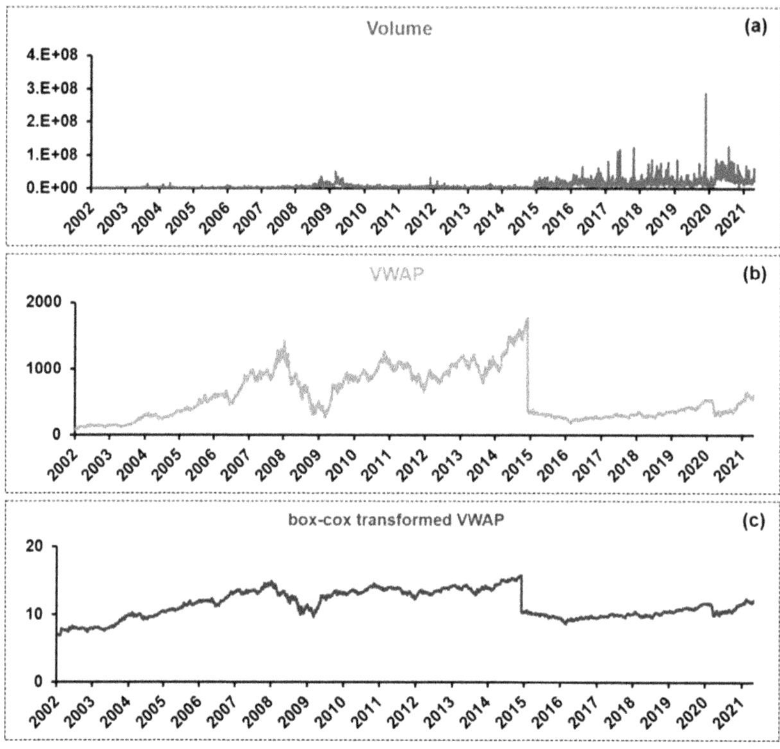

Figure 10.9. Yearly movement of ICICI shares in terms of (a) volume, and (b) VWAP. (c) Box–cox transformed VWAP data for ICICI shares.

shows the processed version of original stock market data, after the execution of the box–cox transformation.

After this, the transformed data was then smoothed to remove the variation that existed between the time steps. This also helped in denoising the time-series data and enabled a better representation to be used for the final predictions. Later, the stationarity of the time series was evaluated using an augmented dicky fuller (ADF) test [51]. In general, a stationary dataset resembles constant statistical measures such as mean, variance, and autocorrelation over a time period. Figure 10.10 represents the statistical outcomes of the ADF test when applied to our data. The ADF statistic was found to be −2.694 529. Subsequently, the p-value for this was 0.074 990, and the critical values were: 1%: −3.432, 5%: −2.862. Hence, with a significant p-value, the null hypothesis was rejected. It should be pointed out that the values were dependent on prior values. Further, the dataset was then decomposed to check for seasonal trends and residuals. However, the data showcased no seasonal behavior.

10.8.3 Computational models and their comparative analysis

The processed data was then utilized for predictive analysis, in which we implemented four algorithms to perform a comparative analysis—auto-regressive

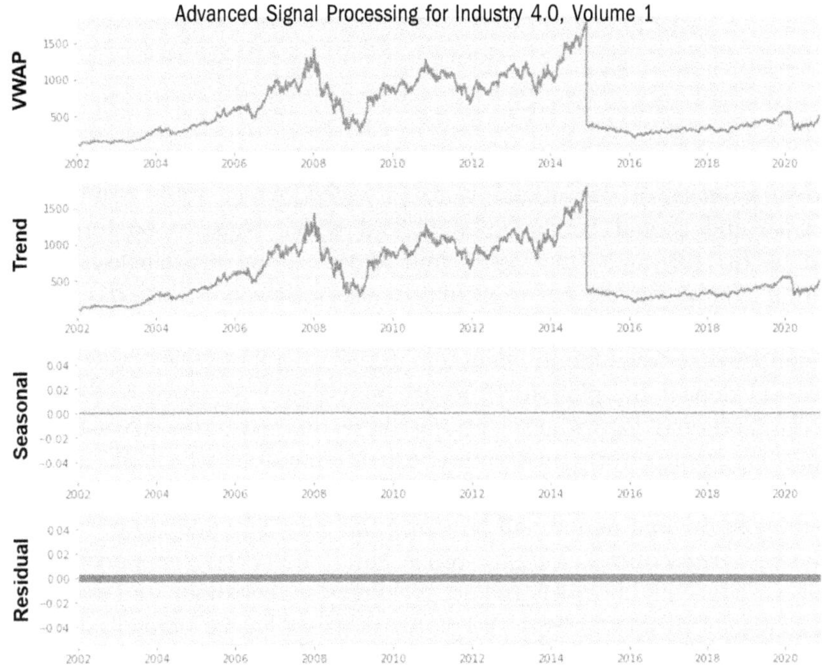

Figure 10.10. Statistical outcomes of ADF test.

integrated moving average (ARIMA), simple artificial neural network (ANN), polynomial regression, and LSTM-based RNN. There are two important elements —one of them is to track the directional changes and the other is to quantify the magnitude of such directional changes. A brief explanation of the used computational models and a comparative evaluation is described below.

First, the ARIMA model was fit to this data. This is a simple yet powerful model for time series forecasting [52]. The ARIMA algorithm aims to look at the variations in values instead of the actual values to establish the final forecasting. For training phase, the fragment of the dataset was kept between March 2000 and December 2019, whereas the dataset between January 2020 and April 2021 was reserved for validation. In the second phase, using a simple ANN model [53], we tried to forecast the values. In this case, the training data was set to have a lag value of 1. The neural network-based model comprised an input layer of five neurons, followed by three hidden layers of size {30, 60, 30}, respectively. In this architecture, the output of one period was modeled to predict the next period and the error was optimized using a back propagation algorithm [54]. Next, we implemented a regression framework with polynomial fitting [55]. Here, linear regression might not be a good choice as the stock pricings rarely follow a linear model and would have high volatility. Additionally, polynomial regression allows the flexibility of using a higher number of dimensions that eventually enables the regression line to bend further and fit better. However, the degree of regression needs to be tuned to obtain a good fit and avoid overfitting. Using a degree of 3, we experimented that the model predicted the stock movement well. However, the differences in these predictions were more as we traversed the timeline. The fourth technique that was used to estimate the movement in ICICI stocks was based on a deep learning strategy. A simple network was implemented with one neuron LSTM. The LSTM approach was incorporated into an RNN structure [6], which involved the implementation of gates to understand that when a series of data times were provided, which data points had more significance and should be remembered and which were to be forgotten. This enabled us to look at a time series and leave out inconsistencies for the final calculation. Hence, if a one-day spike in price was followed by three days of loss then that one-day inconsistent spike was ignored.

A comparative evaluation between these methodologies was performed. Figure 10.11 depicts the plots for forecasted/predicted vs. actual VWAP values of these computational algorithms. Among these methods, the ANN-based model showed a very poor performance while predicting the stock values. Because, as in an ANN, the output of one period was modeled to predict the next period—any error that is there in the first period will flow into the next one, making it comparatively difficult for error control. On the other hand, however, the polynomial-based regression model could properly predict the directional changes, and we can observe a good degree of mismatch from the plot. Additionally, the prediction plot for the ARIMA model reflects a very good match with the actual values, while reducing the directional errors at different instances of time. Finally, the LSTM-based RNN model showcases the best result (among these algorithms) in terms of tracking the actual trend of the data.

Moreover, performances of these algorithms were examined through the measurement of root mean square error (RMSE), mean absolute error (MAE), and directional accuracy. The scores are tabulated in table 10.1. Both RMSE and MAE values were normalized with respect to their individual mean and standard deviation. It is apparent that where the other algorithms reflect large deviations and poor accuracy, the LSTM-based RNN model provides the highest directional accuracy (97.86%) with reduced RMSE (0.0465) and MAE (0.0369) scores. Therefore, using a deep learning framework, a better outcome can be achieved in comparison to the other approaches.

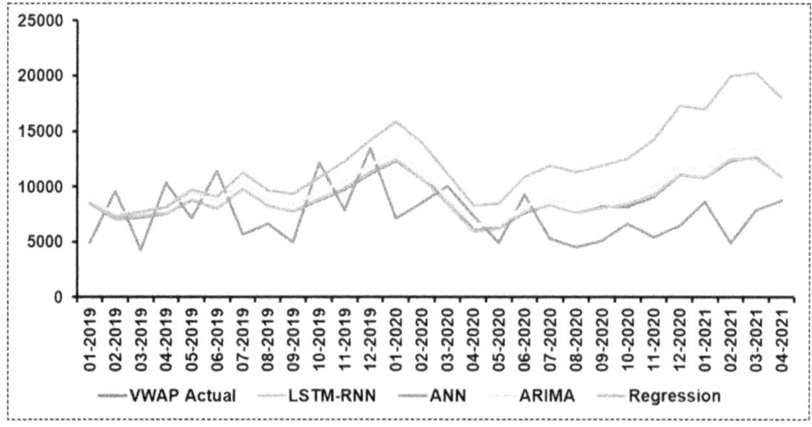

Figure 10.11. Performances of the used computational models, with respect to forecasting the VWAP actual values.

Table 10.1. Comparative analysis between computational models.

Algorithms	RMSE	MAE	Directional accuracy
ARIMA	0.7666	0.5463	91.73%
ANN	1.0622	0.9499	40.82%
Polynomial regression	0.9685	0.8319	89.94%
LSTM-RNN	0.0465	0.0369	97.86%

10.9 Conclusion

Deep learning techniques usually require data and preprocessing with a well-designed computational model to function at its optimal level. Unlike conventional machine learning methodologies, the functional model of the design plays a paramount role. Simultaneously, added computations like regularization and models to tackle overfitting issues can make it more complex compared to traditional learning strategies. Although the overall training time is comparatively higher due to resource-heavy computations, deep learning frameworks are far more enriched with domain knowledge. Interestingly, once the training is completed and deployed in a process, resource requirement minimizes by a significant amount. Depending on the field, it has a deterministic impact on the accuracy of the right prediction. Deep learning-based models are usually in the rear equations that need to be properly designed with the right constants in order for it to work. The weights and bias values, which dominate the core of any deep learning model, are dependent on both low bias and high variance in the data. However, in real cases that is rarely the issue. Data always tend to have sampling bias, missing values, and requires cleaning. Besides, features may require normalization and may need encoding, which also differs depending on the kind of available data. As research progresses in this field, innovative layers with different functions are getting introduced, allowing diverse problem domains to be explored using deep learning. However, a deep learning-based model considers that the data-trend should not change, which makes it difficult in the long run, as the data distribution and inter-relationships between features may alter over time in a real-life scenario. Hence, to maintain a significant overall performance in an industrial process, such deep learning-based frameworks will require periodic observation and scaling-up of the algorithm. As deployment of such processes becomes more mainstreamed, the industry will be able to understand the best approach to solve these issues in an economical and optimal fashion.

References

[1] Kumbure M M, Lohrmann C and Porras P L J 2022 Machine learning techniques and data for stock market forecasting: a literature review *Expert Syst. Appl.* **197** 116659

[2] Sincere M 2022 *Understanding Stocks* 3rd edn (New York: McGraw-Hill)

[3] Gala J 2020 *Guide to Indian Stock Market: Basics of Stock Market for Beginners* (Mumbai: Buzzingstock Publishing House)

[4] Portillo N C 2022 A deep learning state-based market microstructure approach for the price movement prediction task *Doctoral Dissertation* (Chihuahua: Universidad Autónoma De Chihuahua)

[5] Yimian Y 2022 Data analysis on the computer intelligent stock prediction model based on LSTM RNN and algorithm optimization *Proc. IEEE Int. Conf. on Electrical Engineering, Big Data and Algorithms (EEBDA)* pp 480–5

[6] Banik S, Sharma N, Mangla M, Mohanty S N and Shitharth S 2022 LSTM based decision support system for swing trading in stock market *Knowl.-Based Syst.* **239** 107994

[7] Vanshu M, Thakan S and Malik A 2022 Modeling and forecasting the volatility of NIFTY 50 using GARCH and RNN models *Economies* **10** 102

[8] Tong L and Shi Y 2022 Innovation of the component GARCH model: simulation evidence and application on the Chinese stock market *Mathematics* **10** 1903

[9] Chun-Hao C, Lai W H, Hung S T and Hong T P 2022 An advanced optimization approach for long–short pairs trading strategy based on correlation coefficients and Bollinger bands *Appl. Sci.* **12**(3) 1052

[10] Wang J, Rostoker C and Wagner A 2009 A high performance pair trading application *Proc. IEEE Int. Symp. on Parallel and Distributed Processing* pp 1–8

[11] Ramos-Requena J P, Trinidad-Segovia J E and Sánchez-Granero M A 2017 Introducing Hurst exponent in pair trading *Physica* A **488** 39–45

[12] Huang C F, Hsu C J, Chen C C, Chang B R and Li C A 2015 An intelligent model for pairs trading using genetic algorithms *Comput. Intell. Neurosci.* **2015** 16

[13] Bisi L 2022 *Algorithms for risk-averse reinforcement learning, PhD Thesis* Politecnico di Milano

[14] Kim S H, Park D Y and Lee K H 2022 Hybrid deep reinforcement learning for pairs trading *Appl. Sci.* **12** 944

[15] Chen W J, Yao J J and Shao Y H 2022 Volatility forecasting using deep neural network with time-series feature embedding *Econ. Res. -Ekon. Istraž.* **11** 1–25

[16] Niraula G P 2022 Effects of government's policy in stock price: a case of nepse *Jambura Sci. Manage.* **4** 60–7

[17] Adebayo T S, Saint Akadiri S and Rjoub H 2022 On the relationship between economic policy uncertainty, geopolitical risk and stock market returns in South Korea: a quantile causality analysis *Ann. Financ. Econ.* **17** 2250005

[18] Dipashree P, Patil S, Patil S and Arora S 2022 Financial forecasting of stock market using sentiment analysis and data analytics *5th International Conference on Intelligent Sustainable Systems (ICISS 2022)* pp 423–30

[19] Yingmei B 2022 Application of deep learning and big data in marketing evaluation *Proc. Int. Conf. on Multi-Modal Information Analytics* pp 267–73

[20] Lin W 2022 Digital reform of the education industry in the post-epidemic era *Int. J. Manage. Educ. Hum. Dev.* **2** 233–7

[21] Sharma Y, Pandey S and Raheja R 2022 Machine learning: assisting modern education *Int. Res. J. Mod. Eng. Technol. Sci.* **4** 21–9

[22] Önay K F and Esnaf S 2022 Machine learning approach and model performance evaluation for tele-marketing success classification *Int. J. Bus. Anal.* **9** 1–18

[23] Zhang Z, Lim B and Zohren S 2021 Deep learning for market by order data *Appl. Mathe. Finance* **28** 79–95

[24] Adrian M, Capatina A, Cristea D S, Munteanu D, Micu A E and Sarpe D A 2022 Assessing an on-site customer profiling and hyper-personalization system prototype based on a deep learning approach *Technol. Forecast. Soc. Change* **174** 121289

[25] Giri C and Chen Y 2022 Deep learning for demand forecasting in the fashion and apparel retail industry *Forecasting* **4** 565–81

[26] Loureiro A L, Miguéis V L and da Silva L F 2018 Exploring the use of deep neural networks for sales forecasting in fashion retail *Decis. Support Syst.* **114** 81–93

[27] Lalwani P, Mishra M K, Chadha J S and Sethi P 2022 Customer churn prediction system: a machine learning approach *Computing* **104** 271–94

[28] Bristy B N 2022 *Customer churn analysis and prediction, MSci Thesis* United International University, Bangladesh http://dspace.uiu.ac.bd/handle/52243/2325

[29] Pekel O E and Ozcan T 2022 A novel deep learning model based on convolutional neural networks for employee churn prediction *J. Forecast.* **41**(3) 539–50

[30] Lee S and Kim D 2022 Deep learning-based recommender system using cross convolutional filters *Inf. Sci.* **592** 112–22

[31] Maditham V, Reddy N S and Kasa M 2022 A multi-preference integrated algorithm (MPIA) for the deep learning-based recommender framework (DLRF) *Int. J. Intell. Comput. Cybern.* **15** 625–41

[32] Xia Y and Tan L 2022 The construction of accurate recommendation model of learning resources of knowledge graph under deep learning *Sci. Program.* **2022** 1010122

[33] Baxevanis A D, Bader G D and Wishart D S 2022 *Bioinformatics* (New York: Wiley)

[34] Savage N 2014 Bioinformatics: big data versus the big C *Nature* **509** S66–7

[35] Sean W, Schreiber J, Noble W S and Pollard K S 2022 Navigating the pitfalls of applying machine learning in genomics *Nat. Rev. Genet.* **23** 169–81

[36] Michał K, Antonowicz S and Karnkowska A 2022 Tiara: deep learning-based classification system for eukaryotic sequences *Bioinformatics* **38** 344–50

[37] Jo J, Oh J and Park C 2020 Microbial community analysis using high-throughput sequencing technology: a beginner's guide for microbiologists *J. Microbiol.* **58** 176–92

[38] Mathieu A, Leclercq M, Sanabria M, Perin O and Droit A 2022 Machine learning and deep learning applications in metagenomic taxonomy and functional annotation *Front. Microbiol.* **13** 1-10

[39] Yunxiao R, Chakraborty T, Doijad S, Falgenhauer L, Falgenhauer J, Goesmann A, Schwengers O and Heider D 2022 Multi-label classification for multi-drug resistance prediction of *Escherichia coli Comput. Struct. Biotechnol. J.* **20** 1264–70

[40] Arango-Argoty G, Garner E, Pruden A, Heath L S, Vikesland P and Zhang L 2018 DeepARG: a deep learning approach for predicting antibiotic resistance genes from metagenomic data *Microbiome.* **6** 1–5

[41] Liu J, Huang J, Zhou Y, Li X, Ji S, Xiong H and Dou D 2022 From distributed machine learning to federated learning: a survey *Knowl. Inf. Syst.* **22** 1–33

[42] Aledhari M, Razzak R, Parizi R M and Saeed F 2020 Federated learning: a survey on enabling technologies, protocols, and applications *IEEE Access.* **8** 140699–725

[43] McMahan B, Moore E, Ramage D, Hampson S and Arcas B A 2017 Communication-efficient learning of deep networks from decentralized data *Artificial Intelligence and Statistics* pp 1273–82

[44] Hilberger H, Hanke S and Bödenler M 2022 Federated learning with dynamic model exchange *Electronics* **11** 1530

[45] Flores M *et al* 2021 Federated learning used for predicting outcomes in SARS-COV-2 patients *Res. Sq.* **1** 1–31

[46] Cost of Data Breach Report 2021. IBM. https://ibm.com/security/data-breach

[47] The State of Cybersecurity Resilience Report 2021. Accenture. https://accenture.com/us-en/insights/security/invest-cyber-resilience

[48] Berman D S, Buczak A L, Chavis J S and Corbett C L 2019 A survey of deep learning methods for cyber security *Information* **10** 122

[49] Hutchins E M, Cloppert M J and Amin R M 2011 Intelligence-driven computer network defense informed by analysis of adversary campaigns and intrusion kill chains *Leading Issues in Information Warfare and Security Research* 1 (Reading: Academic Conferences Limited) p 80

[50] Atkinson A C, Riani M and Corbellini A 2021 The box–cox transformation: review and extensions *Stat. Sci.* **36** 239–55

[51] Hemavathy P and Gurusamy S 2014 Impact of domestic gold prices on Indian stock market indices with special reference to global financial crisis-an empirical study *Manag. Insight.* **10** 1–8

[52] Ariyo A A, Adewumi A O and Ayo C K 2014 Stock price prediction using the ARIMA model *2014 UKSim-AMSS 16th Int. Conf. on Computer Modelling and Simulation* (Piscataway, NJ: IEEE), pp 106–12

[53] Jain A K, Mao J and Mohiuddin K M 1996 Artificial neural networks: a tutorial *Computer* **29** 31–44

[54] Rojas R 1996 The backpropagation algorithm *Neural Networks* (Berlin: Springer), pp 149–82

[55] Tsai C Y, Kim J, Jin F, Jun M, Cheong M and Yammarino F J 2022 Polynomial regression analysis and response surface methodology in leadership research *Leadersh. Q.* **33** 101592

Advanced Signal Processing for Industry 4.0, Volume 1
Evolution, communication protocols, and applications in manufacturing systems
Irshad Ahmad Ansari and Varun Bajaj

Chapter 11

Digitalization in family businesses—a case study in a food industry in Turkey

Güzide Karakuş and Meral Erdirençelebi

All the industrial revolutions have caused significant changes and transformations in business life in the world. Industry 4.0, unlike other revolutions, has developed much faster and has become an opportunity that will provide significant benefits for businesses of all types with the digital transformation tools it offers. Family businesses are businesses that have a leading position in the economic life of countries. Family businesses have to understand, accept and adapt to the global digital change in order to survive. In this context, the aim of this study is to understand the perspectives and application situations of family businesses regarding digital transformation. In the application part of the study, the digital transformation process of the Muratbey Cheese Company, a leading family business in the cheese manufacturing sector in Turkey, is presented as a situation analysis. Semistructured interviews were held with the Digitalization Department Manager and Communications and Business Development Director of the enterprise and the documents they submitted were examined. It was seen that the activities carried out by the business within the scope of digital transformation were beyond the expectations of the authors and could be considered as an example of good practice.

11.1 Introduction

Rapid developments in the field of communication and technology have made transformation necessary for the sustainability of businesses. In this process, activities such as creating a transformation strategy and determining a roadmap, revising business processes, and designing new business models should be implemented for digitalization-based transformation in businesses. In addition to keeping up with the transformation, it should be the main goal for managers who combine digitalization with creativity and innovation to go beyond managing the transformation process and try to create and present the technology of the future in advance. It is of great importance in terms of competitive advantage to quickly overcome the adaptation process to digital transformation, which is seen as the vision of the future in business management, and to realize the transformation. In addition to the various advantages it provides to businesses, unexpected factors such as the COVID-19 pandemic have shown that digital transformation is no longer a choice but a necessity for the sustainability of social life.

In the digital world, which is developing faster than ever with the development of the digital age, businesses are trying to transform rapidly, while the adoption of digital technologies continues to progress rapidly. Industrial Revolutions are events that have led to major changes in the world [1]. Differences in the structure of the three industrial revolutions aside; the fourth industrial revolution, defined as Industry 4.0, requires greater participation of information and communication technologies in production and a larger change in human resources, which is the main element of production [2]. Compared to previous revolutions, in this revolution, broad-based innovations and new technologies are spreading faster in some parts of the world [3]. The main goal of the Industry 4.0 process is to create a more efficient and competitive production infrastructure with the innovations and opportunities brought by digital developments [4]. Within the scope of the system, the integration of information technologies into the work performed: data analysis and management, simple network solutions in cloud-based applications, smart production systems and digital solutions for all operational activities are in question [5]. However, digital transformation consists of the combined effects of various digital innovations that reveal new actors, structures, practices, values, and beliefs that change, threaten, or complement the existing rules of the game in organizations, ecosystems, and industries [6].

Thanks to the opportunities offered by Industry 4.0 after the information economy, such as big data, artificial intelligence, autonomous robots, and sharing economy, innovations have brought with them a society that can solve various social challenges by being incorporated into both different industries and social life. As of 2016, the world has started to prepare for this transformation by pointing to the super-intelligent society with the title 'Society 5.0'. Society 5.0 refers to a new society led by science and technological innovation in order to balance economic development and solve social problems. In this transition period, there is a radical change. Digitalization, which is the most fundamental catalyst of change, transforms societies and forces their subsystems [7] to change strategically and technologically as well as behaviorally. This need for change harbors great opportunities or threats for businesses of all types.

The implementation of Industry 4.0 applications by considering the principles of Society 5.0 and the digitalization of business functions by improving the organizational structure are of vital importance for all businesses. Family businesses, which are the important building blocks of a sustainable economy in the world, must also successfully complete this transition process. Family businesses, which are the type of business in which share ownership and management activities are carried out by family members, are the engine of the Turkish economy as in many countries around the world. According to 2016 European Union reports, 50% of the world's existing enterprises and 90% of existing enterprises in the United States are family-run [8]. According to a study conducted in European countries in 2016, it is seen that family businesses, which constitute 91% of non-public enterprises, have a share of 57% in total employment in the private sector and a share of 55% in total turnover [9]. The same applies to Turkey, where 95% of the businesses operating can be considered family-run enterprises. Only small and medium-sized enterprises-scale family businesses provide 78% of total employment. 58 of the top 100 businesses with the highest export volume in Turkey are family businesses [10]. Therefore, the sustainability of family businesses is of great importance for the sustainability of the world economy.

The fact that digital transformation has the potential to eliminate these losses in today's conditions where the losses in business subsystems and their effects on sustainability are increasingly important is the important source of motivation for this study. Similarly, the success of family businesses, which have an important place in the global economy, remains important in this process. Accordingly, the following questions are sought in this study.

- *What is the current situation in digitalization in family businesses?*
- *What kind of digitalization measures are taken within the scope of business functions in food businesses?*
- *What are the opportunities and obstacles encountered in the digitalization process in family businesses?*

In this study, first of all, the concepts of Industry 4.0, digitalization and digital transformation, the historical development and scope of digitalization, and the main reasons that lead businesses to digital transformation are discussed. Secondly, the general structure of family businesses and their features and perspectives on digital transformation are given. Afterwards, in order to understand the perspectives of family businesses on digital transformation, the digital transformation processes of the Muratbey Cheese Company, which was established in Turkey and is one of the leading food businesses in the country's economy, was revealed with a single exploratory case study, one of the qualitative research methods. Lastly, suggestions are presented in line with the information obtained in the conclusion section.

11.2 Conceptual framework

11.2.1 Industry 4.0

Industry 4.0, the fourth stage of the industrial revolution, is the increasing digitization of products and systems with their interdependence. Thus, the physical world is connected to the virtual world [11]. The digital transformation implemented in product/service value chains in today's business world is driven by the sub-components of Industry 4.0. Key technologies required for Industry 4.0 transformation, such as artificial intelligence, the Internet of Things, machine learning, cloud systems, cyber security, and adaptive robotics, cause radical changes in the business processes of organizations [12]. Undoubtedly, the digitalization of the business with all its functions is not easy. All supply chain functions, from the supply of raw materials to the delivery of the product to the customer (purchasing, warehousing, logistics, planning, production, quality control, sales, and marketing, etc.) and other support processes (human resources, accounting, finance, public relations, etc.) should be evaluated within the scope of digitalization. The digital transformation in which functions and in which order should be determined by making a cost–benefit analysis and considering the related activities. For digitalization, a roadmap should be determined within the scope of the enterprise, and the transition process should be implemented step by step with the project teams to be created. In this process, institutional memory should be created by recording the progress and the repetition of the mistakes should be prevented.

Smart factories that will be implemented in manufacturing systems should not be considered only within the scope of information technology but should be evaluated systematically by taking into account the principles of production management. These systems are resource systems for the solution of many important business problems, which will enable continuous improvement of competitiveness [13] and access to real-time data. Thus, stock level and machine capacity utilization can be compared in real-time, material stock circulation can be controlled, and the system can automatically trigger the order of materials when necessary. Applications such as low material inventory, logistics and transportation costs, shorter delivery times, and optimum routes in distribution are also possible with Industry 4.0 [14].

Digitization steps of Industry 4.0 will provide customer-specific and personalized design, effective planning, effective management in procurement, production, sales and marketing, maintenance, monitoring and resource use in production systems. Practices to be implemented within the scope of Industry 4.0 will enable dynamic, real-time, and rapid handling of factors such as quality, time, flexibility, transparency, efficiency, reliability, cooperation between humans and machines, price, and environmental sustainability at all stages of the value chain [15]. Technologies to be used in the right place will reduce the costs of all processes related to business functions (production, marketing, human resources, etc.), increase capacity and quality [4], and make businesses efficient and productive. All strategic objectives to be achieved in the current conditions and minimization of all the negatives and mistakes that are desired to be overcome can be achieved through digital transformation. Although this picture of the future includes various risks and uncertainties, it is of the nature to be studied.

Digital transformation requires an infrastructure that goes beyond specific innovations and organizations. In the world of digital innovation, these processes are an important research area [6]. Schwab [3] stated that Industry 4.0 contains various difficulties, but this situation can be turned into an opportunity with proactive preparations. In line with the fourth industrial revolution, the values and ethical principles included in the future systems should be defined within the framework of tolerance and respect, the economic, social, and political systems should be restructured, and awareness should be raised in all stakeholders of the society.

As can be seen, Industry 4.0 and the necessity of digitalization that comes with it, the risks and opportunities related to this requirement, expectations, positive and negative effects are the subjects that should be studied by academic institutions and presented to society in order to raise awareness.

11.2.2 Digitalization and digital transformation

Businesses, as a requirement of the current age, have faced a transformation, sometimes voluntarily and sometimes compulsorily, in order to maintain their existence as subsystems of societies. Digitization means not only a paperless business, but also managing, organizing, controlling digital data, and integrating this data into solutions to create sustainable value and new opportunities for business and society [16]. The development in the use of digital technologies takes place in four stages, depending on the structure of the business. Each stage builds on the previous one, and the contributions and importance of these stages are listed below [17]:

1. *Personal computer stage:* This covers the period until the use of the Internet in businesses. It has effects such as process automation, communication support, and process acceleration in businesses.

2. *Internet stage:* Easy access to information, new means of communication (e-mail, etc.), change in business models (e-commerce), different forms of work such as flexible working, working from home, etc.

3. *Social media stage:* As Web 2.0 technology and mobile devices started to be used regularly in daily life, they started to be used by businesses as well. Web 2.0 technology has enabled crowdsourcing business models based on the idea that collective intelligence produces better solutions than individual intelligence and business models in which customers are actively involved in the product development process of enterprises [18]. Customer focus has become the main goal for businesses. Digital store business models have emerged. Thanks to social media, it has become easier to find new markets through flexible business processes and new collaborations (both inside and outside the organization).

   These three phases took place during the third Industrial Revolution, and the fourth phase, the Internet of Things, gave birth to the fourth Industrial Revolution [19].

4. *Internet of Things:* The integrated use of sensor and network technologies has enabled cyber-physical systems in which every object can be monitored and communicated with others. The concept of the smart factory, which emerged in this way, is based on the principle of self-actualization of production with cyber-physical systems. Along with Industry 4.0, not only the production processes of the enterprise, but the entire business value chain is changing. All stakeholders on the business value chain, from suppliers to customers, are connected to each other and the production process and products are also affected by this situation. Thanks to the traceability of the products, businesses can obtain important information about their customers. The ability to manage data on the cloud changes the size of the relationship between customer, supplier, and business. Thanks to the analysis of all data coming from customers and suppliers, it is easy to develop innovative business models and offer personalized products and services.

In the face of this digital revolution, governments are increasingly defining digitalization as a strategic priority and creating large-scale initiatives to promote the digital transformation of science, industry, and society [20]. While this transformation takes place more easily and synergistically in developed countries, some problems may be encountered in developing countries.

The process that today's businesses have to be involved in is called digital transformation, which means the transition to digitalization, and it started with the use of digital technologies within the business [21]. Digital transformation includes the radical changes that occur in society and industries with the use of high-level digital technologies [22]. Digital transformation is defined by Hess *et al* [23] as the radical change resulting from the inclusion of digital technologies in all organizational structures and business processes, by Henriette *et al* [24] as the development of new business models or reshaping existing business models, and by Schwertner [25] as structural changes in working structures with the integration of new digital technologies into all business areas. These changes result in increased efficiency in the value creation process, stronger customer focus, and new marketing and business models. In other words, in all processes with digitalization; efficiency, agility and innovation are targeted (Schulte, 2016) [26]. The digital transformation, which is advancing rapidly with new digital technologies, changes the economic conditions and expectations. The possibility of accessing all kinds of data, regardless of place and time, brings with it the expectation of a trade environment where data and services are at the forefront, not the physical product, the growth of the economy and the increase in productivity. Existing business models are transforming with digitalization, new business models are emerging, and as a result, business value chains and organizations need different structures. However, businesses that can produce, use, or market different digital technologies will be able to adapt to new economic conditions and survive in the digital market in a new challenging competitive environment [19].

In short, digital transformation is a change caused by the use of technology at many levels in the organization, which includes both the use of digital technologies to improve existing processes and the discovery of digital innovation that can potentially transform the business model [27]. The prominent technologies in the era of digital transformation are presented below [28]:

- Artificial Intelligence
- Internet of Things
- Cloud Computing
- Big Data
- Nano Technology
- Robotic
- Virtual Reality
- Smart Factory
- Cybernetic Physical Systems
- 3D Printers
- Blockchain

Digital transformation can influence many or all parts of a business. For this reason, people who are responsible for business management must ensure the development and use of their internal resources at the same time in order to achieve organizational agility, which is a necessary condition for the successful transformation of their organizations [29]. Accordingly, businesses need a digital transformation strategy that includes the stages of planning, executing, auditing and performance evaluation in order to reduce the risks of digital transformation that must take place in their business models, processes, and relationship networks [30]. The digital transformation strategy is actually a roadmap, and it is a very challenging task. The important thing is that this roadmap is implemented with precision. Due to its unique infrastructure, each business can implement different digital transformation scenarios based on different sensitivities and requirements [19].

Digital transformation requires the transition of production from labor-intensive processes to mechanical processes based on information technology. The point to be considered is the acquisition of information, as well as the interpretation and use of this information [31]. Neumeier *et al* [32] argue that organizations need to develop their ability to adapt flexibly to turbulent environments in order to successfully achieve digital transformation. Digital competencies, management style, and market structure are the factors that trigger digital transformation in enterprises [19], since these factors differ for each enterprise, it should be accepted that the digital transformation journey of each enterprise will also differ.

According to Sağlam [31], factors such as the rapid change in consumer purchasing behavior and the increase in pressure on the manufacturer, the increase in the market share of businesses that rapidly complete digital transformation, the expansion of the competitive space of businesses regardless of their size with the development of digital technologies, and the increase in consumers' personalized service and product expectations are the reasons that lead businesses to digitalization. Regardless of their structure and size, all businesses should clearly define their current status and start this journey by creating a digital transformation roadmap in order to be included in the digital transformation process [31].

Another important issue is that the digital transformation process should not be seen as merely purchasing digital technologies and integrating them into processes. This transformation requires a solid organizational change. It should not be forgotten that the digital transformation process consists of the application of information technology, as well as the support of the senior management, the digital transformation strategy, and the change in the organizational structure [29]. In this process, administrative tools such as organizational agility, flexibility, innovation, change management, organizational learning and organizational memory are used effectively. In addition, the decision makers, crisis solvers, and leaders and managers who will follow the implementation of these issues are of great importance.

11.2.3 Family businesses and digital transformation process

Family businesses are defined as businesses in which people belonging to the same family take part in most of the share ownership and management levels, established to provide their own economic livelihood and leave a legacy to the next generations, and are affected by the values, beliefs, and attitudes of the owner family. Family businesses are the intertwining of family and business systems. The family system, by its nature, is dominated by emotions and values, while the business system is dominated by logic and rules. Employees are both family members and parts of the business. Family businesses have characteristics that differ from other businesses in terms of both these structures and their management styles. Family businesses, which have advantages such as quick decision making, benefiting from family funds financially, and focusing on work, also have weaknesses such as authority problems, financial inadequacy, nepotism, role conflicts, and central management in the growth period [33].

Along with having various advantages and disadvantages, family businesses have an important place in a sustainable economy. In fact, 50% of the businesses in the world, 90% of the businesses in the United States of America [8], and about 95% of the businesses in Turkey are family businesses [10].

However, the life span of family businesses that shape the country's economy is generally limited to the length of time the founder stays at work. In such enterprises, the transition between generations is painful and often the second generation is the breaking point. The number of enterprises that have passed to the third generation is quite low [34]. The founder has an important role in establishing common goals, norms, values, attitudes between the family and employees and has a place accepted by employees. It is known that the latent influence of the founder (founder shadow), especially in the second generation, exists [35]. The cultural structure created by the founder has a great influence on the formation of organizational memory. For their sustainability, family businesses must have the ability to adapt to environmental change. It is also of strategic importance that family businesses, which are living beings, can successfully carry the knowledge and skills they have learned in the adaptation process to the future.

As in every organization or business, decision making and choices are among the main management actions in the management of family businesses. In the vast majority of these businesses, family members make important decisions because they are at the top management levels. It is important to make fast, healthy, consistent, and rational decisions for the family business to compete and survive in the market. Taking decisions in this way becomes possible by making use of decision support systems and technologies adapted to the conditions and conditions of the day [28].

The rapid change in digital technologies has a significant impact on business processes such as business conduct, production processes, management and strategic decision-making processes in all industries and businesses as well as in family businesses. Especially with the Internet of Things technology, the data collected with the help of sensors can be stored on cloud computing, and the large data stacks created by analyzing and interpreting with artificial intelligence can be transformed into information, knowledge, and even strategic wisdom [36]. This transformation brings with it important investment requirements, and family members are also the decision makers in investment decisions in family businesses. For this reason, family members need to have sufficient information about digital transformation and its benefits for many investment decisions such as technology, education, human resources, and technical consultancy.

One of the features that distinguish family businesses from other businesses is the matching of corporate reputation with family reputation. Thanks to the social capital of their family members, these businesses do not have difficulty in finding the necessary support when necessary. For this reason, they have the advantage of accelerating the digitalization process by getting support from their social capital in digital transformation.

Family values, size, institutionalization, and similar effects of family businesses are effective in digital transformation processes. In particular, the importance of senior management should not be overlooked. In family businesses, the power of senior management to influence employees up to the lowest levels is higher than in corporate businesses. The most likely risk to digital transformation will be resistance to change. If there is an effective leader/manager at the head of the family business, this resistance can be easily eliminated. It should be noted that when managed correctly, family businesses can perform their transformation processes much more successfully than other companies, but the impact of a transformation project that is not supported by family leaders will be limited [37].

While the founders and the second generation, who are in the shadow of the founders, are more experienced and more effective in transferring information to the next generations, the new generation is much more effective and competent in digital technologies. However, the previous generations' ignorance of the new generation's effectiveness and competence may have negative consequences and may lead to conflicts [38]. According to a study on family businesses in 31 countries in 2017, it was seen that the new generation could not even express their ideas about digitalization. The digitalization process must also be taken into account in the management of intergenerational conflict [37]. The new generation can take on the role of educating other family members about digital technologies. However, family elders need to get used to these changing roles in their relations with young family members. Otherwise, conflict is inevitable.

On the other hand, one of the most important risks for family businesses in the digitalization process is nepotism. Assigning inadequate family members to positions they do not deserve in digital transformation will pose a risk in the sustainability of the business [39]. Institutionalization will undoubtedly have a very important effect in eliminating all these risks.

Anxiety about leaving a legacy and thinking about the future of the next generations can sometimes cause the managers of family businesses to hesitate to take risks, which may cause them to have a negative view of investing in digitalization processes.

For a successful digital transformation, it is necessary to consider the business as a whole and to adopt a collaborative organizational culture that supports change [40]. Organizational values, innovation, willingness to learn, tolerance for failure, risk propensity, entrepreneurial mindset, trust, participation, and cooperation should be emphasized [41]. Considering the advantages of family businesses, an organizational culture should be created that will support innovative, agile, intrapreneur, and learning organizational structures. Institutionalization should be given importance. Undoubtedly, the founding entrepreneur has a great influence on the creation of organizational culture. For sustainability or the survival of the family name, the founder and other family members must accept the necessity of digital transformation and lead the process.

11.3 Research method

Although family businesses constitute the majority of businesses in almost all of the world, the field of family businesses has started to be seen as an academic discipline after the 1990s, but studies on the subject have not been dealt with as comprehensively as other issues in the field of management [42]. Although it has started to attract a little more attention in the world, especially in recent years, it can be said that studies in the field are still not satisfactory [43]. The most important reason for the inadequacy of studies on family businesses is the difficulty in accessing data for these businesses. Family businesses generally tend to keep information confidential, such as their strategies, financial performance, and family relationships [44], because family businesses represent not only a business but also the speciality of a family. In addition, family businesses, unlike other businesses, have a complex structure created by family interactions and business-family interaction. The different and complex features that distinguish family businesses from other businesses increase the richness of potential research topics, on the other hand, reveal the necessity of examining the subject of family businesses more carefully [45].

Recently, digitalization has been added to the difficulties experienced by family businesses, which are of great importance for national economies. In the literature research conducted within the scope of family businesses, it was seen that the focus was on institutionalization, transfer planning, nepotism, and similar issues, and no study on digitalization was found. The importance of digital transformation and family business concepts has been the main motivation for this study. The study was designed on the basis of how digital transformation is perceived and implemented in family businesses and it is thought to contribute to the literature. The main purpose of the research is to describe how the digital transformation process is carried out in family businesses. For this purpose, *a single exploratory case study*, one of the qualitative research methods, was preferred in the study. The exploratory case study allows one to understand a situation with all its clarity [46]. In the research, by using literature research, observation, interview and document review techniques, in which business functions the business has experienced digital transformation, how the process is applied, and the results obtained are defined. Suggestions for further research on digital transformation in family businesses are presented.

Interviews were conducted with the Digitalization Department Manager and Communications and Business Development Director of Muratbey Cheese, using semi-structured interview techniques, through online interviews in December 2021. The interview started with a short introduction on the subject of digital transformation, and a short discussion was held on the importance of the transformation. Afterwards, clear, non-directive questions were asked by stating the purpose of the research. Where necessary, the interviewees presented documents, conveyed their current practices and offered support.

The research questions are structured in four parts. The first part is aimed at understanding the demographic information of the interviewees and the structure of the business. The second part is about understanding where the business sees itself in digital transformation. In the third part, the interviewer was asked what kind of digitalization steps were taken within the scope of which business functions of the enterprise. Finally, the fourth part was carried out to understand the positive or negative effects of being a family business on digitalization. In the discussion section, the findings obtained are evaluated and suggestions for future studies are presented.

11.4 Research results

11.4.1 Information about the business and interviewees

Having stepped into the cheese industry with a small cheese shop in Istanbul in 1965, Muratbey opened its first production facility in 1992. The elder brother of the family, who is the Chairman of the Board of Directors, focused on innovative studies aimed at bringing a difference to cheese, which is a traditional flavor, with the support and mentality he received from his engineer identity. The company has taken important development steps over the years in line with its goal of producing quality, innovative, and unique products with the taste of natural milk. Muratbey has demonstrated its success in the global market with many domestic and international awards (quality awards, innovation awards, international consumer-friendly brands, best cheese, the product of the year, etc).

In line with its customer-oriented and innovative vision, the business has over 300 product types as the market leader in Turkey in the Special Cheese category. Focusing on activities that will support public health in the face of the COVID-19 epidemic that started in 2020, Muratbey developed its cheeses enriched with vitamin D to support a healthy life in order to support the immunity of consumers.

The company offers innovative cheese varieties to the world market, which it has designed and produced with the machines it produces. The business, which opened abroad in 2008 and registered trademarks in nearly 60 countries, represents Turkish cheeses in the world market. Muratbey reaches more than 50 000 sales points throughout Turkey and five continents around the world with its production facility with a closed area of 35 000 m^2, approximately 400 employees and a daily milk processing capacity of 700 tons.

Muratbey performs an environmentally friendly production, strives for energy-saving and zero chemical waste in production, and continues to be a pioneer and an example in its sector, not only with the product but also with the technological projects it has developed. It carries out its activities with a holistic approach in line with the principle of 'Sustainable Food for a Sustainable Future'.

The **Digitization Department Manager**, after completing his education in computer technologies and programming, worked as an expert and manager in IT units in different sectors such as education, online sales, software, and technology centers. He has been working in the company since 2017.

The **Communications and Business Development Director** has a bachelor's degree in mechanical engineering and has completed two master's programs in senior management and business and management organization. In his 30 year career, he has worked as a senior manager, consultant, and trainer in a leading company in the food sector in Turkey and focused on global branding. Since 2006, he has been involved in business development, marketing, and communication in the company.

There are four family members in the business. The family member, who serves as the Chairman of the Board of Directors, has an electrical–electronics engineering education and has accomplished pioneering works in the sector by constantly seeking differences in cheese. The second member of the family, who serves as the Vice Chairman of the Board of Directors, is responsible for communication, marketing, and finance functions. Two more family members from the second generation of the family are involved in the business. The first of them completed his undergraduate education in the Department of Electrical and Electronics Engineering and his master's degree in business administration in the United Kingdom. He has been working in the company for 5 years and is responsible for the domestic and international sales department. The last second-generation family member working in the business has completed his undergraduate education in industrial engineering and is working as the Supply Chain Director.

11.4.2 Where does the business see itself in digital transformation?

In this section questions were asked to the interviewees to understand where the business sees itself in digital transformation. In this section, questions used by Klötzer and Pflaum [47] to describe the digital transformation in the supply chain in the manufacturing industry were used.

- How would you describe the phenomenon of digitalization from your company's point of view and when do you think your digitization goal will be achieved?

  The phenomenon of digitalization has two main pillars for our business: R&D and innovation. These two titles are among the activities that Muratbey values the most since its establishment. For this reason, we continue our digitalization efforts uninterruptedly under the leadership of our Chairman of the Board of Directors and senior management. In this direction, studies are carried out with our researchers in our two R&D centers on machine design and production for use in both product development and production processes.

  In our company, it is aimed to create a structure that develops and implements effective policies by analyzing global developments and their own dynamics, not following national technological developments. Contributing to Turkey's digital transformation and development in the sector by using and developing digital technologies is among our main goals.

  In order to create a value chain with high added value and competitive power, artificial intelligence, smart factory applications, autonomous robots, digital marketing, cloud computing, automation production lines, new generation smart sensor technologies and cyber security studies are carried out within our enterprise. In line with the strategic goals of our business, we aim to complete the digital transformation in 2025. However, it would not be correct to say that this is enough in the age of developing and changing industry and technology. Apart from keeping up with the change, we will continue to carry out works that are the pioneers of change in the sector.

– How did the digital transformation process start in your business?

In order to understand the impact of digital transformation in production, it would be useful to review the previous industrial revolutions and the basic technologies that triggered them.

The first industrial revolution took place at the end of the 18th century with the use of water and steam technologies, and the second industrial revolution with the use of electricity in production. The second industrial revolution enabled mass production to be started with the use of the first assembly lines.

The third industrial revolution, on the other hand, provided significant advantages in quality, cost and efficiency with the use of programmable logic controller (PLC) systems, automation, and robots in manufacturing thanks to the developments in electronics and automation technologies. Each of the industrial revolutions has affected the way the manufacturing industry works. Our company also provided a rapid transition to the third Industrial Revolution, and revisions were made in our production processes and technologies in line with quality and efficiency improvement. With the work of our R&D centers and IT department, we have been integrated into the third Industrial Revolution.

Our R&D centers are equipped with all kinds of scientific and technological equipment in order to carry out R&D studies at world standards. The data obtained through our R&D studies allow us to evaluate our existing resources in a concrete way, to produce new products, processes, and services, and to develop strategies that will ensure that these are put into practice in the most effective and efficient way.

In our company, automation is given great importance in the production areas, and investments are made continuously. Investments gained momentum, especially with the increase in demand for differentiated cheese types. In 2014, robot systems started to be used in our process with pick and place robots, and then delta robots used in product filling were included in the system. There is already one palletizer robot for palletizing systems, and our additional robot projects are continuing to further develop the system.

We continue to work on projects for different areas where robots can be used in the production process. Thanks to the process automation system, production is controlled from a single point, while auxiliary facilities can be monitored instantly by the Maintenance-Machine unit. Together with these two systems, a large auto-control unit emerges. In the upcoming period, it is planned to invest in low-cost, space-saving, heat-generating, but equally high-reliable hardware and operating and software systems that will run this hardware.

– How would you define certain/special nuances in terms of development stages in the digitalization process?

We address our digital transformation process under three main themes:
- *Planning a strategic perspective on digitalization/creating a vision.*
- *Determining digitalization requirements in core business processes and realizing the transition.*
- *Digitization of infrastructure.*

11.4.3 What kind of digitalization actions have been taken within the scope of which business functions in the enterprise?

This section aims to understand which digitization actions can be taken in the food industry and what kind of steps are taken in the sample business. The interviewees were asked what kind of digitalization actions were taken in which business functions.

Our company, which always prioritizes consumer expectations, launched two R&D Centers in 2017 for innovative products and production processes. Our R&D centers work in line with the goal of developing innovative and unique products in healthy conditions, increasing product quality and standards by innovating production methods, increasing efficiency, and producing value-added products for a sustainable world by reducing costs. Many pieces of software, machinery, and production equipment used in our facilities are designed and produced by researchers in our own R&D centers.

*Every stage of the production activities in our facilities is monitored by food engineers and quality control engineers with 90 different cameras. Studies are carried out with state-of-the-art methods for quality **production** with zero downtime/zero error. A transition to digital methods has been achieved in order to minimize machine downtimes through planned **maintenance** activities. Fault records are kept and analyzed digitally. In this way, the frequency and causes of failures can be determined, and action can be taken on preventive and planned maintenance. Steam boiler, cooling and ventilation systems are monitored and kept under control 24/7 with the Scada system.*

*Automatic pallet changing systems are used in our **warehouses** in order to speed up the **operation processes** and ensure correct shipment. **Quality control** officers can instantly monitor where the pallet they want is and the temperature values of the region where it is located. With the cold storage area, the cold rooms in the production can be monitored instantly from a single center, and interventions can be made.*

*Identified pallets can be automatically assigned according to the product type in the incoming order with first-in-first-out (FIFO) tracking. Incoming orders can be tracked full-time in the Enterprise Resource Planning (ERP) program through integrated assistant software. By using integrated software, production according to the order can be provided in line with the stock quantities in the warehouses. Thanks to the robotic system, **production** can be monitored regularly, and also the errors made in the processes such as energy consumption, production, and storage are stored and analyzed in the system. Pipes, valves, and cold storage surfaces are scanned with thermal cameras, places with heat loss are detected, and **energy losses** are minimized. In this way, **energy management** can also be better monitored.*

Advantages are also provided in the efficient use of water in production through automation systems. Clean in Place (CIP) systems in the machines are a system used in the food industry to eliminate the pollution that occurs in the lines during production.

*Thanks to automation-based CIP systems, the controlled use and circulation of water and chemicals can be ensured during **line cleaning** (sanitation) before/after the production. With the conductivity meter devices in these fully automatic systems, **savings** are also achieved in the use of chemical products.*

In the value chain, many pieces of automation software and machines are used to provide maximum hygienic conditions in all processes from raw material procurement to production, from weighting to boxing and packaging. Our company has an ERP system and business applications integrated into this system, and business intelligence applications, in which the outputs of the system are followed, are also used effectively.

11.4.4 The impact of being a family business on digitalization

In this section, questions were asked to the interviewees to understand the positive or negative effects of being a family business on digitalization. Since no similar study was found in the literature, the questions were formed in line with the author's own knowledge and experience.

- Does your company have strategic plans for digital transformation?

    In line with the strategic goals of our company, a digital transformation roadmap has been determined. This plan is updated in the annual strategic planning periods. We define our digital transformation vision as being a sustainable business focused on digital marketing as well as smart production and supply chain worldwide in the cheese industry. Contributing to sustainability by minimizing environmental impacts is very important for our business, and we have realized that we can achieve this contribution with digital transformation. In order to improve energy efficiency, all water, electricity, air and steam meters in our company were renewed by switching to digital systems. Energy use in production is carefully monitored in every part of our factories, and all measures are taken to save energy. R&D project plans are also developed with this understanding, and any contribution that can be made to a sustainable world for the future is already modeled.

- Has the infrastructure been created for digital transformation? What are the department-based, service procurement, and collaborative activities?

    In line with the awareness that digital transformation in our business can be achieved with a qualified workforce, infrastructure investments have started with investing in people. While digital competence is sought for newly hired staff, planned training sessions are carried out to improve the digital skills of existing personnel.

    Necessary collaborations are carried out at every point of the value chain and in every required field. For example, mobile applications are developed with collaborations in the field of marketing. Important steps are being taken within the scope of business intelligence ERP integration. Applications are developed to provide sales/warehouse controls with dealers.

- Which of the tools expressed by the Industry 4.0 framework concept (big data, RFID, artificial intelligence, Internet of Things, cloud computing, etc.) are used in your business? In which areas are these tools used?

Almost all of these tools are used within our company. Robots, sensors, and software are used in production, quality control, maintenance, and storage. Control and monitoring are provided by sensors and software in production facilities, and necessary warnings can be made to relevant persons. With the business intelligence reporting software related to customer data, the desired reports are presented to the users in line with their requests. Customer data is analyzed by marketing, and all relevant units can be informed in line with their needs. Data is collected and analyzed, and improvements are made to energy management issues.

Data about production and marketing are stored, analyzed and used by relevant managers in decision-making processes. In the next five years, it is planned to implement applications in line with the transition to cloud computing systems and the inclusion of artificial intelligence in decision-making processes.

– Does your organizational culture support the innovative and learning organization structure?

Innovation plays the most important role in the vision of our company, and our organizational structure has been developed in this direction. Our corporate culture is based on a 360-degree understanding of innovation. Innovation and sustainability are at the center of every work carried out in all of our fields of activity and are carefully emphasized. Considering the advantages of digital-ization in business processes, it is understood that it is necessary to integrate employees into this transformation in order to ensure permanence. In this context, studies are carried out and training sessions are provided to increase the digital competencies of employees. In order to ensure the participation of employees, employee suggestions for digital transformation practices are implemented.

– What is the source of finance for projects developed for digital transformation in your business? (Equity capital, public support programs, use of credit, etc.)

The financial infrastructure of our business is based on solid foundations. The majority of the projects are carried out with equity capital, and loan opportunities and public support programs are also evaluated.

– How would you describe the impact of digital transformation on the progress/development of your company? (Efficiency, agility, flexibility, innovation, employee resistance, increase in bureaucracy, etc.)

Along with digital transformation, improvements are observed in productiv-ity, agility, faster, and more accurate decision making. However, every change and transformation has its own pain. In our company, from time to time, employees may encounter the problem of adapting to change and resistance. However, the phenomenon of continuous development and change in our culture enables us to cope with these challenges.

– Who are the decision makers in the projects developed for digital trans-formation in your business? (Family members, professional managers, other stakeholders, joint decisions, etc.)

Our company is a family business. The final decision makers are the Board of Directors, where family members are in the majority. In addition, working experts, professional managers and consultants have an important place in decision-making mechanisms.

– How do family member managers see the digital transformation? Do they lead the digitalization processes in the company?

The main architects of innovation and transformation in our company are family members themselves. Our Chairman of the Board of Directors leads our innovation and digitalization processes with the experience of graduating from the Department of Electrical and Electronics Engineering. The ability to think analytically, which engineering education brings to itself, gives us a great advantage in accelerating the digital transformation of our business. Our chairman of the board of directors, who takes an active position in all stages of production and R&D processes, closely monitors all projects and provides digital leadership. In fact, in case of resistance by the blue-collar employee in the digitalization processes, he contributes to the processes by making direct interviews and briefings with the employees.

– Are there conflicts between family members on digital transformation?

In our business, digitalization is considered a transgenerational development, regardless of industry and companies. The fact that the Chairman of the Board of Directors and the second generation after him have an engineering education and lead both digitalization and R&D and innovation studies constitute an important driving force. In digitalization efforts in every field, the generation X member senior management assumes the role of thought leadership. There were no differences of opinion among family members of different generations in terms of business outlook and solution development in our company.

– Have senior family members and professional managers received training on digitalization?

Along with having an engineering education, senior family members closely follow digital developments and receive support from expert consultations on the necessary issues. Many studies, including the development of production technologies in our company, digital marketing projects, especially R&D center projects, are carried out with university–industry collaborations and academic consultants.

11.5 Conclusion

All the industrial revolutions have caused significant changes and transformations in business life in the world. Over the years, expectations for traceability, speed, flexibility, cost, and efficiency factors have been developed by manufacturers and consumers, without changing the quality expectation for products and services. Industry 4.0, unlike other revolutions, has developed much faster and has become an opportunity that will provide significant benefits for businesses of all types with the digital transformation tools it offers. With the digital transformation, thanks to the technological tools to be used in all functions in the value chain and the big data to be obtained, customer-specific and personalized products/services will be offered, costs will decrease, and quality, flexibility, timeliness, transparency, efficiency, reliability, and traceability will improve, and the sustainability of the business will be contributed to. Similarly, increasing the corporate reputation and improving organizational memory will also be supported.

The COVID-19 pandemic has increased the awareness of businesses on crisis management and accelerated the digitalization process. Regardless of the sector, all companies have been forced to digitalize in order to maintain their existence. Now, new generation technologies such as big data, artificial intelligence, the Internet of Things and cloud computing, which are expressed with the concept of Industry 4.0, and concepts such as robotics, smart factories, dark factory, which are the successors of those technologies, can be encountered at every point of life. However, besides the opportunities and benefits these new technologies offer, there are also some devastating effects. In the case of environmental uncertainty, enterprises that can read, understand and take positions together with change and transformation, opportunities and threats can gain opportunities and advantages by adapting these technologies to their organizations, while some businesses cannot manage the process successfully. These businesses either waste their resources with unnecessary technology investments or cannot adapt to digital transformation by avoiding technology investments, and they have serious problems in sustainability by being at a disadvantage against their competitors who have successfully accomplished digital transformation.

Family businesses are businesses that have a leading position in the economic life of countries. These businesses, on the one hand, struggle to keep up with the change that is mandatory for every business, on the other hand, they strive to fulfill the requirements of being a family business. Wrong decisions taken by family members hinder the sustainability of family businesses. Family businesses have to understand, accept and adapt to the global digital change. In order to achieve this, they must use digitalization in all business functions (R&D and design, inventory management, production, planning, sales and marketing, human resources, maintenance and all management systems). In addition, they should reconsider and regulate issues involving family members such as the transfer plan, the family constitution.

In the application part of the study, the digital transformation process of Muratbey Cheese Company, a leading family business in the cheese manufacturing sector in Turkey, is presented as a situation analysis. Semi-structured interviews were held with the Digitalization Department Manager and Communications and Business Development Director of the enterprise and the documents they submitted were examined. It was seen that the activities carried out by the business within the scope of digital transformation were beyond the expectations of the authors and could be considered an example of good practice.

In the first part of the interview, the interviewees were asked to introduce themselves and the company. It has been observed that family members work on the board of directors and in key positions in the enterprise, but they have the necessary training and technical equipment in their field. It has been observed that professional managers are employed in the organizational chart and expert personnel and researchers in the R&D centers.

In the second part, an answer was sought for the first research question of the study, 'What is the current situation in digitalization in family businesses?'. In line with the questions posed, it has been reached in the evaluation that the company has taken important steps in digitalization and that they have guided this transformation with their corporate structure. It is clearly understood how institutionalization has a significant impact on the success of digital transformation.

First of all, it is seen that the company has established a unit to carry out digitalization at the institutional level and employs expert personnel in this unit. This is an important step in the implementation, internalization, and sustainability of processes and practices. Training sessions on digitalization in the business were given, meetings were held, and strategies were developed. The current situation of the business on digitalization has been revealed by making a SWOT analysis, and opportunities and threats for the future have been defined. The business has prepared a digital transformation roadmap and transformed the digital transformation strategy and action plan into a written pose. In line with this roadmap, strategies and all steps to be taken under the dimensions of suppliers, marketing, software and technologies, people and security, and infrastructure have been defined and presented to the relevant parties.

The human factors, which are one of the most important issues in the digital transformation process in business, have not been ignored. The title of employment and cost of qualified personnel has been defined as a threat in the SWOT analysis and studies have been initiated to overcome this threat. With training programs and suggestion development systems, it was tried to develop the digital skills of the personnel and to internalize the subject.

In the third chapter, an answer has been sought to the question 'What kind of digitalization steps are taken within the scope of business functions in food businesses?'. From the responses received, it was concluded that digitalization is very important for effective management and efficiency in all processes in food businesses. It was asked in which rings of the value chain in the enterprise digitalization steps were taken. It has been observed that robots, tracking systems, business intelligence applications, and software are used in purchasing, storage, production, quality control, facility management, maintenance, marketing, and energy management processes in the enterprise. The company precisely defined the benefits it gained after each digital transformation step and developed applications by taking new digitalization steps to increase this benefit. Thus, it has gained great advantages in terms of cost reduction, reduction in errors, faster decision making, reduction of downtime and waste, and energy savings. Digital transformation provides important tools and benefits for every business function in food businesses.

In the last section, an answer was sought to the question 'What are the opportunities and obstacles encountered in the digitalization process in family businesses?'. The results received in the questions about understanding the effect of being a family business on digital transformation surprised the authors. In the literature, there are generational conflicts in change and transformation in family businesses. While generation Z accepts and implements digitalization more easily, generation X can take a more traditional approach, insist on old methods and resist transformation. However, the situation in the company we examined does not support this assumption.

The Board of Directors of the business consists of family members, and decisions are made here. The Chairman of the Board of Directors is the elder brother of the family and closely monitors all processes in the business. He closely follows innovation and digital transformation, leads the way in implementing new technologies suitable for business structures, and even intervenes in situations where resistance is encountered and persuades the staff. This situation facilitates the work of professional managers and enables them to act more freely in digital transformation. The reason behind the attitude of the Chairman of the Board of Directors is that he has received an engineering education. He developed his digital leadership skills, directed the next generation in this direction, and prevented generational conflict and favoritism, creating an innovative and learning organizational culture. This has provided a very important advantage for the business and prevented it from falling behind its competitors in the global market. As Bretschneider *et al* [37] stated, when managed correctly, family businesses can achieve transformation processes much more successfully than other companies. However, the underlying reason for this situation is education and digital leadership, and it is inevitable for family businesses that fulfill the requirements in this context to achieve success in sustainable digital transformation.

11.6 Future research directions

Within the scope of the study, how digital transformation develops in family businesses and what kind of opportunities or obstacles are encountered were investigated. Within the scope of the subject, a situation analysis was carried out in a leading company operating in the food sector in Turkey. Since there is no study in the literature on digitalization in family businesses, the study is an important source. One of the most important limitations of the study is that the study was conducted on a single enterprise. Considering the topicality of digital transformation and its importance for a sustainable economy, it is recommended to carry out research with different methods in different sectors and to present application examples in order to understand and disseminate the issue.

Another limitation of the study is that it was conducted on a large-scale family business that has achieved institutionalization. The good educational background of the family members involved in business management has enabled them to act consciously and decisively in digitalization. However, it is known that the number of small and medium-sized enterprises in the economies of countries is high, and these enterprises are mostly family businesses. Therefore, it is not possible to generalize the results obtained from the study. In future studies, it is recommended to carry out similar studies in companies of different sizes and structures and to conduct research that will guide companies in their digitalization journey.

References

[1] Carvalho R M D C 2017 Industry 4.0—is Portugal prepared for the future? *Doctoral Dissertation*, International Business Master, the School of Technology and Management of the Polytechnic Institute of Leiria, Portugal

[2] Günaydın D 2018 Türkiye'de dördüncü sanayi devrimini beklerken: Çerkezköy organize sanayi bölgesi'nde bir araştırma *Inst. of Bus. Adm. Man. J.* **29** 73106

[3] Schwab 2016 *Klaus the Fourth Industrial Revolution* (Geneva: World Economic Forum)

[4] Dengiz O 2017 Endüstri 4.0: Üretimde kavram ve algı devrimi *Mak. Tas. ve İm. Der.* **15** 38–45

[5] Karabacak N and Aras N 2017 KOBİ'lerin malzeme aktarma ve depolama sistemlerine yönelik Endüstri 4.0 uygulamalarında karşılaşılan güçlükler ve çözüm önerileri *Mak. Tas. ve İm. Der.* **17** 39–45

[6] Hinings B, Gegenhuber T and Greenwood R 2018 Digital innovation and transformation: an institutional perspective *Inf. Organ.* **28** 52–61

[7] Birkinshaw J and Mol M 2006 How management innovation happens *MIT Sloan Manag. Rev.* **47** 81–8

[8] Deloitte 2019 Aile Şirketlerinde Sürdürülebilir Başarının Anahtarları. https://www2.deloitte.com/content/dam/Deloitte/tr/Documents/risk/aile-sirketlerinde-surdurulebilir-basarinin-anahtarlari.pdf (Accessed on 12 June 2021).

[9] Cravotta S and Grottke M 2019 Digitalization in German family firms—some preliminary insights *J. Evol. Stud. Bus.* **4** 1–25

[10] EY (2020), Aile şirketleri değişime nasıl hazırlanıyor? Aile şirketlerinde yeni normal. https://assets.ey.com/content/dam/ey-sites/ey-com/tr_tr/pdf/2020/09/ey-turkiye-aile-sirketlerinde-yeni-normal.pdf (Accessed on 08 August 2021).

[11] Leyh C, Bley K, Schäffer T and Forstenhäusler S 2016 SIMMI 4.0—a maturity model for classifying the enterprise-wide it and software landscape focusing on Industry 4.0 *2016 Federated Conf. on Computer Science and Information Systems (FedCSIS)* (Piscataway, NJ: IEEE), September pp 1297–302

[12] Sarvari P A, Ustundag A, Cevikcan E, Kaya I and Cebi S 2018 Technology roadmap for Industry 4.0 *Industry 4.0: Managing the Digital Transformation* (Cham: Springer), pp 95–103

[13] Lee J, Jun S, Chang T W and Park J 2017 A smartness assessment framework for smart factories using analytic network process *Sustainability* **9** 794

[14] Sert E 2020 Dijital Dönüşümde Endüstri 4.0 Değer Zinciri Temel Faaliyetler Analizi: Otomotiv Sektörü için Model Önerisi (Doktora Tezi), İstanbul Üniversitesi, Sosyal Bilimler Enstitüsü, İşletme Anabilim Dalı, İstanbul.

[15] Deloitte A G 2015 Industry 4.0: challenges and solutions for the digital transformation and use of exponential technologies www2.deloitte.com/content/dam/Deloitte/ch/Documents/manufacturing/ch-en-manufacturing-industry-4-0-24102014.pdf.

[16] Parida V 2018 Digitalization. J Frishammar and Å Ericson *Adressing Societal Challenges* (Luleå: Luleå University of Technology), pp 23–38

[17] Klein M 2019a *İşletme 4.0 Kapsamında Şirket 2.0 – İşletme Süreçlerinde Sosyal Yazılım Kullanımı* (Ankara: Nobel Akademik Yayıncılık)

[18] Bächle M A 2016 *Wissensmanagement mit Social Media* (Berlin/Boston, MA: Walter de Gruyter)

[19] Klein M 2020 Digital transformation scenarios of businesses—a conceptual model proposal *Electron. J. Soc. Sci.* **19** 997–1019

[20] Nalbantoğlu C B 2021 COVID-19 sürecinin dijital dönüşüme etkileri *Balk. Near East. J. Soc. Sci.* **7** 13–8

[21] Klein M 2019 Geniş Kapsamlı Dijital Dönüşüm Yaklaşımı—Dünya 4.0 *Dijital Dönüşüm Trendleri* ed E S Bayrak Meydanoğlu, M Klein and D Kurt (İstanbul: Filiz Kitabevi)

[22] Majchrzak A, Markus M L and Wareham J 2016 Designing for digital transformation: lessons for information systems research from the study of ICT and societal challenges *MIS Quart* **40** 267–77

[23] Hess T, Matt C, Benlian A and Wiesböck F 2016 Options for formulating a digital transformation strategy *MIS Q. Exec.* **15** 123–39

[24] Henriette E, Feki M and Boughzala I 2015 The shape of digital transformation: a systematic literature review *MCIS 2015 Proc.* **10** 431–3

[25] Schwertner K 2017 Digital transformation of business *Trakia J. Sci.* **15** 388–93

[26] Schulte M A 2016 Digital transformation in the manufacturing industry, Industry 4.0: from vision to reality, IDC White Paper.

[27] Berghaus S and Back A 2017 Disentangling the fuzzy front end of digital transformation: activities and approaches. Association for Information Systems. ICIS 2017 Proc. 4.4. http://aisel.aisnet.org/icis2017/PracticeOriented/Presentations/4

[28] Çark Ö 2020 Aile İşletmelerinde Stratejik Karar Alma Süreçlerinde Dijital Dönüşümün Etkisi *Aile İşletmelerinde Stratejik Yönetim* ed Yılmaz Osman and Kılıç Funda 1st edn (İstanbul: Kriter Yayınevi)

[29] Özmutlu S Y 2021 Organizasyonların dijital dönüşüm sürecine kısa bir bakış ed S Sönmez ed Ü M Yıldırım *Sosyal Bilimlerde Akademik Araştırma ve Derlemeler* 2 (Ankara: Duvar Yayınları)

[30] Kofler T 2018 Digitale transformation in Unternehmen. N Höhne, D Méndez, K B Zimmer, ZD.B Digital Dialogue Positions Papier.

[31] Sağlam M 2021 İşletmelerde geleceğin vizyonu olarak dijital dönüşümün gerçekleştirilmesi ve dijital dönüşüm ölçeğinin Türkçe uyarlaması *İst. Tic. Ün. Sos. Bil. Der.* **20** 395–420

[32] Neumeier A, Wolf T and Oesterle S 2017 The manifold fruits of digitalization-determining the literal value behind. *Proc. 13th Int. Conf. on Wirtschaftinformatik.* Switzerland, February 2017, pp 484–98

[33] Güleş H K, Arıcıoglu M A and Erdirençelebi M 2013 *Aile İşletmeleri: Kurumsallaşma, Sürdürülebilirlik, Uyum* 1st edn (Ankara: Gazi Kitabevi)

[34] Kırım A 2007 *Aile Şirketlerinin Yönetimi, Sistem Yayıncılık: 293* 4th edn (İstanbul: Sistem Yayıncılık A.Ş)

[35] Davis P S and Harveston P D 1999 In the founder's shadow: conflict in the family firm *Fam. Bus. Rev.* **12** 318

[36] Çark Ö, Yıldız İ and Karadeniz A T 2019 Sanayi 4.0 kapsamında işletmeler açısından büyük veri *Int. J. Multidiscip. Stud. Innov. Technol.* **3** 114–20

[37] Bretschneider U, Heider A K, Rüsen T A and Hülsbeck M 2020 *Practical Guide Digitalisation Strategies in Family Businesses on Specific Digitalisation Approaches for Business Families and Family Businesses* (Witten: Witten Institute for Family Business (WIFU))

[38] Erdirençelebi M and Ertürk E 2018 Türkiye'de nepotizm konusunda hazırlanan çalışmalara yönelik bir inceleme. Osmaniye Korkut Ata Üniversitesi İktisadi ve İdari Bilimler Fakültesi İşletme Bölümü, *18th Int, Business Congress, 2–4 May* pp 2344–59

[39] Tosun C 2020 Aile şirketlerinde dijital dönüşüm *Fam. Bus. Rev.* 12–4 https://www2.deloitte.com/content/dam/Deloitte/tr/Documents/deloitteprivate/FBR-Temmuz-2020.pdf (Accessed on 07 June 2021).

[40] Haffke I, Kalgovas B and Benlian A 2017 The transformative role of bimodal IT in an era of digital business *50th Hawaii Int. Conf. on System Sciences* pp 5460–9

[41] Osmundsen K, Iden J and Bygstad B 2018 Digital transformation: drivers, success factors, and implications *MCIS 2018 Proc.* **37** 1–15 https://aisel.aisnet.org/mcis2018/37

[42] Bird B, Welsch H, Astrachan J H and Pistrui D 2002 Family business research: the evolution of an academic field *Fam. Bus. Rev.* **15** 337–50

[43] Benavides-Velasco C A, Quintana-García C and Guzmán-Parra V F 2013 Trends in family business research *Small Bus. Econ.* **40** 41–57

[44] Handler W C 1989 Methodological issues and considerations in studying family businesses *Fam. Bus. Rev.* **2** 257–76

[45] Craig J B and Salvato C 2012 The distinctiveness, design, and direction of family business research: insights from management luminaries *Fam. Bus. Rev.* **25** 109–16

[46] Yin R K 2018 *Case Study Research and Applications: Design and Methods* 6th edn (Los Angeles, CA: SAGE)

[47] Klötzer C and Pflaum A 2017 Toward the development of a maturity model for digitalization within the manufacturing industry's supply chain, Proc. 50th Hawaii Int. Conf. on System Sciences. http://hdl.handle.net/10125/41669

Further reading

Alavi S, Wahab D A, Muhamad N and Shirani B A 2014 Organic structure and organisational learning as the main antecedents of workforce agility *Int. J. Prod. Res.* **52** 6273–95

Bogner E, Voelklein T, Schroedel O and Franke J 2016 Study based analysis on the current digitalization degree in the manufacturing industry in Germany *Procedia CIRP* **57** 14–9

Kuusisto M 2017 Organizational effects of digitalization: a literature review *Int. J. Organ. Theory Behav.* **20** 341–62

Shahi C and Sinha M 2020 Digital transformation: challenges faced by organizations and their potential solutions *Int. J. Innov. Sci.* **13** 17–33

Key terms and definitions

Family business: Businesses that are owned by people belonging to the same family in most of the share ownership and management levels, established for the purpose of providing their own economic livelihood and leaving a legacy to the next generations, and are affected by the values, beliefs and attitudes of the owner family.

Digitization/digital transformation: It is the structural changes that occur with the inclusion of digital technologies in all organizational structures and business processes.

Industry 4.0: It is the fourth Industrial Revolution, which includes many new automation systems, data management and new production technologies.

Nepotism: It is the employment of workers in organizations not based on education, success, or merit, but by blood ties, favoritism, or discrimination.

Leadership: A person who benefits the organization, makes radical changes in the ongoing tradition, and carries the responsibility for managing the environment with decisions and practices based on intuition, intelligence and knowledge.

Sustainability: It is to act by considering environmental, economic and social parameters in all processes of businesses in order to be permanent.

IOP Publishing

Advanced Signal Processing for Industry 4.0, Volume 1
Evolution, communication protocols, and applications in manufacturing systems
Irshad Ahmad Ansari and Varun Bajaj

Chapter 12

Automatic identification of finger movements for industrial robotic applications using electromyogram signals

Anurodh Kumar, Amit Vishwakarma and Varun Bajaj

The evolution of a modern human–machine interface (HMI) has consistently been a fascinating exploration subject in the domain of rehabilitation, in which electromyogram (EMG) signals to play a vital role. EMG control is an up-to-date technique perturbed with the sensing, handling, recognition, and application of EMG signals to command man-aiding robots or restoration appliances. Finger therapy is the challenging aspect of physical therapy and post-stroke rehabilitation. It is tougher to analyze the EMG signal from finger gestures than it is from hand gestures. This chapter presents the automated identification of finger movement using 16 channels. A tunable Q-factor wavelet transform (TQWT) is used for the automated identification of finger movements using EMG signals. TQWT is a structured method for the decomposition of EMG signals into sub-bands. Various features, namely integrated EMG (IEMG), simple square integral (SSI), modified mean absolute value-type 1 (MAV1), waveform length (WL), variance (VAR), average amplitude change (AAC), and Tsallis entropy (TE) are obtained from the sub-bands. Obtained features are fed to the ensemble tree, decision tree, and support vector machine classifier for the identification of finger movements. The ensemble bagged tree (EBT) classifier achieved a classification accuracy of 93.40%, sensitivity of 93. 94%, precision of 93.16%, specificity of 96.84%, F-1 score of 0.9354, and area under the curve of 0.99. The proposed method may be utilized for a hand-robotic prosthesis or exoskeleton for rehabilitation in medical industry applications.

12.1 Introduction

Electromyogram (EMG) signals could be utilized for biomedical, clinical applications, and recent human–machine interfaces (HMI). EMG signals are acquired from the muscles and they measure the electrical activity of muscles. A prosthetic finger can be controlled by using this muscular activity. It is a complicated signal and is controlled through the neurological system. EMG signals depend on the anatomical and physiological properties of muscles. The amplitude of the EMG signal ranges from 0 to 10 mV [1]. An EMG signal is nonstationary and its statistical properties change over time. EMG can be utilized in a variety of applications. It is utilized clinically for the screening of neuromuscular and neurological problems. Additionally, it is utilized in a variety of research labs, including those concerned with neuromuscular, motor control, biomechanics postural control, movement disorders, and physical therapy. Biomedical engineering is increasingly gaining attention for the ability to identify EMG signals utilizing reliable and efficient approaches.

Humans do several daily life activities with their hands, which are essential tools for interacting with the surroundings and manipulating things. Particularly, hand function may be severely affected by stroke, which has an impact on a few hundred per lakh (100,000) of the world's population, and is the major root of motor disability [2]. Moreover, approximately 80% of the sufferers have upper limb disabilities after a stroke [3]. The patient's freedom is substantially impacted by the difficulty of doing various tasks, which can also result in long-term disability. After a stroke, rehabilitation treatment is a typical treatment and is assumed to be crucial to recovery. For instance, between 3 and 6 months following a stroke incident, about 70% of motor function is recovered [4]. The recovery process in the upper extremities may be slower than in the lower extremities, in part because of the complexity of the movements in the upper extremities [5, 6]. Hand therapies often involve repetitive task-specific training, to address finger muscular weakness and spasticity, and sensorial loss of finger muscles during isolated motions [7].

Robotic-assisted therapy may be able to aid in the restoration of motor ability, according to recent developments in hand rehabilitation [8]. In particular, it might raise the therapy's frequency and intensity, provide each motor assistance control and create an objective assessment of the patient's improvement [9]. To address the purpose used in [9], end-effector, and exoskeleton robotic systems have mostly been developed [10]. The end-effector system gives external aid to the movement by exerting forces at the digits' ends [11]. The exoskeleton systems utilize the fingers and hand while allowing direct control of hand joints. Mechanical or electrical signals can be used to control an exoskeleton's control system. Mechanical signals comprise voice commands, torques, and finger or joint positions. The system's response may also be influenced by electrical signals coming from the body. The intention and motor control of the patient are revealed by EMG and electroencephalography (EEG) signals [12, 13]. Invasive and non-invasive electrodes can be used to acquire EMG signals; the latter is commonly utilized in rehabilitation. Surface electrodes in both forearm muscles can be used to partially capture the intention to extend and flex the fingers [14]. But, the EMG signal is heavily influenced by noise due to its low amplitude and interference from other muscle activity, necessitating high-quality analog recording and sophisticated digital processing techniques. Additionally, many DOFs should be controlled simultaneously and proportionally to control the exoskeleton [15]. In particular, rather than focusing on dexterous finger movements, many proposed methods emphasize the rehabilitation of broader hand movements. Additionally, most of the suggested techniques that employ EMG data for exoskeleton control need multichannel systems. Comparing the effectiveness of robotic-assisted rehabilitation to traditional therapy has not yet been validated, and further studies are required. Researchers have put up several methods to date for identifying and analyzing hand movements using EMG signals. The ANFIS-based system is used to classify finger movement [16].

Various approaches for feature extraction and several classifiers have been utilized for the recognition of finger movements using EMG signals [17]. A support vector machine (SVM) classifier has been utilized for the evaluation of finger movement in rehabilitating stroke patients [18]. Six features, namely mean absolute value, Willison amplitude, waveform length, variance, mean frequency, and median frequency have been utilized for the classification of individual finger movement [19]. In [19], extracted features have been given as input to SVM, k nearest neighbor, and artificial neural network classifiers for the performance comparison. An 8-channel armband and a 64-channel high-density grid are wrapped around the forearm and back of the hand, respectively, for the recognition of finger movement [20]. The fractional Fourier transform technique has been utilized for the classification of 10 different finger movements using EMG signals [21]. Four machine learning (ML)-based classifiers have been used to predict finger movements of a prosthetic hand utilizing EMG signals. Out of four classifiers, the XGBoost classifier, provided a better prediction accuracy than other classifiers, namely, decision tree, random forest, and k-NN [22]. An embedded system has been proposed for multivariate thumb and finger movement classification utilizing electroencephalogram signals for upper limb prosthesis control [23].

Time variant linear discriminant technique has been employed for decoding finger movement and hand gestures for brain–computer interfaces [24, 25]. The differential enhanced signal feature has been developed for the classification of upper limb movement using EMG signals [26]. Time–frequency distribution methodology has been proposed for the discrimination of finger movement using EMG signals for prosthetic control [27]. Particle swarm and ant colony optimization techniques have been utilized for suitable feature selection to recognize finger movement using multichannel EMG signals [28]. The empirical mode decomposition method has been employed to recognize hand movement utilizing EMG signals [29, 30]. For real-time EMG-based hand gesture identification, an embedded solution has been presented. The authors focus on the system's multi-level architecture, integrating the hardware and software elements to create a wearable gadget for real-time gesture recognition [31–33]. A nonnegative matrix factorization-based technique has been utilized to identify finger movement using EMG signals [34]. A transfer learning technique has been employed for the classification of hand gestures [35]. Spectral collaborative representation-based classification is proposed for hand gesture recognition using myoelectric signals [36]. A myoelectric control method depending on independent component analysis has been utilized for identifying different movements with a limited number of sensors [37]. The adaptive AdaBoost algorithm has been utilized for the recognition of nine different types of gestures [38]. A low-cost, simple, and fast system has been developed for the recognition of four gestures with a single-channel EMG signal [39]. A 2-channel EMG system has been presented for hand gesture recognition using an SVM classifier [40, 41]. A tunable Q-factor wavelet transform (TQWT)-based filter bank has been developed for the classification of hand gesture recognition using EMG signals [42]. The variational mode decomposition method has been proposed for the identification of hand movements [43]. In [43], the authors also utilized the composite permutation entropy index method. For hand gesture recognition, a nonlinear multiscale maximal Lyapunov exponent has been utilized [44]. A flexible analytic wavelet transform methodology has been proposed to identify various physical actions using EMG signals [45].

The TQWT technique has been utilized to classify ten different actions using EMG signals [46]. For the classification of hand movements, the dimensionality reduction technique (principal component analysis) has been employed, and a linear discriminant analysis classifier is utilized for hand gesture recognition [47–49]. The wavelet packet coefficients have been filtered sparsely for the identification of finger movements using EMG signals [50]. A minimum redundant feature selection method has been utilized for the classification of ten different finger movements utilizing 2-channel EMG [51]. The feature set is chosen from a wavelet-denoised myoelectric signal at four stages of decomposition with the least amount of mutual information redundancy [51]. A logarithmic spectrogram-based graph signal technique has been developed for the classification of multi-class hand grasp [52]. A fuzzy c-means clustering algorithm has been proposed for the classification of hand gestures utilizing EMG signals [53]. A deep convolutional neural network has been proposed to classify hand movements utilizing myoelectric signals [54]. The iterative feature extraction methodology is developed using ternary patterns and discrete wavelets to recognize hand movements using EMG signals [55]. Although all the previously reported techniques in the literature provided good results. But, these methods have some limitations such as features selection, decomposition techniques used, selection of an appropriate classifier, etc. The remaining parts of this chapter are arranged as follows. Section 12.2 presents the methodology. The results are described in section 12.3, and section 12.4 presents this chapter's conclusion.

12.2 Methodology

This section depicts the dataset, TQWT, features extraction, and classification methods used in this chapter. The EMG signals are decomposed into sub-bands utilizing TQWT. Various time domain- and entropy-based features such as integrated EMG (IEMG), modified mean absolute value-type 1 (MAV1), simple square integral (SSI), waveform length (WL), variance (VAR), average amplitude change (AAC), and Tsallis entropy (TE) are obtained from sub-bands. The obtained features are given to various classifiers to identify nine different finger movements. A flow diagram of the proposed method is shown in figure 12.1.

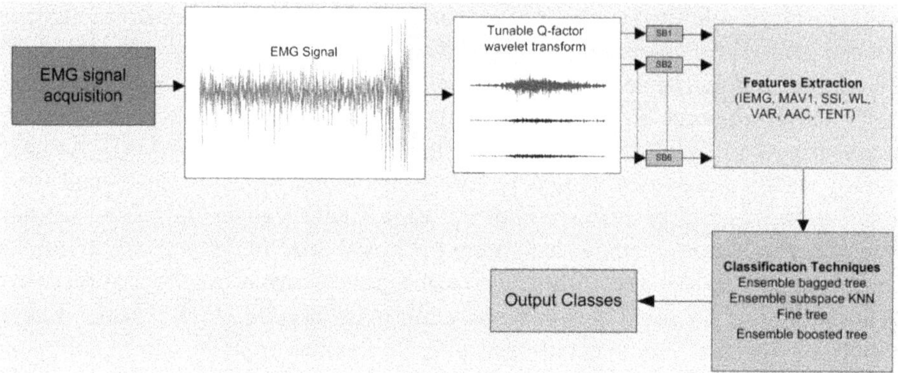

Figure 12.1. Flow diagram of the proposed methodology.

12.2.1 Dataset

The EMG dataset is taken from the Ninapro database website (http://ninaweb.hevs. ch/). This database is intended to serve as a reference point for decoding finger position from EMG measurements. The details of the collection of the dataset are described in [56]. Each session consists of a 140 min recording of 12 subjects. Out of 12 subjects, 10 were able-bodied (all right-hand dominant) and two were right-hand trans-radial amputees. Before their amputations, both amputees were right-handed dominant. All subjects had a median age of 26.5 years. A Delsys Trigno IM Wireless EMG system's 16 active double-differential wireless sensors were used to record the muscle activity. The sampling rate used for the sEMG signals was 1111 Hz. Upsampling all signals to 2 kHz was done. During the signal acquisition, each subject was asked to repeat nine movements utilizing bilateral mirrored movements. Each subject had to repeat the instructed movements following visual cues that are movies displayed on the monitor. Each movement took between 6 and 9 s to complete, and there were 3 s of rest in between each set of trials. Each repetition started with the subject holding their fingers in the resting position. They then gradually moved toward the target posture as it was displayed on the screen and returned to the resting position before the trial ended. The following actions were included for this chapter: 1 shows thumb flexion or extension, 1 indicates thumb abduction or adduction, 2 shows index finger flexion or extension, 3 represents middle finger flexion or extension, 4 shows little finger flexion or extension, 5 shows index pointer, 6 shows cylindrical grip, 7 shows lateral grip, and 8 indicates tripod grip. Figure 12.2 shows the reshaped EMG signal using 16 channels.

12.2.2 TQWT

It is a wavelet transform with a configurable Q-factor. TQWT is frequently utilized for oscillatory and non-oscillatory signals [57]. It is flexible due to its configurable input parameters. It has three important parameters: Q-factor (Q), oversampling rate (r), and the various levels of decomposition (J). These input parameters can be modified empirically [58]. The wavelet's oscillations are regulated by the Q-factor.

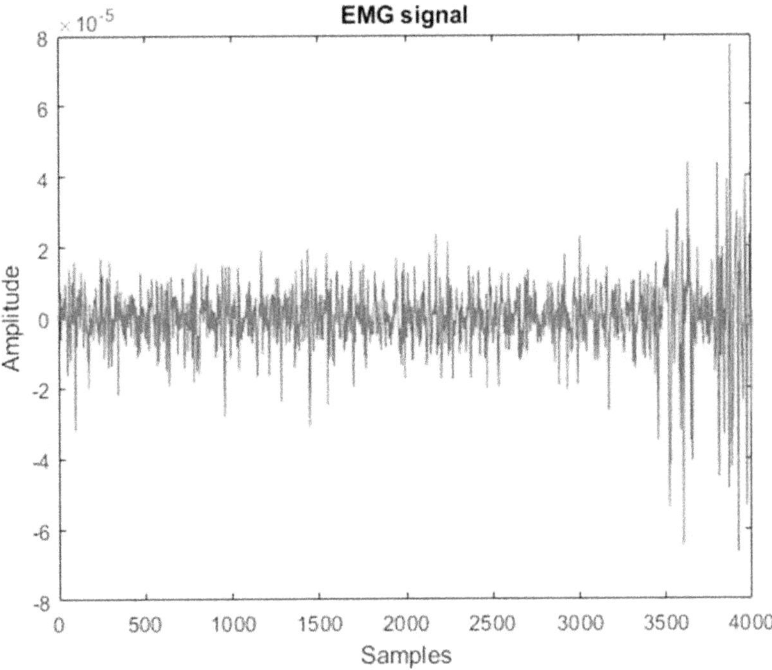

Figure 12.2. Reshaped EMG signal.

For oscillatory signals, a high value of Q is selected such that the real wavelet oscillates more and has a narrower frequency response in comparison to its center frequency. Therefore, more decomposition levels are needed to cover the same range of frequency. A low value of Q is ideally suited for a signal with minimal oscillations [58]. For this chapter, Q is adjusted such that the primary wavelet has a broader frequency response and fewer oscillations. Hence, fewer decomposition levels are needed to cover the frequency spectrum. The Q is given as the proportion of fundamental frequency f_0 to the bandwidth BW as equation (12.1)

$$Q = \frac{f_0}{BW}. \tag{12.1}$$

r controls extra ringing in the wavelet for good time localization without altering the wavelet's shape. The impact of increasing r for a fixed Q is an increase in the crossover between nearby frequency responses. Therefore, to cover the same frequency range, higher values of r require more levels, while lower values of r require fewer levels [57, 59]. TQWT decomposes signals using an iterative filter bank structure with 2-channels, where the number of filter banks is represented by J. The source signal $x(n)$ is split into many high-pass sub-band signals and one low-pass sub-band at each level, with sampling frequencies of αfs and βfs, respectively, where α and β are the scaling parameters. J-level decomposition has $J + 1$ sub-bands. The configuration of filters for decomposing an input signal is shown in figure 12.3(a). d1 (n) represents low-pass sub-band signal and d2(n) represents high-pass sub-band

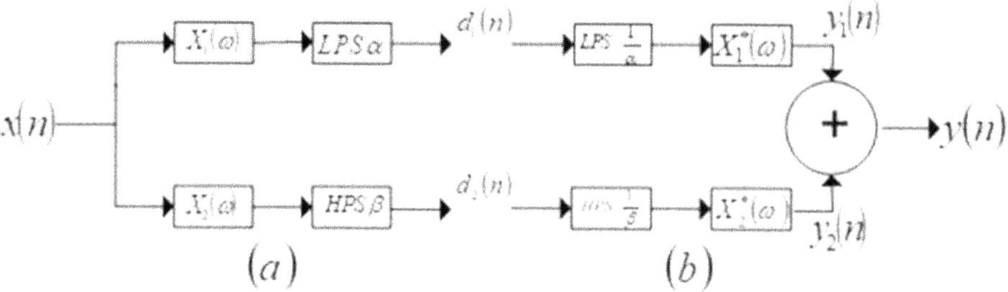

Figure 12.3. Single-level TQWT analysis and synthesis filter bank.

signal. The reconstruction filter depicted in figure 12.3 (b) is suitable for reconstructing the original signal. The frequency response of $X_1^j(\omega)$ and $X_2^j(\omega)$ is given below [57] as equations (12.2) and (12.3)

$$X_1^j(w) = \begin{cases} \prod_{m=0}^{j-1} X_1\left(\dfrac{w}{\alpha^m}\right) & |w| \leqslant \alpha^j \pi \\ 0 & \alpha^j \pi < |w| \leqslant \pi \end{cases} \tag{12.2}$$

$$X_2^j(w) = \begin{cases} X_2\left(\dfrac{w}{\alpha^{i-j}}\right) \prod_{m=0}^{j-1} X_1\left(\dfrac{w}{\alpha^m}\right) & (1-\beta)\alpha^{j-1}\pi \leqslant |w| \leqslant \alpha^{j-1}\pi \\ 0 & \omega \in [-\pi, \pi] \end{cases} \tag{12.3}$$

To avoid oversampling and achieve proper reconstruction, scaling parameters also should satisfy the following relations: $0 < \alpha < 1$, $0 < \beta \leqslant 1$, and $\alpha + \beta => 1$.

The parameters selected for TQWT determine how well it performs in extracting useful information about changes in finger movements from EMG signals. Equations (12.4) and (12.5) explain the relationship between the input parameters and scaling factors.

$$Q = \frac{2-\beta}{\beta} \tag{12.4}$$

and

$$r = \frac{\beta}{1-\alpha} \tag{12.5}$$

The maximum number of decomposition levels (J_{ma}) for an input signal of length N can be calculated as equation (12.6),

$$J_{ma} = \frac{\log\left(\dfrac{\beta N}{8}\right)}{\log\left(\dfrac{1}{\alpha}\right)} \tag{12.6}$$

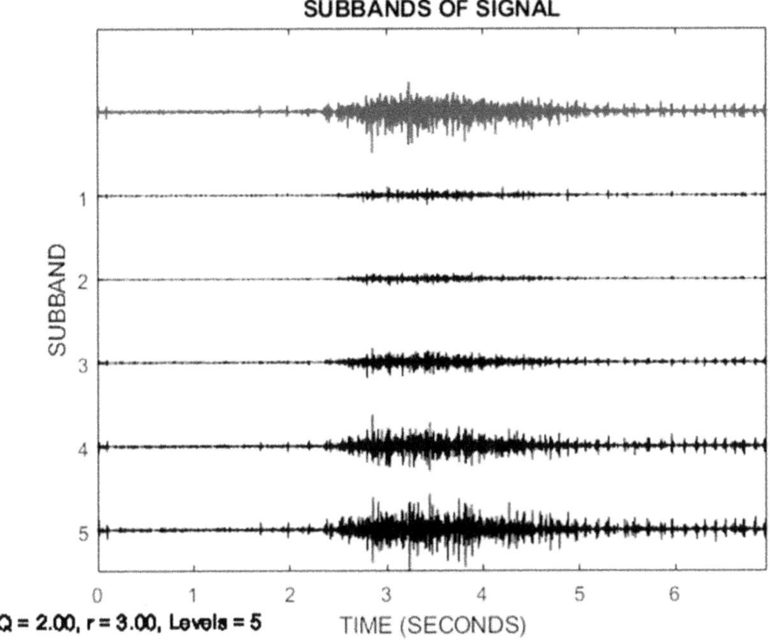

Figure 12.4. Sub-bands for EMG signal using TQWT.

For this chapter, $Q = 2$, $r = 3$, and $J = 5$ are selected. The advantages of using TQWT are

1. Direct depiction in the frequency domain is obtained by the filters due to their computing efficiency and rational transfer functions.
2. Many wavelet transforms, apart from the continuous wavelet transform, cannot regulate their Q-factor [58].
3. TQWT has been frequently utilized to examine multiple physiological signals [60, 61].

The sub-bands obtained from the EMG signal using TQWT are shown in figure 12.4.

12.2.3 Feature extraction

In this chapter, the TQWT approach is used to extract features from the sub-bands of the decomposed EMG signal. Seven features are obtained from the sub-bands and they are given as equations (12.7)–(12.13) [45, 62].

1. IEMG: It is formulated as the sum of absolute values of EMG.

$$IEMG \sum_{n=1}^{N} | x_n |$$ (12.7)

2. MAV1: It is an expansion of MAV and is described as

$$MAV1 \; = \frac{1}{N} \sum_{n-1}^{N} w_n \; |\, x_n \,|$$

(12.8)

$$w_n \; = \begin{cases} 1, & \text{if } 0.25\,N \leqslant n \leqslant 0.75N \\ 0.5, & \text{or else} \end{cases}$$

where w_n represents the weighted window function and is used for increased robustness.

3. SSI: It is the sum of the squared EMG signal amplitude values.

$$\mathrm{SSI} = \sum_{n=1}^{N} x_n^2$$

(12.9)

4. WL: It is described as EMG waveform's total length over the time segment.

$$\mathrm{WL} = \sum_{n=1}^{N-1} |x_{n+1} - x_n|$$

(12.10)

5. VAR: The prediction of the squared deviation from its mean is known as a variance. It is formulated as

$$VAR \; = \frac{1}{N-1} \sum_{n-1}^{N} (x_n - \mu)^2$$

(12.11)

6. AAC: It is roughly identical to WL feature, apart from its wavelength being averaged. It is expressed as

$$AAMC \; \frac{1}{N} \sum_{n-1}^{N-1} |\, x_{n+1} - x_n \,|$$

(12.12)

7. TE: Mathematically, it can be formulated as

$$T = \frac{\dfrac{1}{N} \displaystyle\sum_{n-1}^{N-1} p^q(x_n)}{q-1}$$

(12.13)

where $q > 0$.

12.2.4 Classification methods

This chapter utilized three different classifiers to identify finger movements using EMG signals. An ensemble tree classifier utilized a bagged tree, boosted tree, and subspace kNN kernel. The decision tree classifier utilized medium tree, fine tree, and coarse tree kernels. The linear and polynomial kernel are used in the SVM classifier.

Ensemble tree: This classifier is developed using a complete binary tree. It is based on the ensemble principle, meaning that the stronger learner is always produced from a group of different weak learners. Multiple decisions are integrated into this classifier to achieve optimal classification results.

1. **Decision tree**: It is a supervised ML algorithm that uses a set of instructions to make decisions. In this classifier, each leaf node depicts the output, whereas internal nodes indicate the characteristics of a certain input feature. It is used to separate a complex decision-making process into a set of easier decisions. The types of given input features determine how well the decision tree classifier performs.

2. **SVM classifier**: It is a supervised classification technique that uses a line to distinguish between two different categories. In SVM, two classes can be distinguished using a hyperplane and a linear separator. SVM is a reliable prediction approach that relies on a framework for statistical learning. Polynomial and linear kernel functions are types of kernel functions utilized in SVM. SVM has been used in many cases such as face detection, image classification, text and hypertext classification, handwriting detection, and so on.

12.2.5 Performance metrics

The proposed methodology's effectiveness is evaluated by computing performance metrics, namely, sensitivity (*Sen*), accuracy (*Acc*), precision (*Pre*), specificity (*Spe*), area under curve (*Auc*), and *F-1* score given as equations (12.14)–(12.18) [63–66].

Accuracy: It is the proportion of all correctly identified events to all events. It can be formulated as

$$Acc = \frac{TN + TP}{TP + FN + TN + FP} \tag{12.14}$$

Precision: It is defined as the proportion of truly recognized positive samples to all positive samples that have been classified. It can be computed as

$$Pre = \frac{TP}{TP + FP} \tag{12.15}$$

Sensitivity: It is described as the proportion of truly recognized actual positive events. It can be formulated as

$$Sen = \frac{TP}{TP + FN} \tag{12.16}$$

Specificity: It is described as the proportion of truly recognized actual negative events. It can be given as

$$Spe = \frac{TN}{TP + FP} \tag{12.17}$$

where *TN*, *TP*, *FN*, and *FP* are true negative, true positive, false negative, and false positive events, respectively.

***F-1* score**: It is defined as the harmonic mean of precision and sensitivity. It can be formulated as,

$$F-1 \text{ score} = \frac{2 \times Pre \times Sen}{Pre + Sen} \tag{12.18}$$

Auc: It has a single value that measures the overall effectiveness of a classifier. Its value lies between 0.5 and 1. The value 0.5 shows how efficiently a random classifier performs and 1 represents how efficiently a perfect classifier performs.

12.3 Results

This chapter utilizes TQWT and various classification methodologies for the identification of finger movements. There was a total of 12 subjects present, out of which 10 belong to able-bodied and 2 belong to right-hand amputees. There was a total of 324 signals from nine different classes. Each signal is reshaped into the size of 4000 samples. Each signal is divided into a total of 32 chunks, each of which provides input to the TQWT. The TQWT is tuned using the following parameter values: $Q = 2$, $r = 3$, and $j = 5$. These parameters were selected empirically. Various time domain and entropy-based features such as IEMG, MAV1, SSI, WL, VAR, AAC, and TE are obtained from the sub-bands. These obtained features are given as input to a different classifier. A 10-fold cross-validation method is utilized to evaluate the effectiveness of the proposed approach. The classification accuracy of an ensemble classifier using three kernels is shown in table 12.1. Classification accuracy is achieved by utilizing boosted tree, bagged tree, and Subspace k-NN. The maximum accuracy is obtained by using an ensemble bagged tree classifier for SB-2, which is 93.40%.

Table 12.2 depicts the classification accuracy of the decision tree and SVM classifier. The fine tree (FT) kernel in the decision tree yields the maximum classification accuracy of 66.40% for SB-1. The medium tree (MT) kernel in the decision tree achieved maximum accuracy of 63.24% for SB-2. The coarse tree kernel provides the highest classification accuracy of 61.40% for sB-2, using the decision tree. Utilizing the SVM classifier, the linear kernel gives the highest classification accuracy of 61.88% for SB-2. For the SVM classifier, the fine Gaussian (FG) kernel provides maximum classification accuracy of 65.36% for SB-1. The medium Gaussian (MG) kernel yields the highest classification accuracy of 63.44% for SB-2. Finally, the coarse Gaussian (CG) kernel yields the highest classification accuracy of 61.78% for SB-2. Out of all the classifiers utilized in this chapter, the highest classification accuracy is provided using an ensemble bagged

Table 12.1. Accuracy of ensemble classifier.

SB	BoostedTree (%)	BaggedTree (%)	Subspace k-NN (%)
SB-1	58.10	91.10	81.20
SB-2	63.40	93.40	81.30
SB-3	56.70	92.10	81.70
SB-4	53.60	92.80	81.60
SB-5	58.30	93.10	81.80
SB-6	54.00	93.10	81.20

Table 12.2. Classification accuracy of decision tree and SVM classifier.

SB	Decision Tree (%) FT MT CT	Support Vector Machine (%) Linear Gaussian FG MG CG
SB-1	66.40 62.42 60.38	61.74 65.36 63.28 61.56
SB-2	65.40 63.24 61.40	61.88 65.24 63.44 61.78
SB-3	65.40 63.14 61.08	60.92 64.08 62.68 60.84
SB-4	62.50 62.84 60.96	59.96 64.54 62.28 59.98
SB-5	64.50 62.38 61.24	60.48 63.72 61.88 60.36
SB-6	61.00 60.48 59.16	59.84 62.84 61.56 60.10

tree classifier. Therefore, performance metrics for it are evaluated and are described in table 12.3.

Table 12.3 shows that *Acc* of 93.40%, *Pre* of 93.16%, *Sen* of 93.94%, *Spe* of 96.84%, *F-1 score* of 0.9354, and *Auc* of 0.99 is obtained for SB-2 using the ensemble bagged tree classifier. The confusion matrix obtained for each sub-bands is shown in figures 12.5(a)–(f) for the ensemble bagged tree classifier.

From figure 12.5(b), it can be observed that the ensemble bagged tree for SB-2 more accurately identifies the movements of the finger. The effectiveness of a classification model is depicted by the receiver operating characteristic curve. Figure 12.6 depicts the receiver operating characteristic curve for each sub-band for the ensemble bagged tree classifier. Figure 12.6 shows the receiver operator characteristics of the SB-2 area under the curve as 0.99 which outperformed other sub-bands.

Hand function may be severely affected by stroke, which is the cause of motor disability. Moreover, approximately eighty percent of the sufferers have upper limb disabilities after a stroke. After a stroke, rehabilitation treatment is a typical

Table 12.3. Performance metrics for ensemble bagged tree classifier.

SB	Acc (%)	Pre (%)	Sen (%)	Spe (%)	F-1 score	Auc
SB-1	91.10	90.80	91.64	93.26	0.9120	0.84
SB-2	93.40	93.16	93.94	96.84	0.9354	0.99
SB-3	92.10	91.80	92.24	94.64	0.9200	0.86
SB-4	92.80	92.36	92.56	95.36	0.9246	0.87
SB-5	93.10	92.64	92.68	96.08	0.9266	0.86
SB-6	93.10	91.88	92.84	95.88	0.9236	0.82

treatment and is assumed to be crucial to recovery. The proposed method may be helpful to control a prosthetic arm with higher accuracy, which can be used for industrial applications with more reliability. Based on the above findings, the proposed technique can extract more appropriate features for the recognition of various types of finger movements using electromyogram signals. Hence, the proposed technique may be utilized for a hand-robotic prosthesis or exoskeleton for rehabilitation in medical industry applications. The limitation of utilizing the proposed method is the proper selection of input parameters and selection of features. The proposed method's performance may be improved by optimizing these parameters and appropriate features could be selected using a suitable algorithm. Proper implementation and incorporation may be one of the challenges for industries to move on to new technologies in quick progression. For industries, new technology should be able to integrate with existing technology to fully realize its potential. The other challenge for industries may be inadequate training of health workers for the new technology.

12.4 Conclusion

Finger therapy is the most challenging part of physical therapy and post-stroke rehabilitation. It is more difficult to analyze the EMG signal from finger gestures than it is from hand gestures. This is due to the proximity of the muscles that connect to particular fingers. This chapter presents the automated identification of nine different finger movements. The EMG signals are decomposed through TQWT using five-level decomposition. The obtained sub-bands are used to extract various time domain- and entropy-based features. Obtained features are given as input to various classifiers. The highest identification accuracy of 93.40% is achieved using an ensemble bagged tree classifier. A new strategy for the automatic identification of finger movements can be implemented using the proposed method using an EMG signal. In the future, the identification accuracy of finger movements can be improved with the utilization of more suitable features and the selection of an appropriate classifier. The proposed method may be utilized for a prosthesis limb for rehabilitation in medical industry applications.

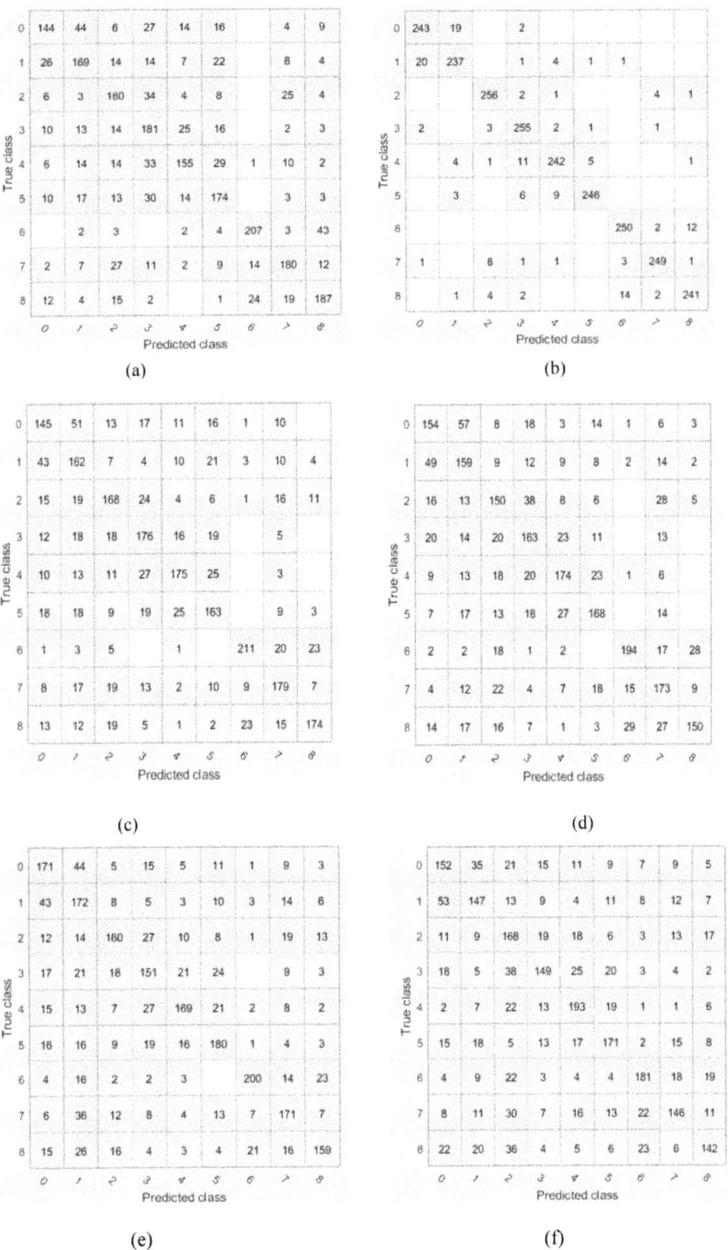

Figure 12.5. Confusion matrix for ensemble bagged tree of six sub-bands SB-1 to SB-6, (a) to (f), respectively.

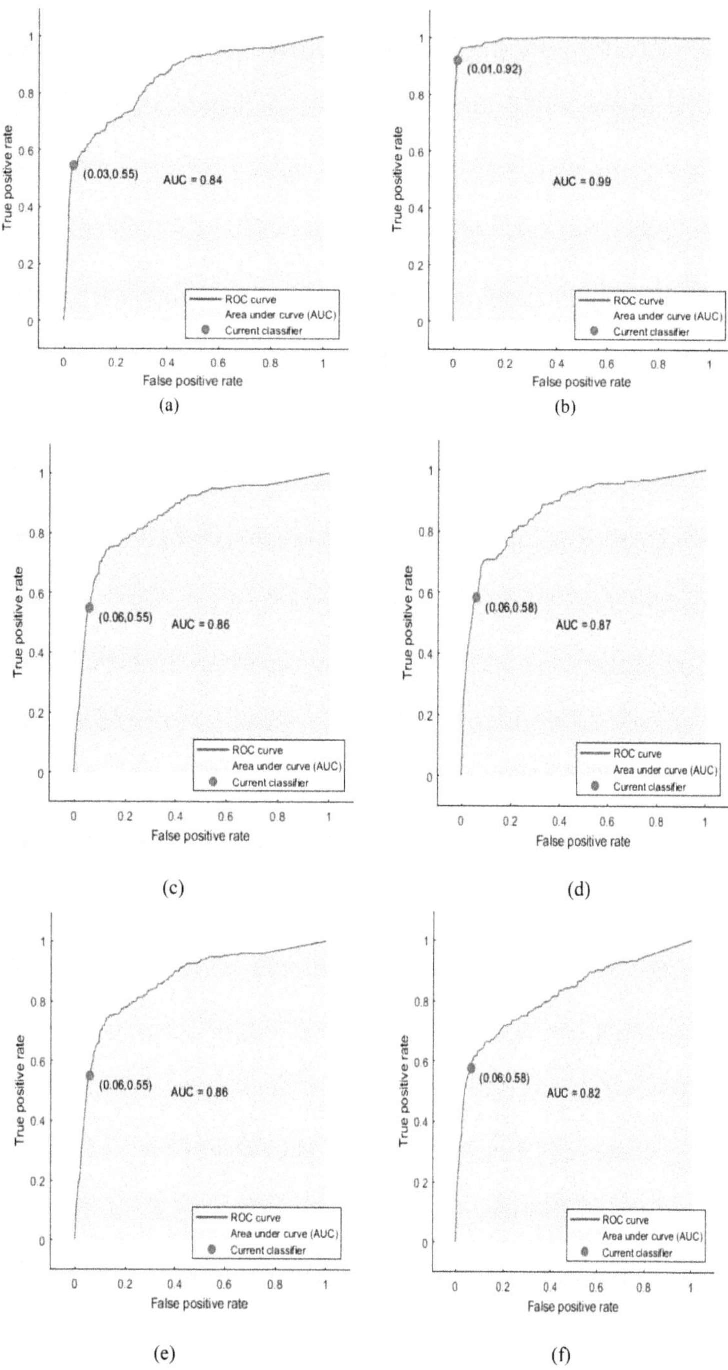

Figure 12.6. Receiver operating characteristics for ensemble bagged tree of six sub-bands SB-1 to SB-6, (a) to (f), respectively.

References

[1] Reaz M B I, Hussain M S and Mohd-Yasin F 2006 Techniques of EMG signal analysis: detection, processing, classification and applications *Biol. Proced. Online* **8** 11–35

[2] Benjamin E J, Virani S S, Callaway C W, Chamberlain A M, Chang A R and Cheng S *et al* 2018 Heart disease and stroke statistics—2018 update: a report from the American Heart Association *Circulation* **137** e67–e492

[3] Lawrence E S, Coshall C, Dundas R, Stewart J, Rudd A G and Howard R *et al* 2001 Estimates of the prevalence of acute stroke impairments and disability in a multiethnic population *Stroke* **32** 1279–84

[4] Stinear C M 2017 Prediction of motor recovery after stroke: advances in biomarkers *Lancet Neurol* **16** 826–36

[5] Lee K B, Lim S H, Kim K H, Kim K J, Kim Y R and Chang W N *et al* 2015 Six-month functional recovery of stroke patients: a multi-time-point study *Int. J. Rehabil. Res.* **38** 173

[6] Sivan M, O'Connor R J, Makower S, Levesley M and Bhakta B 2011 Systematic review of outcome measures used in the evaluation of robot-assisted upper limb exercise in stroke *J. Rehabil. Med* **43** 181–9

[7] Raghavan P 2007 The nature of hand motor impairment after stroke and its treatment *Curr. Treat. Options Cardiovasc. Med.* **9** 221–8

[8] Turolla A 2018 An overall framework for neurorehabilitation robotics: implications for recovery *Rehabilitation Robotics* (Amsterdam: Elsevier), ch 2 pp 15–27

[9] Lambercy O, Dovat L, Yun H, Wee S K, Kuah C and Chua K *et al* 2010 Robotic assessment of hand function with the HapticKnob *Proc. 4th Int. Convention on Rehabilitation Engineering and Assistive Technology* pp 1–4

[10] Yue Z, Zhang X and Wang J 2017 Hand rehabilitation robotics on poststroke motor recovery *Behav Neurol.* **2017** 1–21

[11] Schabowsky C N, Godfrey S B, Holley R J and Lum P S 2010 Development and pilot testing of HEXORR: hand EXOskeleton rehabilitation robot *J. Neuroeng. Rehabil.* **7** 1–16

[12] Balasubramanian S, Garcia-Cossio E, Birbaumer N, Burdet E and Ramos-Murguialday A 2018 Is EMG a viable alternative to BCI for detecting movement intention in severe stroke? *IEEE Trans. Biomed. Eng.* **65** 2790–7

[13] Lu Z, Tong K, Zhang X, Li S and Zhou P 2018 Myoelectric pattern recognition for controlling a robotic hand: a feasibility study in stroke *IEEE Trans. Biomed. Eng.* **66** 365–72

[14] Polygerinos P, Galloway K C, Sanan S, Herman M and Walsh C J 2015 EMG controlled soft robotic glove for assistance during activities of daily living *2015 IEEE Int. Conf. on Rehabilitation Robotics (ICORR)* pp 55–60

[15] Jiang N, Dosen S, Muller K-R and Farina D 2012 Myoelectric control of artificial limbs—is there a need to change focus? [In the spotlight] *IEEE Signal Process. Mag.* **29** 150–2

[16] Caesarendra W, Tjahjowidodo T, Nico Y, Wahyudati S and Nurhasanah L 2018 EMG finger movement classification based on ANFIS *J. Phys.: Conf. Ser.* **1007** 012005

[17] Phukpattaranont P, Thongpanja S, Anam K, Al-Jumaily A and Limsakul C 2018 Evaluation of feature extraction techniques and classifiers for finger movement recognition using surface electromyography signal *Med. Biol. Eng. Comput.* **56** 2259–71

[18] Hamaguchi T, Saito T, Suzuki M, Ishioka T, Tomisawa Y and Nakaya N *et al* 2020 Support vector machine-based classifier for the assessment of finger movement of stroke patients undergoing rehabilitation *J. Med. Biol. Eng.* **40** 91–100

[19] Arteaga M V, Castiblanco J C, Mondragon I F, Colorado J D and Alvarado-Rojas C 2020 EMG-driven hand model based on the classification of individual finger movements *Biomed. Signal Process Control* **58** 101834

[20] Hu X, Song A, Wang J, Zeng H and Wei W 2022 Finger movement recognition via high-density electromyography of intrinsic and extrinsic hand muscles *Sci. Data* **9** 1–12

[21] Taghizadeh Z, Rashidi S and Shalbaf A 2021 Finger movements classification based on fractional Fourier transform coefficients extracted from surface EMG signals *Biomed. Signal Process. Control* **68** 102573

[22] Kumar G, Yadav S, Yogita S and Pal V 2022 Machine learning-based framework to predict finger movement for prosthetic hand *IEEE Sensors Lett.* **6** 1–4

[23] Rashid N, Iqbal J, Javed A, Tiwana M I and Khan U S 2018 Design of embedded system for multivariate classification of finger and thumb movements using EEG signals for control of upper limb prosthesis *BioMed Res. Int.* **2018** 2695106

[24] Gruenwald J, Znobishchev A, Kapeller C, Kamada K, Scharinger J and Guger C 2019 Time-variant linear discriminant analysis improves hand gesture and finger movement decoding for invasive brain–computer interfaces *Front. Neurosci.* **13** 1-18

[25] Qi J, Jiang G, Li G, Sun Y and Tao B 2019 Intelligent human–computer interaction based on surface EMG gesture recognition *IEEE Access* **7** 61378–87

[26] Cene V H and Balbinot A 2018 Using the sEMG signal representativity improvement towards upper-limb movement classification reliability *Biomed. Signal Process. Control* **46** 182–91

[27] Shair E F, Jamaluddin N A and Abdullah A R 2020 Finger movement discrimination of EMG signals towards improved prosthetic control using TFD *Int. J. Adv. Comput. Sci. Appl* **11** 244–51

[28] Purushothaman G and Vikas R 2018 Identification of a feature selection based pattern recognition scheme for finger movement recognition from multichannel EMG signals *Australas Phys. Eng. Sci. Med.* **41** 549–59

[29] Mwata-Velu T, Avina-Cervantes J G, Cruz-Duarte J M, Rostro-Gonzalez H and Ruiz-Pinales J 2021 Imaginary finger movements decoding using empirical mode decomposition and a stacked BiLSTM architecture *Mathematics* **9** 1–14

[30] Sapsanis C, Georgoulas G, Tzes A and Lymberopoulos D 2013 Improving EMG based classification of basic hand movements using EMD *Annu. Int. Conf. IEEE Eng. Med. Biol. Soc.* 5754–7

[31] Benatti S, Casamassima F, Milosevic B, Farella E, Schönle P and Fateh S *et al* 2015 A versatile embedded platform for EMG acquisition and gesture recognition *IEEE Trans. Biomed. Circuits Syst.* **9** 620–30

[32] Liu X, Sacks J, Zhang M, Richardson A G, Lucas T H and Van Der Spiegel J 2017 The virtual trackpad: an electromyography-based, wireless, real-time, low-power, embedded hand-gesture-recognition system using an event-driven artificial neural network *IEEE Trans. Circuits Syst. II Express Briefs* **64** 1257–61

[33] Jiang S, Lv B, Guo W, Zhang C, Wang H and Sheng X *et al* 2018 Feasibility of wrist-worn, real-time hand, and surface gesture recognition via sEMG and IMU sensing *IEEE Trans. Industr. Inform.* **14** 3376–85

[34] Naik G R and Nguyen H T 2015 Nonnegative matrix factorization for the identification of EMG finger movements: evaluation using matrix analysis *IEEE J. Biomed. Health Inform.* **19** 478–85

[35] Côté-allard U, Fall C L, Drouin A, Campeau-lecours A, Gosselin C and Glette K *et al* 2019 Deep learning for electromyographic hand transfer learning *IEEE Trans. Neural Syst. Rehabil. Eng.* **27** 760–71

[36] Boyali A and Hashimoto N 2016 Spectral collaborative representation based classification for hand gestures recognition on electromyography signals *Biomed. Signal Process. Control* **24** 11–8

[37] Naik G R, Al-Timemy A H and Nguyen H T 2016 Transradial amputee gesture classification using an optimal number of sEMG sensors: an approach using ICA clustering *IEEE Trans. Neural Syst. Rehabil. Eng.* **24** 837–46

[38] Liang S, Wang L, Zhang L and Wu Y 2019 Research on recognition of nine kinds of fine gestures based on adaptive AdaBoost algorithm and multi-feature combination *IEEE Access* **7** 3235–46

[39] Tavakoli M, Benussi C and Lourenco J L 2017 Single channel surface EMG control of advanced prosthetic hands: a simple, low cost and efficient approach *Expert Syst. Appl.* **79** 322–32

[40] Tavakoli M, Benussi C, Alhais Lopes P, Osorio L B and de Almeida A T 2018 Robust hand gesture recognition with a double channel surface EMG wearable armband and SVM classifier *Biomed. Signal Process. Control* **46** 121–30

[41] Sezgin N 2019 A new hand finger movements' classification system based on bicoherence analysis of two-channel surface EMG signals *Neural Comput. Appl.* **31** 3327–37

[42] Nishad A, Upadhyay A, Pachori R B and Acharya U R 2019 Automated classification of hand movements using tunable-Q wavelet transform based filter-bank with surface electromyogram signals *Futur. Gener. Comput. Syst.* **93** 96–110

[43] Xiao F, Yang D, Lv Z, Guo X, Liu Z and Wang Y 2020 Classification of hand movements using variational mode decomposition and composite permutation entropy index with surface electromyogram signals *Futur. Gener. Comput. Syst.* **110** 1023–36

[44] Guo Y, Naik G R, Huang S, Abraham A and Nguyen H T 2015 Nonlinear multiscale maximal Lyapunov exponent for accurate myoelectric signal classification *Appl. Soft Comput. J.* **36** 633–40

[45] Sravani C, Bajaj V, Taran S and Sengur A 2020 Flexible analytic wavelet transform based features for physical action identification using sEMG signals *IRBM* **41** 18–22

[46] Chada S, Taran S and Bajaj V 2020 An efficient approach for physical actions classification using surface EMG signals *Health Inf. Sci. Syst.* **8** 3

[47] Alam M S and Arefin A S 2018 Real-time classification of multi-channel forearm EMG to recognize hand movements using effective feature combination and LDA classifier *Bangladesh J. Med. Phys.* **10** 25–39

[48] Crepin R, Fall C L, Mascret Q, Gosselin C, Campeau-Lecours A and Gosselin B 2018 Real-time hand motion recognition using sEMG patterns classification. *Annual Int. Conf. of the IEEE Engineering in Medicine and Biology Society* 2018 July, pp 2655–8

[49] Arozi M, Caesarendra W, Ariyanto M, Munadi M, Setiawan J D and Glowacz A 2020 Pattern recognition of single-channel sEMG signal using PCA and ANN method to classify nine hand movements *Symmetry* **12** 541–59

[50] Bhagwat S and Mukherji P 2020 Electromyogram (EMG) based fingers movement recognition using sparse filtering of wavelet packet coefficients *Sadhana—Acad. Proc. Eng. Sci.* **45** 1–11

[51] Phukan N, Kakoty N M, Shivam P and Gan J Q 2019 Finger movements recognition using minimally redundant features of wavelet denoised EMG *Health Technol.* **9** 579–93

[52] Miften F S, Diykh M, Abdulla S, Siuly S, Green J H and Deo R C 2021 A new framework for classification of multi-category hand grasps using EMG signals *Artif. Intell. Med.* **112** 102005

[53] Jia G, Lam H K, Ma S, Yang Z, Xu Y and Xiao B 2020 Classification of electromyographic hand gesture signals using modified fuzzy c-means clustering and two-step machine learning approach *IEEE Trans. Neural Syst. Rehabil. Eng.* **28** 1428–35

[54] Gupta S H, Sharma A, Mohta M and Rajawat A 2020 Hand movement classification from measured scattering parameters using deep convolutional neural network *Meas. J. Int. Meas. Confed.* **151** 107258

[55] Tuncer T, Dogan S and Subasi A 2020 Surface EMG signal classification using ternary pattern and discrete wavelet transform based feature extraction for hand movement recognition *Biomed. Signal Process. Control* **58** 101872

[56] Krasoulis A, Vijayakumar S and Nazarpour K 2019 Effect of user practice on prosthetic finger control with an intuitive myoelectric decoder *Front. Neurosci.* **13** 891

[57] Selesnick I W 2011 Wavelet transform with tunable Q-factor *IEEE Trans. Signal Process.* **59** 3560–75

[58] Murugappan M, Alshuaib W, Bourisly A K, Khare S K, Sruthi S and Bajaj V 2020 Tunable Q wavelet transform based emotion classification in Parkinson's disease using electro-encephalography *PLoS One* **15** e0242014

[59] Khare S K, Bajaj V and Acharya U R 2021 Detection of Parkinson's disease using automated tunable Q wavelet transform technique with EEG signals *Biocybern. Biomed. Eng.* **41** 679–89

[60] Bajaj V, Taran S, Khare S K and Sengur A 2020 Feature extraction method for classification of alertness and drowsiness states EEG signals *Appl. Acoust.* **163** 107224

[61] Taran S and Bajaj V 2019 Motor imagery tasks-based EEG signals classification using tunable-Q wavelet transform *Neural Comput. Appl.* **31** 6925–32

[62] Phinyomark A, Phukpattaranont P and Limsakul C 2012 Feature reduction and selection for EMG signal classification *Expert Syst. Appl.* **39** 7420–31

[63] Bajaj V and Pachori R B 2012 Classification of seizure and nonseizure EEG signals using empirical mode decomposition *IEEE Trans. Inf. Technol. Biomed.* **16** 1135–42

[64] Hanley J A and McNeil B J 1982 The meaning and use of the area under a receiver operating characteristic (ROC) curve *Radiology* **143** 29–36

[65] Kumar A, Vishwakarma A and Bajaj V 2023 CRCCN-Net: automated framework for classification of colorectal tissue using histopathological images *Biomed. Signal Process. Control* **79** 104172

[66] Gupta K, Bajaj V and Ansari I A 2022 OSACN-Net: automated classification of sleep apnea using deep learning model and smoothed gabor spectrograms of ECG signal *IEEE Trans. Instrum. Meas.* **71** 4002109

IOP Publishing

Advanced Signal Processing for Industry 4.0, Volume 1
Evolution, communication protocols, and applications in manufacturing systems
Irshad Ahmad Ansari and Varun Bajaj

Chapter 13

Data-driven approach to design energy-efficient precoder for QoS-aware MIMO-MRCN system in context of Industry 4.0

Deepak Sahu, Shikha Maurya, Matadeen Bansal and Dinesh Kumar V

Fifth-generation (5G) communication technologies are being used extensively globally and researchers are focusing on the realization of Industry 4.0. Data-driven models for the deployment of energy-efficient wireless networks in the context of Industry 4.0 heavily rely on IoT sensor data or knowledge-based data. Emerging communication networks, i.e., B5G/6G, might be able to provide the necessary information to address the fundamental limitations of the present communication networks as well as the vision of Industry 4.0. Presently, the traditional solutions for various wireless communication problems involve iterative methods or heuristic algorithms, which demand highly complex mathematics. As a result, these solutions either increase the hardware cost and latency rate or give infeasible solutions, which make them inadequate for the real-time realization of Industry 4.0. To overcome these issues, a data-driven model in B5G/6G has been identified as a successful integration of a conventional wireless network into a DL-enabled communication network. In accordance with that, this chapter presents various automatic opportunistic relay selection approaches using deep neural networks for multi-input multi-output multi-relay cognitive networks. The problem is studied to maximize the energy efficiency of the secondary network (i.e., unlicensed network) while ensuring the quality-of-service of both unlicensed and licensed network. Specifically, four different schemes are examined for the joint relay selection and precoder designing at secondary network (SN). DL-enabled scheme achieves the performance similar to that with the conventional relay selection scheme, while exhibiting the minimum computational cost.

13.1 Introduction

The explosive growth of recent technologies in B5G/6G [1] has shown widespread interest in developing the fourth industrial revolution (Industry 4.0) [2]. As per the report of Internet of Things (IoT) analysts [3], more than 30 billion IoT devices will be in human hands and it will reach four times the world population by 2025. A low-cost IoT sensor plays a vital role in the future communication system for achieving the goal of Industry 4.0 [4]. Energy-efficient IoT devices/sensors provide vital information for industrial environments. The deployment and functioning of these devices require more spectrum and energy-efficient systems with much lower computational costs and simple hardware [5]. Traditionally, the designing of these systems demands complex mathematics and inefficient solutions, which usually results in infeasible solutions, high latency rate, and hardware costs. Thus, conventional methods may not be very advantageous for the real-time implementation of an IoT-based system in Industry 4.0 [6].

Nowadays, there is widespread interest in applying machine learning (ML) and deep learning (DL) methods, subsets of artificial intelligence (AI), to solve wireless network designing problems for which either an efficient algorithm does not exist or the algorithmic complexity is prohibitively high [7]. AI approaches have been widely used in various engineering fields such as image processing, pattern recognition, natural language processing, communication, etc [8]. It has the potential to realize the vision of an intelligent fourth industrial revolution by overcoming the limitations of the traditional communication systems [9]. A data-driven model is a promising approach to enhance the utilization of 5G infrastructures [10] and transition into a DL-enabled communication system model. B5G/6G demands a technology that can manage the spectrum scarcity problem for the upcoming industrial-based applications [11]. Cognitive radio (CR) is one of the prominent technologies for solving the spectrum scarcity problem for the upcoming networks by enabling unlicensed IoT users to access the licensed spectrum and utilize the spectrum effectively [12, 13]. Further, the performance of the unlicensed IoT users can be enhanced by utilizing multi-input multi-output (MIMO) [14] and cooperative relaying techniques, which help to provide the link reliability, capacity, and coverage area for the users [15]. This chapter focuses on MIMO-multi relay cognitive networks (MRCNs) to overcome the challenges of Industry 4.0.

CR technology can be used in three modes, i.e., underlay, overlay, and interweave mode [16]. In underlay mode, the unlicensed user can access the licensed spectrum simultaneously with the licensed user until the interference at the licensed user is below the predefined threshold which ensures the quality-of-service (QoS) of the licensed user [17]. In a proposed MIMO-MRCN, the IoT user/node transmits the multiple data streams with the appropriate weights known as the precoders, which assures that the system achieves the maximum energy efficiency (EE) while maintaining the required throughput at the receiver [18]. The conventional methods to solve the energy-efficient precoder designing problem along with relay selection require high computational cost which is not feasible for the real-time industrial applications. In order to avoid these circumstances, a DL-enabled MIMO-MRCN with a data-driven model can be used as it will reduce the computational cost and make it feasible to deploy in the real-time industrial applications. Recently, several researchers have explored relay selection problems using DL for various scenarios of wireless networks in [19–21].

13.2 Related works

In literature, precoder design problems have been extensively studied in various scenarios of wireless communication system, such as MIMO [22, 23], millimeter wave (mm-wave), massive MIMO [24, 25], and MIMO-Cognitive relay network (MIMO-CRN) [26–30]. Authors in [26–30] studied the impact of precoder designing in MIMO-CRN with a single relay node for improving the EE at the unlicensed receiver. Authors in [26] studied the problem of a relay precoder for maximizing the EE at the secondary receiver with consideration of a MIMO-CRN in a non-regenerative scenario with consideration of the QoS constraint. This work is further extended for the joint designing of precoders at source and relay, presented in [27]. Authors in [26, 27] used the iterative method to solve the optimization problem of precoder design, which imposed higher computational cost. The issue of complexity was considered in [28, 29] for problem precoder design, and authors in [28] employed a DL-based model for solving the non-convex problem of designing a relay precoder and reduced the computational cost. Further, [29] approximated the non-convex problem of precoder designing at the source and relay using the data-driven method and robust system presented for the MIMO-CRN system. The performance of these systems can be further improved by employing multiple relays to assist the transmission between source and receiver. In the same direction, papers [31–33] explored the relay selection in the multi-relay scenario for various wireless networks. Authors in [31, 32] studied the relay selection problem for the CR scenario with multiple relays. In addition, [33] studied the outage performance with selection of optimal relay and antenna in a CR environment. There are various relay selection schemes present in the literature. These schemes have generally used traditional optimization methods, which impose a higher computational cost. Therefore, this chapter deals with the designing of a MIMO-MRCN using a data-driven approach. This system tries to provide the computational cost effective solutions, which makes it suitable to implement in real-time scenarios in the context of Industry 4.0. This work explores the various relay selection schemes along with the impact of precoders design to improve the end-to-end EE of SN subjects to the power, interference, and QoS constraints at the secondary source S_{TX} and relays S_{R_i}, respectively. These constraints meet the need of maintaining the QoS of both SN and primary network (PN). The relay selection and precoder designing problem is solved in two phases. The first phase deals with the precoder designing, whereas, in the second phase, the best relay is selected based on the designed precoders and obtained end-to-end EE.

The rest of the chapter is organized as follows: Section 13.3 introduces the system model. Section 13.3.1 describes the problem formulation for designing of precoders and relay selection. The DL-based low complexity approach for precoder designing is studied in section 13.3.2. Numerical results are presented in section 13.4. Finally, conclusions are drawn in section 13.5. A list of important notations used in the rest of the chapter is given in Table 13.1.

13.3 System model

This chapter considers a MIMO-MRCN, illustrated in figure 13.1. It includes two networks, PN and SN, where PN is the licensed user and consists of a transmitter

Table 13.1. Important notations.

Notation	Description
P_{TX}	Primary transmitter
P_{RX}	Primary receiver
S_{TX}	Secondary transmitter
S_{R_i}	ith secondary relay
S_{RX}	Secondary receiver
A^T	Transpose of matrix A
$\mathrm{tr}(A)$	Trace of matrix A
$P_{S_{TX}}$	Power transmitted by S_{TX}
$I_{S_{max}}$	Interference at P_{RX} caused by the transmission of SN
$I_{R_{i max}}$	Interference at P_{RX} caused by the transmission of S_{R_i}
$P_{S_{max}}$	Maximum available transmit power at S_{TX}
$P_{R_{i max}}$	Maximum available transmit power at S_{R_i}
P_{R_i}	Power transmitted by S_{R_i}
$E\{A\}$	Expectation of A
I_x	Identity matrix of size $x \times x$
$\mathbb{C}$	Complex space
$\mathbb{R}$	Real space

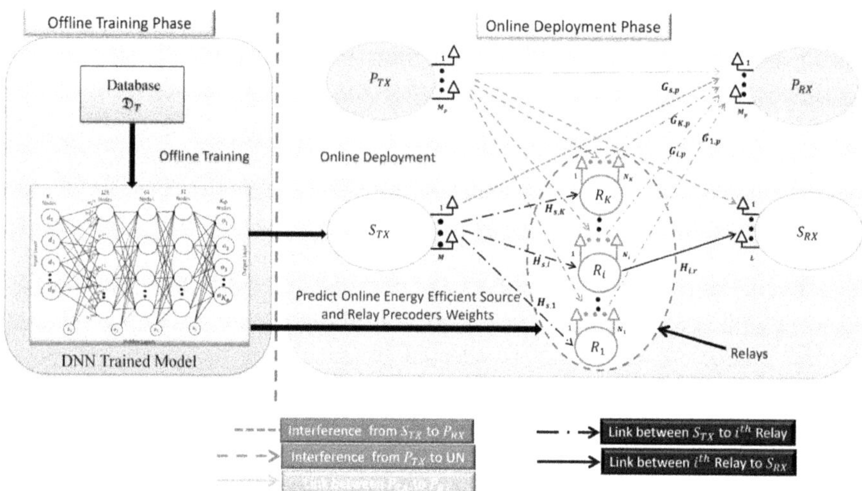

Figure 13.1. System model of a DL-enabled MIMO-MRCN.

P_{TX} and a receiver P_{RX} equipped with m_p antennas. The SN works in an underlay mode, which includes a transmitter S_{TX}, a receiver S_{RX}, and K amplify-and-forward relays ($S_{R_i}, \forall i \in 1, 2, \ldots, K$). Let, m_s, m_d, and m_r denote the number of antennas at the S_{TX}, S_{RX}, and S_{R_i}, respectively. The channel matrices from S_{TX} to S_{R_i} and from

S_{R_i} to S_{RX}, are denoted by the $\mathbf{H}_{SR_i} \in \mathbb{C}^{(m_r \times m_s)}$ and $\mathbf{H}_{R_iD} \in \mathbb{C}^{(m_d \times m_r)}$, respectively. While the interference channel matrices from S_{TX} and S_{R_i} to P_{RX} are represented by the $\mathbf{G}_{SP} \in \mathbb{C}^{(m_p \times m_s)}$ and $\mathbf{G}_{R_iP} \in \mathbb{C}^{(m_p \times m_r)}$, respectively. It is assumed that the S_{TX} and S_{RX} have the knowledge of channel state information (CSI) of all the links due to the feedback system between SN and PN [34]. In the SN, all relays operate in half duplex mode. Among all relays, one relay is selected to transmit the information from S_{TX} to S_{RX}. Thus, SN requires two time-slots for end-to-end transmission. In the first time-slot, S_{TX} transmits the data stream $\mathbf{s}$ of length $d_u \times 1$ after multiplying it with a precoding matrix $U_i \in \mathbb{C}^{m_s \times d_u}$, where $d_u = \min\{m_s, m_r, m_d\}$ and $\mathbb{E}\{\mathbf{ss}^H\} = I_{d_u}$. Received signal at S_{R_i} (ith relay) is denoted as follows:

$$\mathbf{y}_{R_i} = \mathbf{H}_{SR_i}\mathbf{U}_i\mathbf{s} + \mathbf{n}_{R_i}. \tag{13.1}$$

Further, S_{R_i} forwards the multiplied signal $\mathbf{y}_{R_i}$ with the relay precoding matrix $\mathbf{Q}_i \in \mathbb{C}^{(m_r \times m_r)}$ to S_{RX}. The received signal at S_{RX} is given as follows:

$$\mathbf{y}_D = \mathbf{H}_{R_iD}\mathbf{Q}_i\mathbf{y}_{R_i} + \mathbf{n}_D \tag{13.2}$$

$$\mathbf{y}_D = \mathbf{H}_{R_iD}\mathbf{Q}_i\mathbf{H}_{SR_i}\mathbf{U}_i\mathbf{s} + \mathbf{H}_{R_iD}\mathbf{Q}_i\mathbf{n}_{R_i} + \mathbf{n}_D \tag{13.3}$$

$$\mathbf{y}_D = \mathbf{H}_{R_iD}\mathbf{Q}_i\mathbf{H}_{SR_i}\mathbf{U}_i\mathbf{s} + \mathbf{n}_1. \tag{13.4}$$

Here, $\mathbf{n}_1 = \mathbf{H}_{R_iD}\mathbf{Q}_i\mathbf{n}_{R_i} + \mathbf{n}_D$ represents the total noise plus interference caused by the PN to the SN with zero mean and covariance of $\mathbf{Z}_D = \sigma_{n_{R_i}}^2\mathbf{H}_{R_iD}\mathbf{Q}_i\mathbf{Q}_i^H\mathbf{H}_{R_iD}^H + \sigma_{n_D}^2\mathbf{I}_{m_d}$ in both time slots, where n_{R_i} and n_D are the complex white Gaussian noise vectors having zero means with variances of $\sigma_{n_{R_i}}^2\mathbf{I}_{m_r}$ and $\sigma_{n_D}^2\mathbf{I}_{m_d}$, respectively.

In the MIMO-MRCN, signal transmits in the two time slots, S_{TX} to S_{R_i} and S_{R_i} to S_{RX}. Therefore, the average power consumption at the SN is given as follows [27]:

$$P_{Total} = \frac{1}{2}\left(P_{S_{TX}}/\rho_{S_{TX}} + P_{R_i}/\rho_{R_i} + P_{ckt}\right), \tag{13.5}$$

where $P_{S_{TX}}$ and P_{R_i} is the transmit power at S_{TX} and S_{R_i}, respectively, represented by

$$P_{S_{TX}} = \text{tr}(\mathbf{W}_i), \tag{13.6}$$

$$P_{R_i} = \text{tr}\left(\mathbf{Q}_i\mathbf{H}_{SR_i}\mathbf{W}_i\mathbf{H}_{SR_i}^H\mathbf{Q}_i^H + \sigma_{n_{R_i}}^2\mathbf{Q}_i\mathbf{Q}_i^H\right). \tag{13.7}$$

These transmit powers should be less than the maximum available transmit power at source and relay, which is denoted by $P_{S_{max}}$ and $P_{R_{imax}}$, respectively. $\rho_{S_{TX}}$ and ρ_{R_i} represent the power amplifier efficiencies of S_{TX} and S_{R_i}, respectively. The quantity P_{ckt} is the total circuit power consumed by the SN.

The achievable rate of the SN can be represented mathematically as follows [35]:

$$\text{Rate}_i = \frac{1}{2}\log_2\left[\det\left\{\mathbf{I} + \mathbf{H}_{R_iD}\mathbf{Q}_i\mathbf{H}_{SR_i}\mathbf{U}_i\mathbf{U}_i^H\mathbf{H}_{SR_i}^H\mathbf{Q}_i^H\mathbf{H}_{R_iD}^H\mathbf{Z}_D^{-1}\right\}\right]. \tag{13.8}$$

This paper finds the best relay selection scheme along with the impact of precoder designing in relay selection. Both relay selection and precoder designing is done with the aim to maximize the end-to-end EE of the SN. Now, the EE of SN can be defined as the ratio of average rate to the total consumed power, denoted as

$$\eta_i = \frac{\text{Rate}_i}{P_{\text{total}}}$$

$$\eta_i = \frac{\frac{1}{2}\log_2\left[\det\left\{\mathbf{I} + \mathbf{H}_{R_iD}\mathbf{Q}_i\mathbf{H}_{\text{SR}_i}\mathbf{U}_i\mathbf{U}_i^H\mathbf{H}_{\text{SR}_i}^H\mathbf{Q}_i^H\mathbf{H}_{R_iD}^H\mathbf{Z}_D^{-1}\right\}\right]}{\frac{1}{2}\left(P_{S_{\text{TX}}}/\rho_{S_{\text{TX}}} + P_{R_i}/\rho_{R_i} + P_{ckt}\right)}. \tag{13.9}$$

Further, it can be simplified as follows:

$$\eta_i = \frac{\log_2\left[\det\left\{\mathbf{I} + \mathbf{H}_{R_iD}\mathbf{Q}_i\mathbf{H}_{\text{SR}_i}\mathbf{U}_i\mathbf{U}_i^H\mathbf{H}_{\text{SR}_i}^H\mathbf{Q}_i^H\mathbf{H}_{R_iD}^H\mathbf{Z}_D^{-1}\right\}\right]}{\left(P_{S_{\text{TX}}}/\rho_{S_{\text{TX}}} + P_{R_i}/\rho_{R_i} + P_{ckt}\right)}. \tag{13.10}$$

Additionally, during transmission, SN causes interference to PN, which are given as

$$I_{S_{\text{TX}}} = \text{tr}\left(\mathbf{G}_{\text{SP}}\mathbf{W}_i\mathbf{G}_{R_iP}^H\right), \tag{13.11}$$

$$I_{R_i} = \text{tr}\left(\mathbf{G}_{R_iP}\mathbf{Q}_i\mathbf{H}_{\text{SR}_i}\mathbf{W}_i\mathbf{H}_{\text{SR}_i}^H\mathbf{Q}_i^H\mathbf{G}_{R_iP}^H + \sigma_{n_{R_i}}^2\mathbf{G}_{R_iP}\mathbf{W}_i\mathbf{G}_{R_iP}^H\right), \tag{13.12}$$

where $\mathbf{W}_i = \mathbf{U}_i\mathbf{U}_i^H$. The interference power at PN in both time slots, i.e., $I_{S_{\text{TX}}}$ and I_{R_i}, should be less than a predefined threshold value $I_{S_{\max}}$ and $I_{R_{i\max}}$, respectively.

In order to find joint precoders at S_{TX} and S_{R_i}, i.e., $\mathbf{W}_i$ and $\mathbf{Q}_i$, an energy efficiency maximization problem is devised subject to the QoS, interference, and transmit power constraints. Here, the QoS constraint for SN is used in the form of a minimum rate requirement. Mathematically, the problem can be written as

Problem 13.1:

$$\max_{(\mathbf{W}_i),(\mathbf{Q}_i)} \frac{\log_2\left[\det\left\{\mathbf{I} + \mathbf{H}_{R_iD}\mathbf{Q}_i\mathbf{H}_{\text{SR}_i}\mathbf{W}_i\mathbf{H}_{\text{SR}_i}^H\mathbf{Q}_i^H\mathbf{H}_{R_iD}^H\mathbf{Z}_D^{-1}\right\}\right]}{\left(P_{S_{\text{TX}}}/\rho_{S_{\text{TX}}} + P_{R_i}/\rho_{R_i} + P_{ckt}\right)}, \tag{13.13}$$

subject to

$$\text{tr}\left(\mathbf{G}_{\text{SP}}\mathbf{W}_i\mathbf{G}_{R_iP}^H\right) \leqslant I_{S_{\max}}, \tag{13.14}$$

$$\text{tr}(\mathbf{W}_i) \leqslant P_{S_{\max}}, \tag{13.15}$$

$$\text{tr}\left(\mathbf{G}_{R_iP}\mathbf{Q}_i\mathbf{H}_{\text{SR}_i}\mathbf{W}_i\mathbf{H}_{\text{SR}_i}^H\mathbf{Q}_i^H\mathbf{G}_{R_iP}^H + \sigma_{n_{R_i}}^2\mathbf{G}_{R_iP}\mathbf{W}_i\mathbf{G}_{R_iP}^H\right) \leqslant I_{R_{i\max}}, \tag{13.16}$$

$$\text{tr}\left(\mathbf{Q}_i\mathbf{H}_{\text{SR}_i}\mathbf{W}_i\mathbf{H}_{\text{SR}_i}^H\mathbf{Q}_i^H + \sigma_{n_{R_i}}^2\mathbf{Q}_i\mathbf{Q}_i^H\right) \leqslant P_{R_{i\max}}, \tag{13.17}$$

$$\frac{1}{2} \log_2\left[\det\left\{ \mathbf{I} + \mathbf{H}_{R_iD}\mathbf{Q}_i\mathbf{H}_{SR_i}\mathbf{W}_i\mathbf{H}_{SR_i}^H\mathbf{Q}_i^H\mathbf{H}_{R_iD}^H\mathbf{Z}_D^{-1} \right\} \right] \geqslant \text{Rate}_{\min}, \tag{13.18}$$

$$\forall = 1, 2, \ldots, K.$$

Problem 13.1 consists interference and power constraints at the S_{TX} and S_{R_i}, where $P_{\max}$ and $I_{\max}$ denote the maximum transmitting power and interference at the S_{TX} and S_{R_i}, respectively. Further, this problem also considers the QoS constraints at the SN, which ensures that the minimum rate requirement at each node of the SN is achieved. Further, the problem 13.1 is converted into the minimization problem represented as follows:

Problem 13.2:

$$\max_{(\mathbf{W}_i),(\mathbf{Q}_i)} \frac{\left(P_{S_{TX}}/\rho_{S_{TX}} + P_{R_i}/\rho_{R_i} + P_{ckt}\right)}{\log_2\left[\det\left\{ \mathbf{I} + \mathbf{H}_{R_iD}\mathbf{Q}_i\mathbf{H}_{SR_i}\mathbf{W}_i\mathbf{H}_{SR_i}^H\mathbf{Q}_i^H\mathbf{H}_{R_iD}^H\mathbf{Z}_D^{-1} \right\} \right]}, \tag{13.19}$$

subject to

$$\text{tr}\left(\mathbf{G}_{SP}\mathbf{W}_i\mathbf{G}_{R_iP}^H\right) \leqslant I_{S_{\max}}, \tag{13.20}$$

$$\text{tr}(\mathbf{W}_i) \leqslant P_{S_{\max}}, \tag{13.21}$$

$$\text{tr}\left(\mathbf{G}_{R_iP}\mathbf{Q}_i\mathbf{H}_{SR_i}\mathbf{W}_i\mathbf{H}_{SR_i}^H\mathbf{Q}_i^H\mathbf{G}_{R_iP}^H + \sigma_{n_{R_i}}^2\mathbf{G}_{R_iP}\mathbf{W}_i\mathbf{G}_{R_iP}^H\right) \leqslant I_{R_{i\max}}, \tag{13.22}$$

$$\text{tr}\left(\mathbf{Q}_i\mathbf{H}_{SR_i}\mathbf{W}_i\mathbf{H}_{SR_i}^H\mathbf{Q}_i^H + \sigma_{n_{R_i}}^2\mathbf{Q}_i\mathbf{Q}_i^H\right) \leqslant P_{R_{i\max}}, \tag{13.23}$$

$$\frac{1}{2} \log_2\left[\det\left\{ \mathbf{I} + \mathbf{H}_{R_iD}\mathbf{Q}_i\mathbf{H}_{SR_i}\mathbf{W}_i\mathbf{H}_{SR_i}^H\mathbf{Q}_i^H\mathbf{H}_{R_iD}^H\mathbf{Z}_D^{-1} \right\} \right] \geqslant \text{Rate}_{\min}, \tag{13.24}$$

$$\forall = 1, 2, \ldots, K.$$

It may be noted that problem 13.2 is a non-convex problem with respect to the two variables W_i and Q_i for which the global optimum solution is difficult to calculate. Hence, problem 13.2 can be simplified and divided into two sub-problems 13.3 and 13.4 as follows:

Problem 13.3:

$$\max_{(\mathbf{Q}_i)} \frac{\left(P_{S_{TX}}/\rho_{S_{TX}} + P_{R_i}/\rho_{R_i} + P_{ckt}\right)}{\log_2\left[\det\left\{ \mathbf{I} + \mathbf{H}_{R_iD}\mathbf{Q}_i\mathbf{H}_{SR_i}\mathbf{W}_i\mathbf{H}_{SR_i}^H\mathbf{Q}_i^H\mathbf{H}_{R_iD}^H\mathbf{Z}_D^{-1} \right\} \right]}, \tag{13.25}$$

subject to

$$\text{tr}\!\left(\mathbf{G}_{R_iP}\mathbf{Q}_i\mathbf{H}_{\text{SR}_i}\mathbf{W}_i\mathbf{H}_{\text{SR}_i}^{H}\mathbf{Q}_i^{H}\mathbf{G}_{R_iP}^{H} + \sigma_{n_{R_i}}^{2}\mathbf{G}_{R_iP}\mathbf{W}_i\mathbf{G}_{R_iP}^{H}\right) \leqslant I_{R_{i_{\max}}}, \tag{13.26}$$

$$\text{tr}\!\left(\mathbf{Q}_i\mathbf{H}_{\text{SR}_i}\mathbf{W}_i\mathbf{H}_{\text{SR}_i}^{H}\mathbf{Q}_i^{H} + \sigma_{n_{R_i}}^{2}\mathbf{Q}_i\mathbf{Q}_i^{H}\right) \leqslant P_{R_{i_{\max}}}, \tag{13.27}$$

$$\frac{1}{2}\log_2\!\left[\det\!\left\{\mathbf{I} + \mathbf{H}_{R_iD}\mathbf{Q}_i\mathbf{H}_{\text{SR}_i}\mathbf{W}_i\mathbf{H}_{\text{SR}_i}^{H}\mathbf{Q}_i^{H}\mathbf{H}_{R_iD}^{H}\mathbf{Z}_D^{-1}\right\}\right] \geqslant \text{Rate}_{\min}, \tag{13.28}$$

$$\forall = 1, 2, \ldots, K.$$

Problem 13.4:

$$\max_{(\mathbf{W}_i)} \frac{\left(P_{S_{\text{TX}}}/\rho_{S_{\text{TX}}} + P_{R_i}/\rho_{R_i} + P_{ckt}\right)}{\log_2\!\left[\det\!\left\{\mathbf{I} + \mathbf{H}_{R_iD}\mathbf{Q}_i\mathbf{H}_{\text{SR}_i}\mathbf{W}_i\mathbf{H}_{\text{SR}_i}^{H}\mathbf{Q}_i^{H}\mathbf{H}_{R_iD}^{H}\mathbf{Z}_D^{-1}\right\}\right]}, \tag{13.29}$$

subject to

$$\text{tr}\!\left(\mathbf{G}_{\text{SP}}\mathbf{W}_i\mathbf{G}_{R_iP}^{H}\right) \leqslant I_{S_{\max}}, \tag{13.30}$$

$$\text{tr}(\mathbf{W}_i) \leqslant P_{S_{\max}}, \tag{13.31}$$

$$\frac{1}{2}\log_2\!\left[\det\!\left\{\mathbf{I} + \mathbf{H}_{R_iD}\mathbf{Q}_i\mathbf{H}_{\text{SR}_i}\mathbf{W}_i\mathbf{H}_{\text{SR}_i}^{H}\mathbf{Q}_i^{H}\mathbf{H}_{R_iD}^{H}\mathbf{Z}_D^{-1}\right\}\right] \geqslant \text{Rate}_{\min}, \tag{13.32}$$

$$\forall = 1, 2, \ldots, K.$$

Problems 13.3 and 13.4 deal with a single variable either $\mathbf{W}_i$ or $\mathbf{Q}_i$ assuming another variable considers as a constant. Problem 13.3, designs $\mathbf{Q}_i$ while considering $\mathbf{W}_i$ as a constant. Similarly, problem 13.4 designs $\mathbf{W}_i$, while considering $\mathbf{Q}_i$ as a constant. Here, problems 13.3 and 13.4 are convex and can be solved using the interior point method. However the joint precoders can be designed by solving problems 13.3 and 13.4 in iterative manner [27] as shown in figure 13.2.

13.3.1 Relay selection and precoder design schemes

This section introduces four different schemes for relay selection and precoder designing for secondary nodes. The proposed schemes for joint relay selection and precoder designing are carried out in two phases. One phase is for precoder designing and second is for relay selection. Detailed explanation of the proposed four schemes are given below.

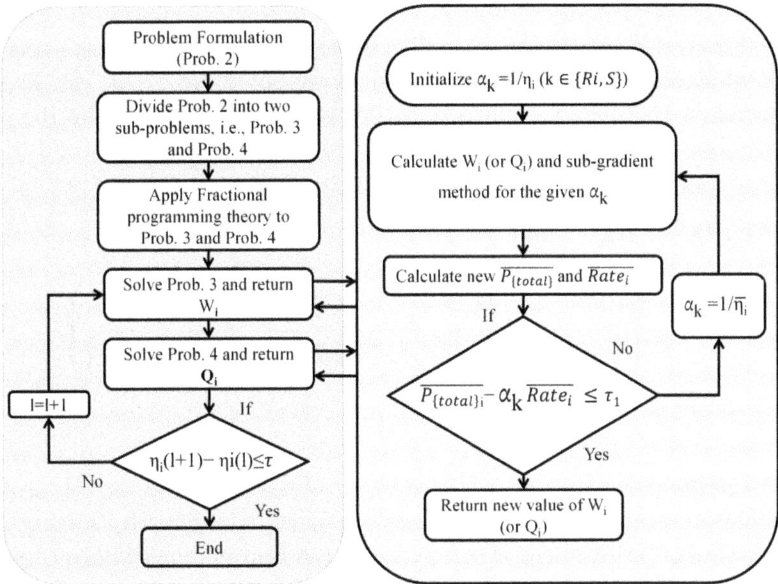

Figure 13.2. Iterative algorithm.

13.3.1.1 Opportunistic scheme

In the first phase of this scheme, source and relay precoders are jointly designed by solving the given problems 13.3 and 13.4. These problems are solved iteratively using the algorithm given in figure 13.2. In the second phase, the best relay S_{R_i} is selected based on the maximum end-to-end achieved EE at S_{RX}, which can be mathematically presented as

$$\left(S_{R_{i^*}},\ \mathbf{W}_{i^*},\ \mathbf{Q}_{i^*}\right) = \arg \min_{i\in\{1,2,\dots,K\}}$$

$$\left(\frac{\left(P_{S_{TX}}/\rho_{S_{TX}} + P_{R_i}/\rho_{R_i} + P_{ckt}\right)}{\log_2\left[\det\left\{\mathbf{I} + \mathbf{H}_{R_iD}\mathbf{Q}_i\mathbf{H}_{SR_i}\mathbf{W}_i\mathbf{H}_{SR_i}^{H}\mathbf{Q}_i^{H}\mathbf{H}_{R_iD}^{H}\mathbf{Z}_D^{-1}\right\}\right]}\right). \tag{13.33}$$

13.3.1.2 Reactive scheme

The first phase of this criteria is to design the S_{TX} precoder while considering the fixed precoder at all relay nodes. For this scheme, problem 13.4 is solved. In the second phase of this scheme, the best relay S_{R_i} is selected based on the maximum end-to-end EE at SN, calculated while using the designed precoder. Mathematically, it can be represented as:

$$\left(S_{R_{i^*}},\ \mathbf{W}_{i^*}\right) = \arg \min_{i\in\{1,2,\dots,K\}}$$

$$\left(\frac{\left(P_{S_{TX}}/\rho_{S_{TX}} + P_{R_i}/\rho_{R_i} + P_{ckt}\right)}{\log_2\left[\det\left\{\mathbf{I} + \mathbf{H}_{R_iD}\mathbf{Q}_i\mathbf{H}_{SR_i}\mathbf{W}_i\mathbf{H}_{SR_i}^{H}\mathbf{Q}_i^{H}\mathbf{H}_{R_iD}^{H}\mathbf{Z}_D^{-1}\right\}\right]}\right). \tag{13.34}$$

13.3.1.3 Proactive scheme

In this criterion, the optimal energy-efficient precoder weights at each relay $(\mathbf{Q}_i)$ is designed, while considering the fixed precoder weight at S_{TX}. This can be done by solving problem 13.3 while assuming that $\mathbf{W}_i$ is known. After finding the precoder weights at relays, the best relay $S_{R_i^*}$ is selected based on the maximum end-to-end EE from S_{TX} to S_{RX}. Relay selection criteria can be mathematically presented as:

$$(S_{R_{i^*}}, \mathbf{Q}_{i^*}) = \arg \min_{i \in \{1,2,\ldots,K\}} \left(\frac{\left(P_{S_{\text{TX}}}/\rho_{S_{\text{TX}}} + P_{R_i}/\rho_{R_i} + P_{ckt} \right)}{\log_2 \left[\det \left\{ \mathbf{I} + \mathbf{H}_{R_iD} \mathbf{Q}_i \mathbf{H}_{\text{SR}_i} \mathbf{W}_i \mathbf{H}_{\text{SR}_i}^H \mathbf{Q}_i^H \mathbf{H}_{R_iD}^H \mathbf{Z}_D^{-1} \right\} \right]} \right) \tag{13.35}$$

13.3.1.4 Random scheme

Here, both S_{TX} and S_{R_i} precoder matrices are jointly designed by solving problem 13.2. Afterwards, the relay S_{R_i} is selected in a random manner without taking into consideration the obtained end-to-end EE.

13.3.2 A DL-based low complexity approach for precoder designing

In order to implement the computationally cost-effective scheme for relay selection while considering the impact of precoders at the S_{TX} and S_{R_i}, a data-driven approach can be considered using DL, where problem 13.1 is approximated using the properties of a deep neural network (DNN) and on the basis of extracted features, the solution can be predicted in a fraction of a second without the need of any complex mathematical framework. For the given problem 13.1, the channel features $(\mathbf{d})$ (acts as input of the DNN) are mapped through a nonlinear function to the optimal precoder weights $(\mathbf{q})$ (output of DNN). Mathematically, it can be represented as:

$$f: \mathbf{d} \in \mathbb{R}^{\eta} \rightarrow \mathbf{q}^* \in \mathbb{R}^{2j_\phi}, \tag{13.36}$$

The schematic representation of the input vector $(\mathbf{d})$ and output vector $(\mathbf{q})$ is given in figures 13.3 and 13.4, respectively. Furthermore, using the above mapping function, the DNN can perform the channel features extraction according to the communication environment and predict the nearly optimal precoder weights $\mathbf{W}_i$ and $\mathbf{Q}_i$ with minimal error and computational cost.

In the DL-enabled scheme, the channel gains between all the links are considered for the DNN input. Since these MIMO channel matrices are complex, therefore, first, we pre-process it and decompose into a simple form via singular value decomposition (SVD) as DNNs process real numbers only [36]. The SVD to the complex channel matrices [37] can be represented as follows:

$$\mathbf{H}_{\text{SR}_i} = \mathbf{U}_{H_{\text{SR}_i}} \mathbf{S}_{H_{\text{SR}_i}}^{1/2} \mathbf{V}_{H_{\text{SR}_i}}^H, \tag{13.37}$$

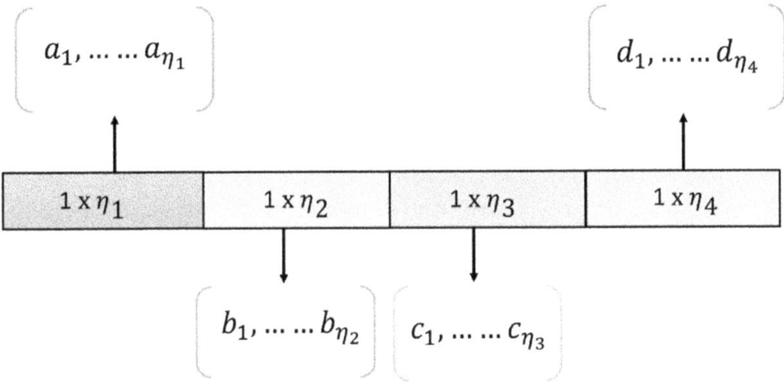

Figure 13.3. DNN input vector.

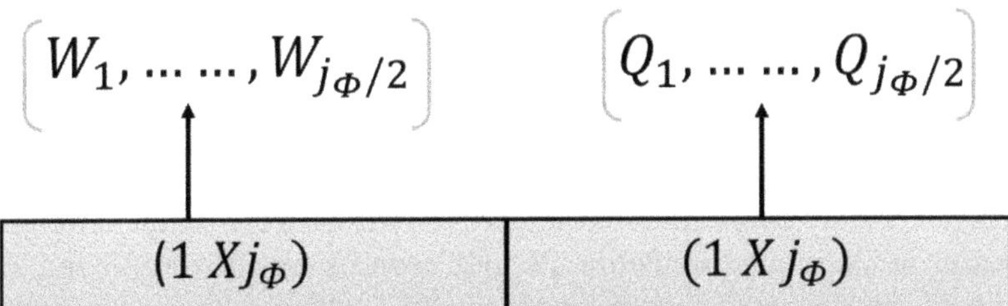

Figure 13.4. DNN output vector.

$$\mathbf{H}_{R_iD} = \mathbf{U}_{H_{R_iD}}\mathbf{S}_{H_{R_iD}}^{1/2}\mathbf{V}_{H_{R_iD}}^{H}, \tag{13.38}$$

$$\mathbf{G}_{\text{SP}} = \mathbf{U}_{G_{\text{SP}}}\mathbf{S}_{G_{\text{SP}}}^{1/2}\mathbf{V}_{G_{\text{SP}}}^{H}, \tag{13.39}$$

$$\mathbf{G}_{R_iP} = \mathbf{U}_{G_{R_iP}}\mathbf{S}_{G_{R_iP}}^{1/2}\mathbf{V}_{G_{R_iP}}^{H}, \tag{13.40}$$

where $\mathbf{U}_{H_{\text{SR}_i}}$, $\mathbf{V}_{H_{\text{SR}_i}}^{H}$, $\mathbf{U}_{H_{R_iD}}$, $\mathbf{V}_{H_{R_iD}}^{H}$, $\mathbf{U}_{G_{\text{SP}}}$, $\mathbf{V}_{G_{\text{SP}}}^{H}$, $\mathbf{U}_{G_{R_iP}}$, and $\mathbf{V}_{G_{R_iP}}^{H}$ are unitary matrices. Further, diagonal matrices of singular values can be represented as

$$\mathbf{S}_{H_{\text{SR}_i}} = \text{diag}(a_1, a_2 \cdots a_{\eta_1}), \tag{13.41}$$

$$\mathbf{S}_{H_{R_iD}} = \text{diag}(b_1, b_2, \cdots b_{\eta_2}), \tag{13.42}$$

$$\mathbf{S}_{G_{\text{SP}}} = \text{diag}(c_1, c_2 \cdots c_{\eta_3}), \tag{13.43}$$

$$\mathbf{S}_{G_{R_iP}} = \text{diag}(d_1, d_2, \cdots d_{\eta_4}), \tag{13.44}$$

where η_1, η_2, η_3, and η_4 refer as the rank of H_{SR_i}, H_{R_iD}, G_{SP}, and G_{R_iP}, respectively. Finally, input features can be expressed as follows:

$$\mathbf{d} = \left[a_1, \ldots, a_k, \ldots, a_{\eta_1}, b_1, \ldots, b_k, \ldots b_{\eta_2}, c_1, \ldots c_{\eta_3}, d_1, \ldots d_{\eta_4} \right]^T \in \mathbb{R}^\eta. \quad (13.45)$$

where $\eta = (\eta_1 + \eta_2 + \eta_3 + \eta_4)$. SVD of $\mathbf{Q}_i$ are given as

$$\mathbf{Q}_i = \mathbf{V}_{H_{R_iD}} \mathbf{S}_{Q_i}^{1/2} \mathbf{U}_{H_{SR_i}}^H, \quad (13.46)$$

where, $\mathbf{S}_{Q_i} = \mathrm{diag}(s_1, s_2 \cdots s_{j_\phi})$ are the diagonal matrices with singular values. Further, the eigenvalue decomposition (EVD) of $\mathbf{W}_i$ is represented as follows:

$$\mathbf{W}_i = \mathbf{S}_{H_{SR_i}} \mathbf{S}_{W_i} \mathbf{V}_{H_{SR_i}}^H \quad (13.47)$$

where $\mathbf{S}_{W_i} = \mathrm{diag}(g_1, g_2 \cdots s_{j_\phi})$. Thus, the equivalent output of the DNN can be represented as

$$\mathbf{q}^* = \left[s_1, s_2 \cdots s_{j_\phi}, g_1, g_2 \cdots g_{j_\phi} \right]^T \in \mathbb{R}^{2j_\phi}, \quad (13.48)$$

where j_ϕ refers to the rank of $\mathbf{q}^*$, i.e., $j_\phi = \min(\eta_1, \eta_2)$.

13.3.3 DNN architecture

In this chapter, a fully connected multilayer DNN is used for offline data generation and training, as illustrated in figure 13.5. It consists of an input layer, L hidden layers, and a output layer. The input layer proceeds the input vector given in (13.37) with η neurons to the first hidden layer. Let x_l denote the output of the l_{th} hidden layer, which can be expressed as [36]:

$$x_l = f_l(\mathbf{W}_l x_{l-1} + \mathbf{B}_l), \quad (13.49)$$

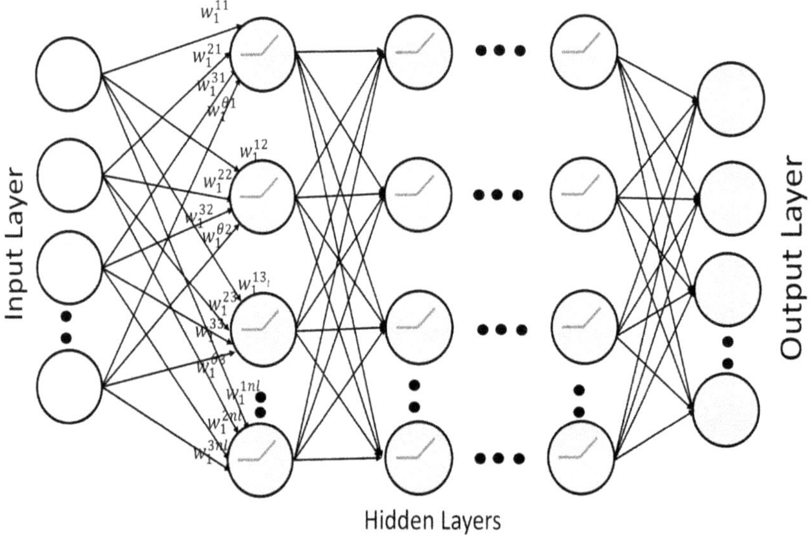

Figure 13.5. Fully connected DNN.

where $\mathbf{W}_l$ and $\mathbf{B}_l$ denote the weight and bias matrix at the l_{th} layer. In order to add nonlinearity we use a rectified linear unit (ReLU) function at the hidden layer, mathematically represented as:

$$f_l = \begin{array}{ll} x_l & x_l > 0 \\ 0 & x_l \leqslant 0. \end{array} \tag{13.50}$$

The ReLU function helps to avoid the vanishing gradient problem during the entire training process. In the training process, the parameters ($\theta \in \mathbf{W}, \mathbf{B}$) are updated iteratively until the threshold falls below the threshold value ϵ, which can mathematically be written as follows [38]:

$$|\theta[t] - \theta[t-1]| \leqslant \epsilon. \tag{13.51}$$

Gradient descent with momentum (GDM) optimizer use for tuning the parameter, it helps to accelerate gradient vectors in the right directions [39]. After the successful execution of the training process, the trained model is tested over the fresh test dataset. When the tested model has achieved the desired performance, it is known as a fit or well-trained model. Mean square error (MSE) validates the performance of the trained DNN. It calculates the loss between the actual and predictive solutions, which can mathematically be expressed as:

$$\mathcal{L}_{\text{mse}} = \frac{1}{|\mathcal{D}_T|} \sum_{k \in \mathcal{D}_T} \left\| \mathbf{x}_{L+1}^{(k)} - \mathbf{q}^{*(k)} \right\|_2^2. \tag{13.52}$$

13.3.4 Dataset generation, training, and deployment phase

In this section, we focuses on the offline dataset generation, training process, and online deployment for the architecture of DNN presented earlier. Schematic diagram of all the phases is presented in figure 13.6. In the offline dataset generation phase, $\mathcal{D}_T = 10\,000$ samples of the optimal relay precoder weights are calculated for different channel realizations. In each realization, random channels are generated and the corresponding precoder weights are calculated via a conventional scheme [27] and stored in the dataset. Further, the total dataset $\mathcal{D}_T$ is split randomly for the training and testing process of the DNN. In the training process, the parameters ($\theta \in \mathbf{W}, \mathbf{B}$) are updated iteratively until the threshold falls below the tolerable value ($\epsilon = 0.001$). Gradient descent with momentum (GDM) optimizer use for tuning the parameter. The GDM is preferred over the other methods such as stochastic gradient descent (SGD) optimization method because it has to accelerate gradient vectors in the right directions that leads to faster convergence toward the global minimum. On the other hands SGD method can veer off in the wrong direction due to the frequent updates.

Finally, in the online deployment phase, the well-trained DNN can be used for automatic relay selection with the impact of predicted energy-efficient precoders for

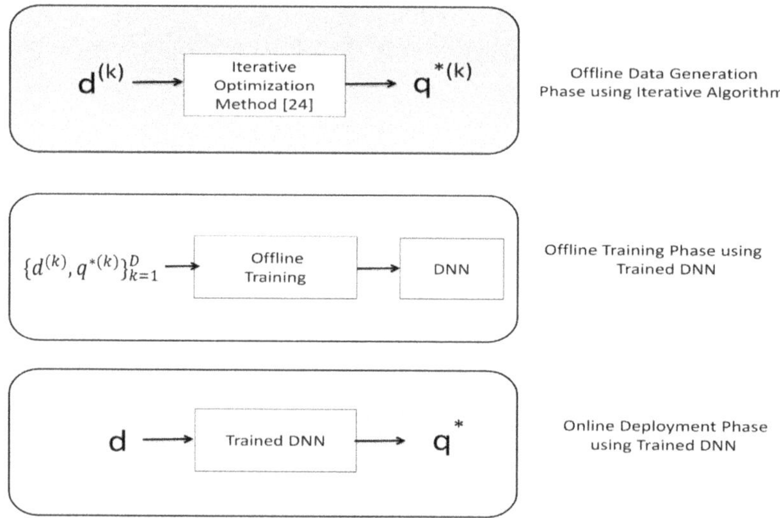

Figure 13.6. Schematic diagram of offline data generation, training, and online deployment phase.

enhancing the end-to-end EE of MIMO-MRCN with low complexity. Furthermore, it can be used to implement a real-time scenario in the context of Industry 4.0.

13.4 Numerical results

This section depicts an experimental analysis of the DL-enabled scheme for relay selection with impact of precoder designing over the conventional relay selection schemes via MATLAB simulation. The proposed DL-enabled scheme considers all system parameter settings the same as the conventional scheme used in [27] during the dataset generation $\mathcal{D}_T$. In the data generation phase, the channel parameters are independent and identically distributed (i.i.d.) circularly symmetric complex Gaussian variables having zero means with variances of 1 for $\mathbf{H}_{SR_i}$ and $\mathbf{H}_{R_iD}$; variances of 0.3 for $\mathbf{G}_{SP}$ and $\mathbf{G}_{R_iP}$. The parameter settings are summarized in table 13.2. A total of 10 000 samples are generated during the offline data generation phase. Out of the total dataset, 8000 samples are used for training and 2000 samples are used for the testing. The proposed scheme uses the DNN architecture of 5 layers with 3 hidden layers for training and testing. All hyper parameters are selected empirically, which is summarized in table 13.3. A GDM optimizer has been used to train all the networks. All simulation results are plotted for an average of 1000 channels. At last, the efficacy of the proposed DL-enabled scheme is compared with the conventional relay selection schemes on the same parameter settings. Figure 13.7 shows the comparison of all the proposed schemes along with the variation of the EE with respect to the total transmit power. One can see that the EE of all the schemes increases with the transmit power, reaches the peak, and saturates afterwards. This behavior occurs because as we increase the transmit power level, secondary nodes can consume more power to improve the data rate, which also results in the increase of EE. However, after a certain point if we increase the transmit power, nodes

Table 13.2. Simulation parameters [27].

Simulation parameters	Settings
m_s	2
m_r	2
m_d	2
m_p	2
P_{ckt}	3 dBm
$I_{S_{\max}} = I_{R_{i_{\max}}} = I_{S_{\max}}$	-1 dBW
$P_{S_{\max}} = P_{R_{i_{\max}}} = P_{\max}$	10 dBW
$\sigma^2_{n_{R_i}} = \sigma^2_{n_D}$	1
$\rho_{S_{\mathrm{TX}}}$	1
ρ_{R_i}	1
$\mathrm{Rate}_{\min}$	0.4
Variances of $\mathbf{G}_{\mathrm{SP}}$ and $\mathbf{G}_{R_iD}$	0.3
Variances of $\mathbf{H}_{\mathrm{SR}_i}$ and $\mathbf{H}_{R_iP}$	1

Table 13.3. DNN hyperparameter settings.

Parameters	Values
Number of hidden layers	3
Number of neurons in 1st hidden layer	128
Number of neurons in 2nd hidden layer	64
Number of neurons in 3rd hidden layer	32
Batch size	2000
Epochs size	1000
Minimum gradient	0.000 01
Learning rate	0.001
Optimizer	Gradient descent with momentum (GDM)
Nonlinear activation function	Rectified Linear unit (ReLU)
Loss function	Means Square Error (MSE)

cannot utilize more power due to the limitation caused by the interference constraint. It results in the saturation of the achieved EE. In addition, we can also analyze and compare the performance of all schemes using figure 13.7. As can be seen, the opportunistic scheme and DL-enabled scheme provide better EE than any other schemes. It is by the fact that both schemes designed S_{TX} and S_{R_i} precoders as well as selected the best relay according to the maximum obtained EE. The DL-enabled scheme provides similar performance to the opportunistic scheme with lower computational cost. The other two proposed schemes, i.e., proactive and

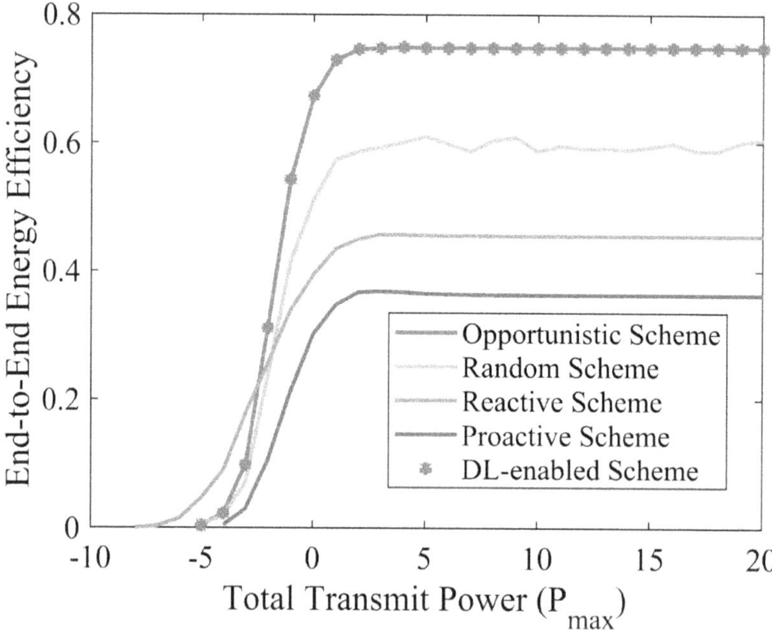

Figure 13.7. EE versus P_{max} (dB) by various relay selection schemes with $m_s = m_r = m_d = m_p = 2$, $I_{R_{i_{max}}} = -1$ (dBW), $P_{S_{max}} = 10$.

reactive schemes, also select the best relay according to maximum obtained EE, however, in both schemes only one precoder is designed (i.e., either $\mathbf{W}_i$ or $\mathbf{Q}_i$), which resulted in the decreased EE. Finally, the performance of the fourth scheme, i.e., the random scheme, lies between these three schemes. This scheme designed both precoders $\mathbf{W}_i$, and $\mathbf{Q}_i$, thus, it achieves better end-to-end EE than the proactive and reactive schemes. However, due to the random selection of relay for transmission, the achieved performance is less than the opportunistic scheme. Figure 13.8 examines the MSE performance for the proposed DNN on the two different datasets i.e., $\mathcal{D}_{T_1}$ and $\mathcal{D}_{T_2}$. Here, DNN1 and DNN2 use the same DNN architectures and hyperparameter settings during the training and testing trained, except the dataset. Here, DNN1 uses the $\mathcal{D}_{T_1} = 10\,000$ samples for training and testing; out of the total samples, 8000 samples are used for the training and 2000 samples are used for testing. However, DNN2 uses the $\mathcal{D}_{T_1} = 8000$ samples for training and testing; out of the total samples, 6400 samples are used for the training and 1600 samples are used for testing. Here, the entire dataset $\mathcal{D}_{T_1}$ and $\mathcal{D}_{T_2}$ passes through the 1000 epochs. One epoch refers to an iteration in which the DNN passes a dataset from the forward to backward pass. From figure 13.8, it is observed that the DNN1 performed better than the DNN2 in terms of MSE. Therefore, 10 000 samples are considered for the offline training and testing. Moreover, the training and testing errors of DNN1 are closely matched to each other and saturated down with epoch. It shows that the proposed DNN i.e., DNN1, is well trained with good fits and ready for the implementation in the online deployment phase for predicting the precoder weights.

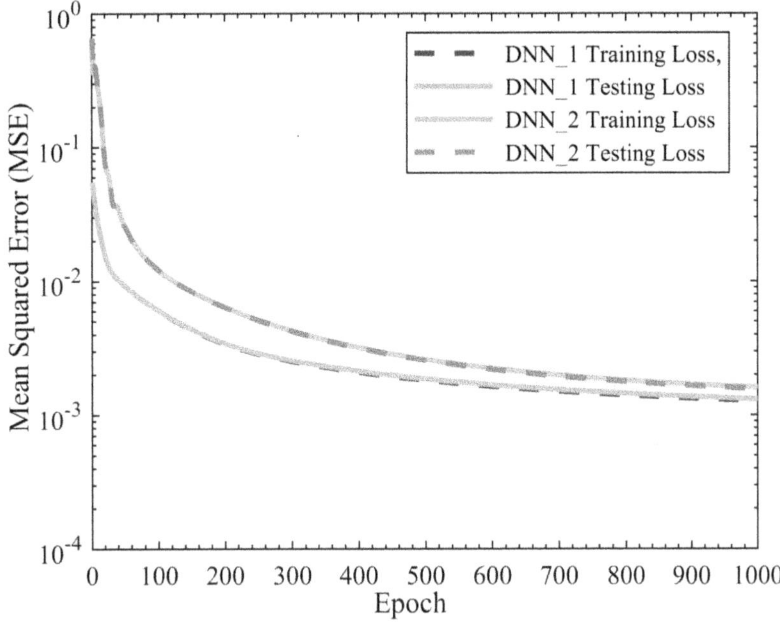

Figure 13.8. MSE versus epochs for different number of data samples with same DNN architecture.

13.5 Conclusion

In this chapter, DL-enabled data-driven model using DNN has been studied for MIMO-MRCN for an automatic relay selection while considering the impact of energy-efficient precoders at the source and relays. The different relay selection schemes were investigated with the objective to maximize the end-to-end EE of SN subject to different sets of constraints to satisfy the QoS of both licensed and SN. The schemes were analyzed in terms of the achieved EE. Numerical results have shown that the DNN-based relay selection scheme give performance similar to that with the conventional relay selection scheme in terms of maximum end-to-end EE for the SN, while exhibiting the minimum computational cost as well as MSE. It shows that the proposed scheme can be a good candidate to implement in the online deployment phase for predicting the precoder weights in the context of Industry 4.0.

References

[1] Dogra A, Jha R K and Jain S 2021 A survey on beyond 5G network with the advent of 6G: architecture and emerging technologies *IEEE Access* **9** 67512–47

[2] Sasiain J, Sanz A, Astorga J and Jacob E 2020 Towards flexible integration of 5G and IIoT technologies in Industry 4.0: a practical use case *Appl. Sci.* **10** 7670

[3] Lueth K L, Retrieved from state of the IoT 2020: 12 billion IoT connections, surpassing non-IoT for the first time. https://iot-analytics.com/state-of-the-iot-2020-12-billion-iot-connections-surpassing-non-iot-forthe-first-time/

[4] Rao S K and Prasad R 2018 Impact of 5G technologies on Industry 4.0 *Wirel. Pers. Commun.* **100** 145–59

[5] Morocho-Cayamcela M E and Lee H *et al* 2019 Machine learning for 5G/B5G mobile and wireless communications: potential, limitations, and future directions *IEEE Access* **7** 137184–206

[6] Lin C-C, Deng D-J, Chen Z-Y and Chen K-C 2016 Key design of driving Industry 4.0: joint energy-efficient deployment and scheduling in group-based industrial wireless sensor networks *IEEE Commun. Mag.* **54** 46–52

[7] Hussain F, Hassan S A and Hussain R *et al* 2020 Machine learning for resource management in cellular and IoT networks: potentials, current solutions, and open challenges *IEEE Commun. Surv. Tutor.* **22** 1251–75

[8] Huang H *et al* 2020 Deep learning for physical-layer 5G wireless techniques: opportunities, challenges and solutions *IEEE Wirel. Commun.* **27** 214–22

[9] Simeone O 2018 A very brief introduction to machine learning with applications to communication systems *IEEE Trans. Cogn. Commun. Netw.* **4** 648–64

[10] Dai L, Jiao R and Adachi F *et al* 2020 Deep learning for wireless communications: an emerging interdisciplinary paradigm *IEEE Wirel. Commun.* **27** 133–9

[11] Ahmad W S H M W *et al* 2020 5G technology: towards dynamic spectrum sharing using cognitive radio networks *IEEE Access* **8** 14460–88

[12] Goldsmith A, Jafar S A and Maric I *et al* 2007 Breaking spectrum gridlock with cognitive radios: an information theoretic perspective *IEEE Proc.* **97** 894–914

[13] Sahu D and Trivedi A 2014 A Bayesian approach using M-QAM modulated primary signals for maximizing spectrum utilization in cognitive radio *Int. Conf. on Advances in Computing, Communications and Informatics (ICACCI)* 1183–7

[14] Deligiannis A, Daniyan A, Lambotharan S and Chambers J A 2018 Secrecy rate optimizations for MIMO communication radar *IEEE Trans. Aerosp. Electron. Syst.* **54** 2481–92

[15] Chen X, Chen H-H and Meng W 2014 Cooperative communications for cognitive radio networks-from theory to applications *IEEE Commun. Surv. Tutor.* **16** 1180–92

[16] Akyildiz I F *et al* 2006 NeXt generation/dynamic spectrum access/cognitive radio wireless networks: a survey *Comput. Netw.* **50** 2127–59

[17] Haykin S 2005 Cognitive radio: brain-empowered wireless communications *IEEE J. Sel. Areas Commun.* **23** 201–20

[18] Cover T and Thomas J 2006 *Elements of Information Theory* 2nd edn (New York: Wiley)

[19] Nguyen T V, Tran T N, Shim K, Huynh-The T and An B 2021 A deep-neural-network-based relay selection scheme in wireless-powered cognitive IoT networks *IEEE Internet Things J.* **8** 7423–36

[20] Su Y, Lu X, Zhao Y, Huang L and Du X 2019 Cooperative communications with relay selection based on deep reinforcement learning in wireless sensor networks *IEEE Sens. J.* **19** 9561–9

[21] Guo Y-Y, Yang J, Tan X-L and Liu Q 2022 An energy-efficiency multi-relay selection and power allocation based on deep neural network for amplify-and-forward cooperative transmission *IEEE Wireless Commun. Lett.* **11** 63–6

[22] Zhu X *et al* 2020 Deep learning-based precoder design in MIMO systems with finite-alphabet inputs *IEEE Commun. Lett.* **24** 2518–21

[23] Zhang X and Vaezi M 2021 Multi-objective DNN-based precoder for MIMO communications *IEEE Trans. Commun.* **69** 4476–88

[24] Elbir A M and Papazafeiropoulos A K 2020 Hybrid precoding for multiuser millimeter wave massive MIMO systems: a deep learning approach *IEEE Trans. Veh. Technol.* **69** 552–63

[25] Huang H, Song Y and Yang J *et al* 2019 Deep-learning-based millimeter-wave massive MIMO for hybrid precoding *IEEE Trans. Veh. Technol.* **68** 3027–32

[26] Maurya S and Bansal M 2017 Energy efficient precoder design for non-regenerative MIMO-CRN *IEEE Wirel. Commun. Lett.* **6** 646–9

[27] Maurya S, Bansal M and Trivedi A 2019 Joint source and relay precoder design for energy-efficient MIMO-cognitive relay networks *IET Commun.* **13** 2226–34

[28] Sahu D, Maurya S, Bansal M and Dinesh Kumar V 2022 Deep learning based energy-efficient relay precoder design in MIMO-CRNs *Phys. Commun.* **50** 1–8

[29] Sahu D, Maurya S, Bansal M and Dinesh Kumar V 2022 Data-driven approach to design energy-efficient joint precoders at source and relay using deep learning in MIMO-CRNs *Trans. Emerg. Telecommun. Technol.* **33** e4454

[30] Maurya S and Bansal M 2021 Joint relay selection and precoder designing for energy-efficient MIMO-CRNs *5th Conf. on Information and Communication Technology (CICT)* 1–5

[31] Li Q, Zhang Q, Feng R, Luo L and Qin J 2013 Optimal relay selection and beamforming in MIMO cognitive multi-relay networks *IEEE Commun. Lett.* **17** 1188–91

[32] Yan P, Zou Y, Ding X and Zhu J 2020 Energy-aware relay selection improves security-reliability tradeoff in energy harvesting cooperative cognitive radio systems *IEEE Trans. Veh. Technol.* **69** 5115–28

[33] Das P 2020 Outage performance of cognitive relay networks with optimal relay and antenna selection *National Conf. on Communications (NCC)* 1–6

[34] Perez-Romero J, Sallent O, Agusti R and Giupponi L 2007 A novel on-demand cognitive pilot channel enabling dynamic spectrum allocation *2nd IEEE Int. Symp. on New Frontiers in Dynamic Spectrum Access Networks* 46–54

[35] Telatar E 1999 Capacity of multi-antenna Gaussian channels *Eur. Trans. Telecommun.* **10** 585–95

[36] Goodfellow I, Bengio Y and Courville A 2016 *Deep Learning* (Cambridge, MA: MIT Press) http://www.deeplearningbook.org

[37] Lebrun G, Gao J and Faulkner M 2005 MIMO transmission over a time-varying channel using SVD *IEEE Trans. Wirel. Commun.* **4** 757–64

[38] Loizou N and Richtarik P 2020 Momentum and stochastic momentum for stochastic gradient, newton, proximal point and subspace descent methods *Comput. Optim. Appl.* **77** 653–710

[39] Ning Q 1999 On the momentum term in gradient descent learning algorithms *Neural Netw.* **12** 145–51

9 780750 352482